New Diagnostic Technology: Applications in Animal Health and Biologics Controls

Developments in Biologicals

Vol. 126

This series "Developments in Biologicals" begins with Vol. 102 and is the continuation of IABs symposia series "Progress in Immunobiological Standardization", Vols 1-5, "Immunobiological Standardization", Vols 1-22 and "Developments in Biological Standardization", Vols 23-101.

Organized and published by the International Association for Biologicals (IABs)

KARGER

Basel · Freiburg · Paris · London · New York
Bangalore · Bangkok · Singapore · Tokyo · Sydney

New Diagnostic Technology: Applications in Animal Health and Biologics Controls

Saint-Malo, France
October 3-5, 2005

Volume editors

Philippe Vannier
AFSSA, Ploufragan, France

David Espeseth
IABs, Perkasie, PA, USA

Proceedings of an international conference organized by
French Food Safety Agency (AFSSA)
Institute for Higher Education in Animal Production and Food Industries (ISPAIA)
and
International Association for Biologicals (IABs)
USDA Animal and Plants Health Inspection Service (APHIS)
World Organization for Animal Health (OIE)

58 figures and 50 tables, 2006

KARGER

Basel · Freiburg · Paris · London · New York
Bangalore · Bangkok · Singapore · Tokyo · Sydney

Developments in Biologicals

Library of Congress Cataloging-in-Publication Data

International Conference on New Diagnostic Technology (2005 : Saint-Malo, France)
New diagnostic technology : applications in animal health and biologics controls : Saint-Malo, France, October 3-5, 2005 / volume editors, Philippe Vannier, David Espeseth : proceedings of an international conference organized by French Food Safety Agency (AFSSA) ... [et al.].
p. ; cm. -- (Developments in biologicals ; v. 126)
Includes bibliographical references.
ISBN-13: 978-3-8055-8116-5 (soft cover : alk. paper)
ISBN-10: 3-8055-8116-5 (soft cover : alk. paper)
1. Veterinary medicine--Diagnosis--Congresses. I. Vannier, Philippe. II. Espeseth, David. III. Agence française de sécurité sanitaire des aliments. IV. Title. V. Series.
[DNLM: 1. Animal Diseases--diagnosis--Congresses. 2. Molecular Diagnostic Techniques --veterinary--Congresses. 3. Polymerase Chain Reaction--Congresses. 4. Technology, Medical --methods--Congresses. W1 DE997VKF v.126 2006 / SF 772.6 I61n 2006]
SF771.I58 2005
636.089'6075--dc22

2006022543

Bibliographic Indices. This publication is listed in bibliographic services, including Current Contents® and Index Medicus.

Drug dosage. The authors and the publisher have exerted every effort to ensure that drug selection and dosage set forth in this text are in accord with current recommendations and practice at the time of publication. However, in view of ongoing research, changes in government regulations, and the constant flow of information relating to drug therapy and drug reactions, the reader is urged to check the package insert for each drug for any change in indications and dosage and for added warnings and precautions. This is particularly important when the recommended agent is a new and/or infrequently employed drug.

Worldwide distribution by S. Karger AG, Basel (Switzerland).
Printed in Switzerland by instaprint.
ISBN-10: 3-8055-8116-5
ISBN-13: 978-3-8055-8116-5

Scientific Committee

P. Vannier, President (AFSSA, France)
D. Espeseth (IABs)
D. Gaudry (IABs)
R. Hill (CVB, USA)
M. Lombard (IABs)
A. Schudel (OIE)
D. Senne (NVSL, USA)
C. Jungbaeck (PEI, Germany)
P. Midtlyng (VESO, Norway)
Y. Blanchard (AFSSA, France)

Organisation of the Meeting

Session I	Principles of Novel Techniques: Tests Description and Potentialities Chairman: *Rick Hill,* USA
Session II	Standardisation, Validation and Overcoming the Problems of Molecular Diagnostic Methods Chairman: *Alejandro Schudel,* OIE
Session III	Application in Detection of Pathogens and Screening Programmes Chairman: *Paul Midtlyng,* Norway
Session IV	Certainties and Uncertainties of Agent or Toxin Identification by Molecular Methods Chairman: *Steve Edwards,* United Kingdom
Session V	Examples of Application in Quality Control of Human and Veterinary Vaccines Chairman: *Jacques Léchenet*, France
Conclusions and Recommendations	Chairman: *David Espeseth,* USA

ACKNOWLEDGEMENTS

The International Conference on **New Diagnostic Technology: Applications in Animal Health and Biologics Controls** is grateful to the following organizations for their financial support.

Union Européenne
Région Bretagne
Zoopole Développement
Conseil Général des Côtes d'Armor
AES Laboratoire
Intervet
Fort Dodge
Virbac
Ceva Santé Animale
Merial
Pfizer
Qiagen
Merck Research Laboratories
Tetracore
SIMV

Contents

Vannier P, Espeseth D (eds): New Diagnostic Technology: Applications in Animal Health and Biologics Controls. Dev Biol (Basel). Basel, Karger, 2006, vol 126, p 1.

SESSION I

Principles of Novel Techniques: Tests Description and Potentialities

Chairman: *Rick Hill,* USA

Vannier P, Espeseth D (eds): New Diagnostic Technology: Applications in Animal Health and Biologics Controls. Dev Biol (Basel). Basel, Karger, 2006, vol 126, pp 3-5.

Experiences on the Application of Various Nucleic Acid Amplification Systems for the Improved Detection of Swine Vesicular Disease Virus

M. Hakhverdyan[1], *T.B. Rasmussen*[2], *Å. Uttenthal*[2], *P. Thorén*[1], *S. Belák*[1]

[1] National Veterinary Institute and the Swedish University of Agricultural Sciences, Departments of Virology, OIE Collaborative Center for the Application of Polymerase Chain Reaction Methods for Diagnosis of Viral Diseases in Veterinary Medicine, Uppsala, Sweden

[2] Danish Institute for Food and Veterinary Research, Department of Virology, Lindholm, Kalvehave, Denmark

Introduction

During the last one-and-a-half-decades more than 50 nested classical PCR assays have been developed in our laboratory for detecting a variety of DNA and RNA viruses. Recently, real-time PCR methods (such as TaqMan, Molecular Beacons, and the Primer-Probe Energy Transfer System) have been elaborated for the diagnosis of trans-boundary animal diseases (TADs), such as foot-and-mouth disease (FMD), swine vesicular disease (SVD), vesicular stomatitis (VS), classical swine fever and African swine fever. Conventional and real-time PCR packages have been developed for the multiplex detection of eight important viruses of swine. By introducing nucleic acid extraction and pipetting robotics, together with the multi-channel real-time PCR machines, the diagnostic procedures of TADs have become rapid, robust and automated. Herewith, current data on comparative studies of two real-time PCR systems are presented – a molecular beacon and recently developed novel primer-probe energy transfer (PriProET). Swine vesicular disease virus (SVDV) is used as a model in the study, representing the vesicular cluster of TADs.

Materials and Methods

Twenty SVDV strains, the closely related human enterovirus Coxsackie B5 (CB5), as well as foot-and-mouth disease virus (FMDV) and vesicular stomatitis virus (VSV) were used in the comparative study. A molecular beacon assay was used according the standard procedures. The principle of molecular beacon is well known and has been described in many publications. The PriProET assay, is based on two fluorescently labelled oligonucleotides (1, 2). The basic principle of the assay relies on fluorescence resonance energy transfer (FRET) between donor and acceptor oligonucleotides, which must be in the vicinity of each other. Briefly, one of two primers is labelled with a donor fluorescence dye on the 5´-end, in addition to the hybridisation probe, which is labelled with an acceptor/reporter fluorescence dye on the 3´-end. During the PCR cycles, when the probe anneals to the extended fluorescent primer, energy transfers from donor to acceptor due to their close proximity. Acceptor dye emits fluorescence that can be monitored and quantified at each cycle by real-time PCR instrument.

A conserved region of the SVDV 3D-gene was chosen for the design of primers and probes. The forward primer was labelled with a donor fluorophore (FAM) on the 5´end, whereas the probe was labelled with a reporter fluorophore (Texas Red) on the 3´end. FAM and black hole quencher were used for the molecular beacon probe. An ABI PRISM 7700 Sequence Detector was used for the experiments (Applied Biosystems, USA). Fluorescence was monitored in real-time in fifty-five amplification cycles followed by a melting peak profile of the probe to confirm specific amplification of the target virus. Fluorescent data were analysed on an in-house pre-programmed Excel datasheet.

Results

The molecular beacon probe detected a high number of SVDV strains, but failed to detect some of them. PriProET specificity tests have shown successful amplification of all twenty SVDV strains, despite eight of the SVDV strains showing amplification with shifted melting peaks. CB5, FMDV and VSV gave negative results. For determination of analytical sensitivity, a 10-fold dilution series of a synthetic amplicon was tested. The detection limit was five viral genome equivalents.

Discussion

Since the molecular beacon failed to detect some SVDV strains, this system cannot be recommended as a general diagnostic tool. The PriProET system was previously applied to the detection of FMDV and VSV (1, 2), and it was found to be a powerful general tool for SVDV detection. This means that all three viruses can be detected side-by-side in one PCR assay without any further optimisation of the system, as the thermodynamic profile used for the amplification is the same. Detection of all tested SVDV strains certifies the high specificity of the assay. The PriProET system can detect divergent SVDV strains even if there are 1-2 mutations in the probe region. High sensitivity and tolerance of the assay towards mutations in the probe region increase the chances of amplifying those SVDV strains, which recently have diverged during the course of evolution.

Conclusion

The PriProET method provides a sensitive, specific and high throughput novel tool for the rapid detection of SVDV and for the differentiation of all three viruses in the vesicular disease complex of TADs

Acknowledgements

This study was supported by European Union (Project number 513645).

References

1 Rasmussen TB, Uttenthal A, de Stricker K, Belak S, Storgaard T: Development of a novel quantitative real-time RT-PCR assay for the simultaneous detection of all serotypes of foot-and-mouth disease virus. Arch Virol 2003;148:2005-2021.

2 Rasmussen TB, Uttenthal A, Fernandez J, Storgaard T. Quantitative multiplex assay for simultaneous detection and identification of Indiana and New Jersey serotypes of vesicular stomatitis virus. J Clin Microbiol 2005;43:356-362.

Dr. M. Hakhverdyan, National Veterinary Institute and the Swedish University of Agricultural Sciences, Departments of Virology, SE - 75189 Uppsala, Sweden.
E-mail: mikhayil.hakhverdyan@bmc.uu.se

Vannier P, Espeseth D (eds): New Diagnostic Technology: Applications in Animal Health and Biologics Controls. Dev Biol (Basel). Basel, Karger, 2006, vol 126, pp 7-15.

Detection of Animal Viruses Using Nucleic Acid Sequence-Based Amplification (NASBA)

L.T. Lau[1,2,3], *Y.W.W. Fung*[1,2], *A.C.H. Yu*[1,2,3]

[1] Hong Kong DNA Chips Ltd., Hang Tung Resources Centre, Shau Kei Wan, Hong Kong SAR, China

[2] Neuroscience Research Institute, Key Laboratory of Neuroscience, Peking University and Department of Neurobiology, Peking University Health Science Centre, Beijing, China

[3] Anigen Diagnostics Limited, Hang Tung Resources Centre, Shau Kei Wan, Hong Kong SAR, China

Abstract: As seen in recent avian influenza outbreaks in Asia, prevention is the key to fighting infectious disease successfully. Efficient disease surveillance systems on the basis of molecular diagnostics will help monitor the emergence of viruses in the early stage and thus prompt containment measures can be in place to minimize disease spread. Here we describe and review molecular diagnostics focusing on nucleic acid sequence-based amplification (NASBA) technology in detecting viruses causing animal diseases, such as avian influenza, foot-and-mouth disease, and Newcastle disease. NASBA offers high sensitivity, specificity, accuracy, and speed of availability of results, and NASBA would be the most applicable molecular diagnostics for disease surveillance and control.

INTRODUCTION

Viruses are known to expand their range of hosts by mutation, with humans being unfortunate in becoming haphazard victims of viral mutation when humans and animals co-exist in close proximity. Outbreaks of avian influenza H5N1 infection since 1997 and the emergence of SARS (severe acute respiratory syndrome) in 2003 have raised serious concern of possible animal-to-human transmission of viruses through contact of humans with birds, poultry and other livestock. Many Asian countries have experienced the loss of human lives and devastation to their economy brought about by H5N1 outbreaks, and the need to develop rapid and sensitive diagnostic methods to detect such viruses is vital. Rapid detection tests for viruses will allow timely surveillance of possible sources of diseases and immediate actions could be taken to contain the diseases and prevent outbreaks [1].

Current virus identification methods include conventional serology (such as antigen detection and antibody testing), virus culture (usually considered as the gold standard), and nucleic acid-based testing (NAT) (such as real-time PCR and nucleic acid sequence-based amplification (NASBA)). Blood-based serological methods may reveal some of the infection history of the test subject, but these methods suffer from lack of sensitivity and frequent false positives. Culturing live virus usually provides unambiguous results, but culturing often requires a longer time, and negative culture results do not necessarily mean absence of the virus tested. The shortcomings of these traditional methods may cause delay or mis-interpretation of results and lead to spreading of the viruses. From the experience of recent H5N1 outbreaks, diagnostic methods offering speed, specificity, and high sensitivity will be particularly important to the front line surveillance of infectious diseases. NAT are molecular diagnostic tools that can be used in disease surveillance to achieve early and sensitive detection of viruses. The main NAT streams are the PCR-based technologies, particularly real time PCR, and NASBA-based technologies.

Preventing global outbreaks of infectious diseases should be a mandate for all countries as globalization and advanced transportation systems bring the whole world closer by promoting frequent mobility of people and livestock. Exemplified by the SARS coronavirus spread from China to many parts of the world, governments and laboratories of different countries bear the responsibility to share disease surveillance information and technology available in monitoring disease outbreaks. Prevention is the key to fighting infectious diseases successfully, and efficient disease surveillance systems will help monitor the emergence of viruses and prompt containment measures to minimize disease spread.

Principle of NASBA

NASBA, developed in 1991, is an RNA-based amplification technology [2]. It is a continuous, isothermal, and enzyme-based method for the amplification of nucleic acids [3] which can serve as an alternative to PCR. NASBA uses a mixture of reverse transcriptase, ribonuclease-H, and RNA polymerase and two target-specific primers. The DNA oligonucleotide primers are specially designed with forward primer containing a 5' extension carrying the promoter sequence for bacteriophage T7 DNA-dependent RNA polymerase, and a reverse primer containing a 5' extension with a sequence complementary to a DNA detection probe labelled with a ruthenium-based electro-chemiluminescent (ECL) tag. Amplification takes place with the 5' primer extensions fully incorporated into the amplified products allowing both highly efficient production of complementary RNA template (directed by RNA polymerase) and specific detection by a ruthenium-containing probe. This method of detection requires the use of a chemi-luminenscence detector. Under optimum conditions, it is possible to achieve 10^{12}-fold amplification in 1.5 hours, versus the 10^{9}-fold amplification accomplished by PCR [4].

NASBA has the edge in amplifying specific single-stranded RNA targets without the need for nucleic acid denaturation among common contaminants such as genomic DNA, heparin, EDTA, citrate, haemoglobin, albumin, and lipids [3]. Moreover, the end product of amplification is RNA, which tends to be unstable under normal environmental conditions, thus minimizing the possibility of carry-over contamination of equipment from previous experiments. There are numerous NASBA-based diagnostic tests for viruses such as human immunodeficiency virus type 1 (HIV) [3], cytomegalovirus (CMV) [5], hepatitis C virus [6], and papillomavirus type 16 [7]. Recently, we have proved the applicability of NASBA technology in the surveillance of animal diseases by successful development of NASBA-based methods to detect a number of animal viruses including avian influenza virus (AIV) and foot-and-mouth disease virus (FMDV) [8-11]. The use of this isothermal method renders the purchase of expensive and maintenance-requiring PCR thermocycler (conventional or real-time type) unnecessary, since a simple temperature-calibrated water bath will be able to provide the isothermal environment.

NASBA-Based Animal Virus Detection System

As cross-species viral infections are becoming more common, there is an imperative need for detecting animal viruses to control potential infection in livestock. With the high sensitivity and specificity offered by the NASBA technology, we have successfully developed various methods for AIV [8-10], FMDV [11], Newcastle disease virus, classical swine fever virus, and porcine reproductive and respiratory syndrome virus [unpublished data]. Various public and private laboratories in collaboration with us have started to use our NASBA-based methods for routine testing and surveillance of AIV and FMDV on their farm animals and poultry [personal communications]. The ultra-high sensitivity of the NASBA method to detect AIV H5N1 has been proved to be comparable to virus isolation [12]. The high sensitivity achieved by NASBA may be contributed by the direct incorporation of reverse transcription and highly efficient in vitro transcription by T7 RNA polymerase into one isothermal amplification reaction, whereas the reverse transcription step at a different temperature precedes PCR amplification involving more temperature changes for RT-PCR.

Avian Influenza Virus (AIV)

All AIVs belong to influenza type A, which can be further classified according to the antigenicity of two envelope glycoproteins, haemagglutinin (HA) and neuraminidase (NA). Currently there are 15 subtypes of HA (H1-15) and nine subtypes of NA (N1-9) [8]. H5N1 and H7N2 are the two most dangerous subtypes of AIV strains known in terms of flock mortality and potential for virulence in humans. They belong to the category of highly pathogenic avian influenza (HPAI) with a mortality rate reaching 100%, while low pathogenic avian influenza (LPAI) strains cause less serious infections mainly in the respiratory and intestinal tracts [8].

The conventional method for diagnosis of avian influenza is through virus culture in embryonated fowl's eggs, albeit sensitive but labour-intensive and time-consuming [13]. Virus antigen detection techniques such as enzyme-linked immunoassay and immunofluorescence are relatively fast but less sensitive and less specific than virus isolation. Both RT-PCR and NASBA are highly sensitive and rapid diagnostic methods. We have previously conducted a study to compare the NASBA method with virus culture in the detection of AIV H5N1 strain, demonstrating equivalent sensitivity of the NASBA method to virus culture (Table 1, [12]). The study elegantly shows the applicability of a four-hour NASBA method in detecting H5N1 with equivalent sensitivity to the "gold standard" virus culture approach, which takes days to be completed [12].

In addition to high sensitivity, NASBA is an effective way to test specifically for the presence of H5N1 virus in poultry vaccinated with attenuated AIV [14]. Conventional serology-based detection methods could not distinguish whether the presence of H5N1 antibodies is due to induction by vaccine or from natural immunity after past infection. NASBA detection of H5N1 using throat or cloacal swab samples targets H5N1 nucleic acid (RNA) to reflect the current presence of viral load and therefore is unaffected by vaccination.

Table 1: Comparable sensitivity of NASBA and virus culture in the detection of AIV H5N1.

Sample	**NASBA**[a]	**Virus culture**
10^{-1} dilution	Positive/positive	Positive
10^{-2} dilution	Positive/positive	Positive
10^{-3} dilution	Positive/positive	Positive
10^{-4} dilution	Positive/positive	Positive
10^{-5} dilution	Positive/positive	Positive
10^{-6} dilution	Positive/negative[b]	Positive
Negative Control	Negative	Negative

[a] Replicate analyses were performed on serially diluted aliquots of A/chicken/Hong Kong/1000/97 (H5N1) with known viral titre [12] using NASBA H5N1 method or virus culture.

[b] The limit of detection in these NASBA assays is probably at 10^{-6} dilution.

Foot-and-Mouth Disease Virus (FMDV)

FMDV is a single-stranded RNA virus belonging to the family of *Picornaviridae*, including other members such as swine vesicular disease virus (SVDV), Coxsackie virus, and poliovirus [11]. FMDV can be classified into seven antigenically and genetically distinct serotypes, A, O, C, Asia-1, SAT-1, SAT-2, and SAT-3 [11]. Infection by this virus can be symptomatic or asymptomatic, and for the asymptomatic ones, both vaccinated and unvaccinated animals could be found [15,16]. It is thus important to develop an FMDV detection method to rapidly and sensitively identify genuine FMDV carriers from vaccinated animals [11], or the spread of FMDV amongst livestock will pose enormous economic disaster to the livestock industry.

We have been able to detect the major seven serotypes of FMDV using the NASBA method [11]. Recently, NASBA methods for specifically serotyping FMDV A, O, and Asia 1 have been developed in our laboratory. Table 2 summarizes the high specificity exhibited by NASBA in distinguishing A, O, or Asia 1 serotype, showing no evidence of cross hybridization of the specific primers/probes among these three closely related serotypes. Therefore, the FMDV NASBA method can be applied to generally screen for the presence of FMDV carrier animals (using the universal primers/probe for seven serotypes) and serotyping of A, O, or Asia 1 can be performed to study the prevalence and epidemiology of various FMDV serotypes in the infected regions. The use of FMDV A/O/Asia 1 serotyping NASBA may be particularly useful for tracing the outbreak origin as seen in the recent FMDV outbreak in China.

Table 2: Specificity of FMDV NASBA method for serotyping A, O, and Asia 1[a].

	FMDV A RNA	FMDV O RNA	FMDV Asia RNA
FMDV A NASBA	Positive	Negative	Negative
FMDV O NASBA	Negative	Positive	Negative
FMDV Asia 1 NASBA	Negative	Negative	Positive

[a] RNA samples extracted from FMDV A, O, and Asia 1 were subjected to NASBA amplification using primers/probes specific to A, O, or Asia 1.

Newcastle Disease Virus (NDV)

Newcastle disease is a contagious and widespread avian disease that affects most species of birds. It is caused by the avian paramyxoviridae serotype 1 (APMV1), also known as the Newcastle disease virus (NDV). Its negative-sense RNA genome contains six genes encoding an RNA-directed RNA polymerase, haemagglutinin-neuraminidase protein, fusion protein, matrix protein, phosphoprotein, and nucleoprotein from 5' to 3' direction [17]. NDV is classified according to its virulence. Lenogenic strains display low virulence, mesogenic isolates display intermediate virulence, while velogenic ones are the most virulent [17]. Velogenic strains are further classified as neurotropic or viscerotropic depending on the virus's clinical manifestation. Unvaccinated flocks infected with velogenic strains of NDV often have 100% mortality rate, and this is a threat to commercial poultry industries as a huge number of infected poultry might be decimated with an NDV outbreak. Turkeys, pigeons, and caged birds such as cockatiels, budgies, and amazons are also prone to infection. The Office International des Epizooties (OIE) has in fact designated the velogenic strains as List A reportable viral agents, requiring federal reporting, strict quarantine measures and trade embargoes [18].

We have once again applied the NASBA technology to detect NDV isolates rapidly and reliably of a broad genetic lineage [manuscript submitted]. NDV NASBA shares similar sensitivity to virus isolation and demonstrates high specificity only to NDV strains.

Classical Swine Fever Virus (CSFV) and Porcine Reproductive and Respiratory Syndrome Virus (PRRSV)

In the swine farming industry, two commonly seen swine diseases, classical swine fever and porcine reproductive and respiratory syndrome, can lead to economically significant loss, as seen in reduced pork production and decreased piglet population. These two diseases are highly contagious and have high mortality rates. They are among the most serious forms of porcine illnesses known, and have caused devastation to the economy in countries affected by these diseases. Rapid and sensitive diagnostics, such as NASBA, would allow early detection of these diseases to minimize spreading.

Detection of NASBA amplicons by Enzyme-linked Oligonucleotide Capture (EOC)

NASBA with ECL detection, being highly sensitive, specific, and accurate for nucleic acid amplification, allows rapid diagnosis of viral infections in animals and humans. However, the high cost of the equipment to perform electrochemiluminescent (ECL) detection may make this method less affordable to laboratories with a limited testing budget. In an attempt to enhance the applicability of NASBA technology, an enzyme-linked oligonucleotide capture (EOC) detection method making use of standard ELISA plate readers has been developed. The EOC detection procedure requires no extra equipment investment and would make the highly sensitive NASBA method an affordable technology in many laboratories. In the EOC detection method, NASBA amplicons are immobilized by hybridization to a biotinylated oligonucleotide capture probe bound to a streptavidin-coated surface. Detection is then mediated by a digoxigenin (DIG)-labelled detection probe and an anti-DIG antibody-alkaline phosphatase conjugate, which provides the colorimetric end-product.

The use of common 96-well microtitre plate spectrophotometers allows relatively inexpensive and high throughput analyses of NASBA amplicons, thus providing a viable option for highly sensitive and specific molecular diagnostics without the need to bear high machine costs as seen in real-time PCR equipment. Table 3 shows the equivalent magnitude of detection exhibited by NASBA/ECL and NASBA/EOC methods using primers/probes specific for all FMDV serotypes on FMDV field samples. NASBA/EOC will allow a more versatile application of this highly sensitive detection assay in locations where direct testing of field samples is crucial to monitor livestock crossing country borders, since only simple equipment, such as water bath and ELISA plate reader, are required.

Ease of Standardizing NASBA over PCR Technology

NASBA is a continuous, isothermal (usually at 41°C) and enzyme-based method [2]. NASBA experiments require equipment as simple as a water bath, and there is no need for expensive thermocyclers, which become obsolete when newer models are marketed. Without specialized equipment, the entire NASBA reaction can be performed with a water bath at one temperature, regardless of laboratories, regions, or even laboratory technicians. Because no specialized machine is required for the NASBA, developing countries and budget-restricted laboratories can afford to include NASBA-based molecular diagnosis among their routine disease surveillance detection systems. NASBA is able to deliver accurate, consistent, and reliable results with high sensitivity. NASBA may equal or even surpass real-time PCR in sensitivity; in our numerous experiments with the NASBA method, NASBA has consistently demonstrated a sensitivity level exceeding that of real-time PCR in detecting avian influenza A virus (Table 4). Moreover, because of the isothermal nature, NASBA methods can be standardized easily. On the contrary, although real-time PCR is currently the commonly used molecular method in the detection of various viruses, the amplification phase involves stepwise changes in temperatures. Different brands and types of thermocyclers (laser or white lamp based), primers, probes (various fluorescent labels and quenchers), *Taq* DNA polymerase ("hot-start" or "cold-start"), different reaction additives (passive reference), and different real-time detection chemistry are commonly seen in real-time PCR. Temperature transition rates to achieve denaturation, annealing, and extension also depend on the machines used. All these factors contribute to variations of real-time

Table 3: Applicability of NASBA/EOC method to field samples of FMDV compared to NASBA/ECL method.

Sample number[a]	ECL result	EOC result
400	Positive	Positive
409	Positive	Positive
415	Positive	Positive
418	Positive	Positive
420	Positive	Positive
423	Negative	Negative
426	Negative	Negative
428	Positive	Positive
445	Positive	Positive
461	Negative	Negative
486	Positive	Positive
488	Positive	Positive
Positive control	Positive	Positive
Negative control	Negative	Negative

[a] FMDV RNA field samples (RNA samples obtained from the Institute of Animal Health, UK as part of the collaboration to evaluate the FMDV NASBA method) were subjected to NASBA amplification using standard NASBA protocol. The NASBA amplicon samples were subsequently detected by the ECL or EOC methods. Signals higher than the cut-off values were considered "Positive" whereas those below the cut-off values were considered "Negative". Cut-off value is determined by the mean value of signals obtained from negative controls ± k*S.D. where k is a constant.

Table 4: Comparison of NASBA and real-time PCR in detecting avian influenza A virus (AIV) (all subtypes).

Sample[#]	AIV NASBA	Taqman real-time PCR
10^{-1}	Positive	Positive
10^{-2}	Positive	Positive
10^{-3}	Positive	Positive
10^{-4}	Positive	Positive
10^{-5}	Positive	Negative
10^{-6}	Negative	Negative

[#] An AIV sample of known viral titre was serially diluted 10-fold and extraction was carried out to isolate the AIV RNA. RNA samples were subjected to AIV NASBA [8] and AIV real-time PCR [using a TaqMan real-time PCR method].

PCR at inter-laboratory and intra-laboratory levels. In the event of a disease outbreak, there is the possibility of inconsistent or even contradictory results due to the intrinsic differences in real-time PCR settings, thus impeding testing speed and timely quarantine or treatment actions when tested samples constantly require confirmation by other laboratories. Sample confirmation and inter-laboratory comparisons have been particularly seen recently with frequent H5N1 outbreaks in Southeast Asian regions. For real-time PCR-based nucleic acid testing methods, variations in reaction temperatures, signal-generating chemistry and detection technologies, and use of different reagents, will render standardization of PCR conditions almost unfeasible.

One unique and valuable feature of NASBA is its ability to resist interference from inhibitory substances commonly present in biological samples, especially faeces. Our in-house evaluation has demonstrated that RNA extracted from positive viral material spiked into faeces-containing samples can be detected by NASBA, but not real-time PCR [unpublished data]. Our evaluation indicates that NASBA technology is highly applicable in clinical and veterinary settings when faecal materials are among one of the most accessible types of samples.

The control of the spreading of diseases in Asia has been a difficult task, as public healthcare systems and policies vary among Asian countries and the detection technology is far beyond the current needs [19]. The World Health Organization, Food and Agriculture Organization of the United Nations (FAO), and the Office International des Epizooties (OIE) have been pleading the need for the surveillance, early warning, detection, and reporting of diseases. Despite the wide recognition of real-time PCR technology, the high machine cost usually prohibits its use for nation-wide surveillance and front-line screening. Instead of being the method of first choice, real-time PCR becomes restricted to certain central or reference laboratories and only serves as an alternative method. Also, the intrinsic difficulties in standardizing real-time PCR methods will render this technology less likely to serve the purpose of unified surveillance programmes. NASBA-based technology can provide a good option in controlling and fighting against disease outbreak.

Conclusions

Molecular diagnostic tools for virus detection have been widely developed and used for human diseases. While these techniques are usually less readily available for animals, pioneering research and development in animal virus detection would contribute tremendously to our knowledge and monitoring of modern infectious diseases caused by animal viruses. NASBA, because of its sensitivity, specificity, accuracy, and rapid availability of results, would be a good method of choice for disease surveillance and control.

References

1 Fung YW, Lau LT, Yu AC: The Necessity of Molecular Diagnostics for Avian Flu. Nat Biotechnol 2004;22:267.

2 Compton J: Nucleic Acid Sequence Based Amplification. Nature 1991;350:91-92.

3 Romano JW, van Gemen B, Kievits T: NASBA: a novel, isothermal detection technology for qualitative and quantitative HIV- RNA measurements. Clin Lab Med 1996;16:89-103.

4 Boom R, Sol CJ, Salimans MM, Jansen CL, Wertheim-van Dillen PM, van der Noordaa J: Rapid and Simple Method for Purification of Nucleic Acids. J Clin Microbiol 1990;28:495-503.

5 Blok MJ, Lautenschlager I, Christiaans MHL, van Hooff JP, Goossens VJ, Middeldorp JM, Sillekens P, Ramon A, Hockerstedt K, Bruggeman CA: Nucleic acid sequence based amplification: a new technique for monitoring cytomegalovirus infection in transplant recipients. Transplant Proc 1999;31:308-309.

6 Damen M, Sillekns P, Cuypers THM, Frantzen I, Melser R: Characterization of the quantitative HCV NASBA assay. J Virol Meth 1999;82:45-54.

7 Smits HL, van Gemen B, Schukkink R, van der Velden J, Tjong-A-Hung SP, Jebbink MF, ter Schegget J: Application of the NASBA nucleic acid amplification method in the detection of human papillomavirus typr 16 E6-E7 transcripts. J Virol Meth 1995;54:75-81.

8 Lau LT, Banks J, Aherne R, Brown IH, Dillon N, Collins RA, Chan KY, Fung YW, Xing J, Yu AC: Nucleic acid sequence-based amplification methods to detect avian influenza virus. Biochem Biophys Res Commun 2004; 313:336-342.

9 Collins RA, Ko LS, So KL, Ellis T, Lau LT, Yu AC: Detection of highly pathogenic and low pathogenic avian influenza subtype H5 (Eurasian lineage) using NASBA. J Virol Methods 2002;103:213-225.

10 Collins RA, Ko LS, Fung KY, Chan KY, Xing J, Lau LT, Yu AC: Rapid and sensitive detection of avian influenza virus subtype H7 using NASBA. Biochem Biophys Res Comm 2003;300:507-515.

11 Collins RA, Ko LS, Fung KY, Lau LT, Xing J, Yu ACH: A Method to Detect Major Serotypes of Foot-and-Mouth Disease Virus. Biochem Biophys Res Comm 2002;297:267-274.

12 Shan S, Ko LS, Collins RA, Wu ZL, Chen JH, Chan KY, Xing J, Lau LT, Yu ACH: Comparison of Nucleic Acid-Based Detection of Avian Influenza H5N1 with Virus Isolation. Biochem Biophys Res Comm 2003;302:377-383.

13 Kox NJ, Subbarao K: Influenza. Lancet 1999;354:1277-1282.

14 Swayne DE, Beck JR, Perdue ML, Beard CW: Efficacy of Vaccines in Chickens against Highly Pathogenic Hong Kong H5N1 Avian Influenza. Avian Dis 2001;45:355-365.

15 Burrows R: The Infectivity Assay of Foot-and-Mouth Disease Virus in Pigs. J Hyg (Lond) 1966;64:419-429.

16 Salt JS, Samuel AR, Kitching RP: Antigenic Analysis of Type O Foot-and-Mouth Disease Virus in the Persistently Infected Bovine. Arch Virol 1996;141:1407-1421.

17 Alexander DJ: Newcastle disease and other avian Paramyxoviridae infections, in Calnek BW, Barnes HJ, Beard CW, McDougald LR, Saif YM (eds): Disease of Poultry, 10th edition. Ames IA, Iowa State University Press, 1997, pp 541-569.

18 Office International des Epizootes (OIE): Manual for Animal Disease Reporting to the OIE. World Organization for Animal Health, Paris, 1996.

19 Writing Committee of the World Health Organization (WHO) Consultation on Human Influenza A/H5: Avian Influenza A (H5N1) Infection in Humans. New Engl J Med 2005;353:1374-1385.

Prof. A.C.H. Yu, Neuroscience Research Institute, 38 Xue Yuan Road, Peking University, Beijing 1000083, China.
E-mail: achy@hkdnachips.com; achy@bjmu.edu.cn

Vannier P, Espeseth D (eds): New Diagnostic Technology: Applications in Animal Health and Biologics Controls. Dev Biol (Basel). Basel, Karger, 2006, vol 126, pp 17-21.

Introduction to DNA Micro-Arrays

C. Campbell

Scottish Centre for Genomic Technology and Informatics, University of Edinburgh, Edinburgh, U.K.

INTRODUCTION

DNA micro-array experimental procedures are redefining vast areas of experimental science. DNA micro-arrays function on the basis of molecular recognition between DNA molecules immobilised on a solid surface (probes) and their binding partners (targets) in solution. The fact that this recognition is specific and has high affinity means that several probes can simultaneously recognise their targets from complex mixtures in solution.

While micro-arrays can be created from virtually any type of molecule ranging from small molecules to cells, the most commonly used types are protein and DNA micro-arrays. The idea of immobilising probe molecules to detect targets can be traced back to immunoassay experiments of the 1960s; however this technology is improved greatly by creating a micro-array of such probes since this technique means not only that several experiments can be run in parallel, but also that detection limits are lowered and binding kinetics improved [1]. Similarly, DNA micro-array technology evolved from the use of solid substrates in performing Southern Blot experiments and the multiplexing of this technology has found greater use to date, (mostly in laboratory research) [2], than corresponding protein chips. Current technology allows the immobilisation of over a million DNA probes on an array, for example Affymetrix have developed an array with 1,300,000 spots for measuring gene expression across the entire human genome [3]. Such high density arrays are not always necessary though, since a more focussed approach might interrogate a subset of genes of known interest, requiring the fabrication of a lower density custom array.

This review will focus mainly on the use of DNA micro-array technology for gene expression profiling but there are several published reviews on single nucleotide polymorphism discrimination [4] and protein micro-array technology that might be of help to the interested reader [5-8].

Since its advent, micro-array technology has shown itself to be a powerful tool in the probing of gene activity and system responses to disease. DNA micro-arrays generally contain hundreds or thousands of gene probes which are simultaneously exposed to a target sample. The gene probes are generally selected from cDNA fragments derived by polymerase chain reaction (PCR) or from synthetic oligonucleotides which can either be synthesised and robotically spotted on to a solid support or synthesised in situ using standard photolithographic techniques. Gene-probes can be attached to the solid support by a variety of techniques ranging from physisorption to covalent attachment. Physisorption is procedurally simple and works by pairing areas of complementary charge on the DNA and a charged surface such as poly-l-lysine. Covalent attachment is a slightly more involved process, but the chemical nature of the bond formed imparts a higher stability to the attachment, leading to less signal loss due to leaching. The covalent bond formed might, for example, be between a functional silane surface and a functional group attached to one end of the DNA molecule. An example of a micro-array created using standard photolithographic techniques is shown in Figure 1.

Figure 1 shows a micro-array on which 400,000 gene probes have been immobilised, representing 12,000 human genes. The inset shows 16 gene probes which represent a single gene.

DNA micro-arrays can be used to assay two classes of biomolecule: RNA and DNA. RNA is used for gene expression analysis (transcription profiling) and DNA is used for genotyping. Assaying RNA requires the extraction of RNA from tissue culture or biopsy samples, usually using a commercial RNA extraction kit. It is important to establish the quality of the RNA (by, for example, gel electrophoresis) before starting a micro-array experiment, since a degraded sample may not give reliable results. The RNA transcripts are assayed by converting them enzymatically (via reverse transcription) to DNA. Since this reaction is a "test-tube" process, labelled bases can be incorporated into the synthetic DNA.

Labelling is generally either direct or indirect. Directly labelled DNA usually has a fluorescent dye covalently bound to it. Indirectly labelled DNA generally has a small molecule bound to it that can be recognised by (for example) a fluorescently-labelled antibody in a process analogous to a "sandwich immuno-assay". A variety of labels are available including organic dyes (for fluorescent detection), metal nanoparticles (for resonant light scatter), semiconductor nanoparticles (Quantum Dots for enhanced fluorescent detection) or redox molecules such as ferrocene (for electrochemical detection).

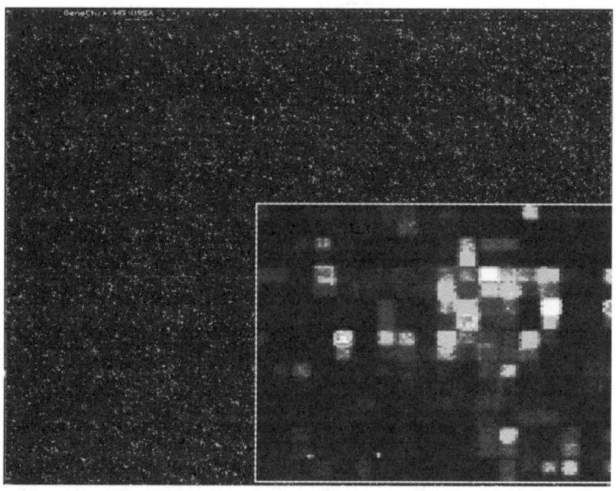

Fig. 1

At this stage in the experiment, it is also possible to incorporate either a linear or non-linear amplification step to improve the sensitivity of detection. Amplification can however sometime skew results and should be used judiciously in a well controlled manner. The labelled DNA is then hybridised to the array of probes and washed to ensure that any labelled DNA that has bound is specifically attached and not (for example) a partial match or non-specifically bound to the underlying array surface.

Through the use of different dyes, multiple samples can be assayed on the same array. For example, an RNA sample from an infected tissue sample may be labelled with cy3, while a sample from healthy tissue is labelled with cy5. These can then be hybridised on the same array and the fluorescence read separately. Such experiments require stringent controls to ensure that any bias, due to dye incorporation differences, is not allowed to skew the data.

To "read" the array, a commercial micro-array scanner is usually used. These range from dedicated instruments which can measure the intensity of only two common dyes, to more versatile instruments which interrogate dyes across the entire visible spectrum. Almost all use confocal optics in combination with either white-light or laser sources and a variety of filters. To obtain meaningful data at this stage, the scanner settings have to be carefully chosen to minimise the number of array features (spots) showing saturated fluorescence and maximise the linear range of detection. Suitable standards have been detailed by Forster et al [9].

At this point you have a picture of a micro-array with up to 1.5 million spots of differing intensity each corresponding to either part of a gene of interest or control spots which can be used for data normalisation. Clearly these data cannot be analysed manually. Significantly, bio-informatic techniques have grown with micro-array technology to enable the meaningful interpretation of micro-array data. Specialised image analysis software recognises which spot is which and calculates its fluorescence; bio-informatic tools can then be used to interrogate and analyse the data for significance. This may involve identifying genes which behave similarly and clustering them into groups. Bio-informatic tools also enable comparison of gene expression between experiments carried out at different times and in different laboratories. However, care must be taken when making such comparisons and the researcher should ensure that all data have been generated with sufficient levels of standardisation and comply with international standards on the Minimum Information about a Micro-array Experiment (MIAME) [10].

While the technical aspects of an array experiment can be challenging and sometimes expensive, it is worth noting that the greatest value is always gained when an experiment is correctly designed in advance of any technical work. Correct experimental design should eventually give data that address the biological question being posed, and have statistical robustness which makes the ultimate analysis of such data much more meaningful [11,12]. Some of the important aspects of experimental design are outlined in Table 1.

Table 1: Experimental issues and challenges.

Sample size	Minimisation of print variability
Statistical design and modelling	Standardisation of sample handling
Multiple testing	Appropriate QC steps
Statistical significance	Standardisation of data handling

Applications

Micro-arrays have found many useful applications, primarily in research. As the technology translates from the laboratory to the clinic there will be an increasingly broad acceptance of micro-arrays as medical diagnostic tools.

In the field of infectious disease, Wilson et al describe a short-oligonucleotide-based approach to the sequence specific identification of 18 pathogens including *Bacillus anthracis, Vibrio cholorae*, and Ebola [13]. Similarly, Wang et al have described a long-oligonucleotide micro-array capable of detecting over 140 respiratory viruses [14]. The authors have validated this technology by using clinical specimens isolated from the human respiratory tract. This extension of micro-array technology clearly shows its clinical use at the point-of-need. The application of micro-array technology in such fields requires high degrees of protocol standardisation and there is a growing movement towards the incorporation of several procedures on to a common platform, thus deskilling and automating the experimental process.

The application of this technology need not be limited to a clinical environment; its applicability is also clear in counter-terrorism applications for detecting viral or bacterial agents. For example, Read et al have used pathogen micro-arrays to determine the origin and mode of anthrax infection in a recent outbreak [15]. Genome-based analysis has thus shown itself to be a useful tool in diagnosing whether an outbreak of infectious disease comes from a bio-warfare source or a more familiar source of infection. This "microbial forensics" approach also has potential use in a clinical setting where outbreaks of infection are becoming an increasingly common complication in hospitals.

The above application examples directly interrogate the genome of infective agents. An interesting variation on the theme is to probe changes in host-gene expression as a response to disease. Host-gene chips have been used as a predictive tool in cancer profiling. Van de Vijver et al used a micro-array format based on a 70-gene prognosis profile to classify a sample of patients with primary breast cancer into those with gene expression signatures characteristic of a good prognosis and those with characteristics of a poor prognosis [16]. This characterisation of patients into low- and high-risk subgroups provides a clinically useful guide by informing decisions regarding the appropriate therapy. This should minimize the risk to the patient from either under-treatment or over-treatment. It is clear that future studies will aim to identify infectious diseases by discrimination of host gene expression signatures in addition to the established method of probing the pathogen genome.

In conclusion, it can be seen that a correctly performed micro-array experiment can give a wealth of data, which the able experimentalist can translate into valuable information. Because of the scale of information generated, the analysis and management of these data sometimes limit micro-array techniques but bio-informatics tools play an increasing role in aiding this process.

Only with the proper management of information and databases can the full potential of micro-array technology be realised. With the correct choice of experimental variables, analysis of the generated data will yield statistically significant results. Micro-arrays have already shown their worth in research and will increasingly find applications in clinical medicine.

REFERENCES

1 Ekins R, Chu FW: Micro-arrays: their origins and Applications. Tribtech 1999;17:217-218.

2 Epstein JR, Epstein JR, Biran I, Walt DR: Fluorescence-based nucleic acid detection and micro-arrays. Anal Chim Acta 2002;469:3-36.

3 www.affymetrix.com.

4 Kwok P-Y: Methods for Genotyping Single Nucleotide Polymorphisms. Annu Rev. Genomics Hum Genet 2001;2:235-258.

5 Melton L: Protein arrays: Proteomics in multiplex. Nature 2004;429:101-107.

6 Tyers M, Mann M: From genomics to proteomics. Nature 2003;422:193-197.

7 Phizicky E, Bastiaens PIH, Zhu H, Snyder M, Fields S: Protein analysis on a proteomic scale. Nature 2003;422:208-215.

8 Liotta LA, Espina V: Encyclopaedia of Diagnostic Gemomics and Proteomics, Marcel Dekker, 2005.

9 Forster T, Roy D, Ghazal P: Experiments using micro-array technology: limitations and standard operating procedures. J. Endocrinol. 2003;178:195-204.

10 www.mged.org.

11 Kerr K, Churchill G: Statistical design and the analysis of gene expression micro-array data. Genetics Research 2001;77:123-128.

12 Churchill G: Fundamentals of experimental design for cDNA micro-arrays. Nat Gen Suppl 2002;32:490-495.

13 Wilson WJ, Strout CL, DeSantis TZ, Stilwell JL, Carrano AV, Andersen GL: Sequence-specific identification of 18 pathogenic microorganisms using micro-array technology. Molecular and Cellular Probes 2002;16:119-127.

14 Wang D, Coscoy L, Zylberberg M, Avila PC, Boushey HA, Ganem D, DeRisi JL: Micro-array-based detection and genotyping of viral pathogens. Proc Natl Acad Sci 2002;99:15687-15692.

15 Read TD, Salzberg SL, Pop M, Shumway M, Umayam L, Jiang LX, Holtzapple E, Busch JD, Smith KL, Schupp JM, Solomon D, Keim P. Fraser CM: Comparative Genome Sequencing for Discovery of Novel Polymorphisms in Bacillus anthracis. Science 2002;296:2028-2033.

16 van de Vijver MJ, He YD, van 't Veer LJ, Dai H, Hart AAM, Voskuil DW, Schreiber GJ, Peterse JL, Roberts C, Marton MJ, Parrish M, Atsma D, Witteveen A, Glas A, Delahaye L, van der Velde T, Bartelink H, Rodenhuis S, Rutgers ET, Friend SH Bernards R: A gene-expression signature as a predictor of survival in breast cancer. N Engl J Med 2002;347:1999-2009.

Dr. C. Campbell, Scottish Centre for Genomic Technology and Informatics, University of Edinburgh, 49 Little France Crescent, Edinburgh, EH25 9RG, U.K.
E-mail: colin.campbell@ed.ac.uk

Vannier P, Espeseth D (eds): New Diagnostic Technology: Applications in Animal Health and Biologics Controls. Dev Biol (Basel). Basel, Karger, 2006, vol 126, pp 23-30.

High Throughput Sequencing and Comparative Genomics of Foot-and-Mouth Disease Virus

C. Carrillo, E.R. Tulman, G. Delhon, Z. Lu, A. Carreno, A. Vagnozzi, G.F. Kutish, D.L. Rock

Plum Island Animal Disease Center, Agricultural Research Service, United States Department of Agriculture, Greenport, NY, USA

Abstract: Despite a basic understanding of many aspects of FMD biology, much information regarding FMDV virulence, host range, and virus transmission remains poorly understood. Here we present how the use of high throughput sequencing for complete genome sequences of foot-and mouth disease virus (FMDV) led to a series of new insights into viral genome sequence conservation and variability, genetic diversity in nature and phylogenetic classification of isolates, including the first complete sequences of the South African Territories type 1 and 3 (SAT1 and SAT3) genomes. Comparative genomic analysis of full-length sequences of FMDV isolates did allow: (i) the identification of highly conserved regulatory or coding regions which are critical for aspects of virus biology as well as novel viral genomic motifs with likely biological relevance; (ii) characterization of the first complete sequences of the SAT1 and SAT3 genomes; (iii) identification of a novel SAT virus lineage genetically distinct from other SAT and Euro-Asiatic lineages; (iv) precise identification of strains circulating around the world for epidemiological and forensic attribution; (v) assessment of mutation and recombination processes as mechanisms equally involved in evolution; (vi) mutation rates, tolerance and constraints of genes and proteins during evolution of FMD viruses during in vivo replication and (vi) support for the hypothesis of a new evolutionary model.

INTRODUCTION

Foot-and-mouth disease (FMD) is a highly contagious disease that affects domestic and wild cloven-hoofed animals [1]. Only certain geographical regions are free of the disease, but still FMD is endemic in many countries of South and Central America, Asia, the Middle East and Africa, posing a serious risk of dissemination. Despite the success of conventional inactivated vaccines in the eradication and control of FMD in several countries, the lack of protection between different serotypes and the presence of the virus in such extended territories constitutes a permanent hazard. As an alternative to the conventional inactivated vaccines, the use of attenuated antigenically marked virus capable of inducing a solid and durable immunity through replication in the animal is highly desirable. However, there is not enough information on the tolerance and constraints of FMDV genes and proteins to manipulate an infectious clone safely and achieve its attenuation without compromising its biology and/or immune response in the natural host. Despite a basic understanding of many aspects of FMD biology, much information regarding FMD virulence, host range, and virus transmission remains poorly understood. The aetiological agent, FMD virus (FMDV), is an Aphthovirus in the *Picornavirus* family [2]. The non-enveloped virus contains a single molecule of single-stranded positive sense, polyadenylated RNA. The 8.5 kb genome consists of a single ORF flanked by two untranslated regions (5'- and 3'- UTRs) that codes for a polyprotein 2,300 to 2,350 amino acids long, which will be cleaved and processed during and after translation by viral proteases to yield a number of intermediates plus the final 12 mature viral polypeptides.

Complete genome sequence analysis of viruses has proved effective and accessible for several micro-organisms. High throughput DNA sequencing has accelerated the availability of information in recent years, but the amount of data available related to animal diseases is still rather limited. In most cases, restricted areas of the viral genome are extensively characterized for routine epidemiological purposes (mostly VP1 and other capsid proteins for FMDV). Insights into viral genome sequence conservation/variation and genetic diversity in nature will probably have an impact on our understanding of FMDV infection, host range, and transmission. There is a pressing need to identify all informative areas of the FMDV genome to improve routine epidemiological analysis and meet the new demands of forensic microbiology. Development of spatial epidemic models to simulate transmission or to assess biosecurity planning and emergency-response preparedness requires better knowledge of FMDV evolution. Prevention and minimization of the impact of foreign infectious diseases on animal health demands the development of new, more powerful molecular techniques of rapid diagnostic methods. Here we present and summarize the advances obtained in the knowledge of FMDV biology, epidemiology and evolution through the use of a rapid full-length sequencing method and computer analysis of more than 150 FMDV genomes from a variety of origins.

Materials and Methods

FMDV isolates

One hundred and three FMDV isolates were obtained from the Animal Plant Health Inspection Service (APHIS) as described [3] representing a broad temporal and geographical distribution. For evolution and transmission experiments, O/Taiwan/97/FMDV was used to start both in vivo and in vitro transmission chains, as described elsewhere [4].

Sample preparation

FMDV RNA isolation, RT-PCR amplification, sequencing and animal experiments are extensively explained elsewhere [3,4].

Sequence analysis

Sequence assembly, multiple alignments and consensus master sequences were performed as described elsewhere [3,4]. Here we include analysis, both at the nucleic acid and protein levels, obtained by using a number of computer programs to calculate substitution rate ratios, statistical evaluation of selective pressures, prediction of secondary structures and a variety of phylogenetic studies and results obtained from BLASTS and motif searches in PROSITE and GenBank databases [3,4].

Results and Discussion

Identification of highly conserved regulatory or coding regions

The 5'-UTRs are organized into a small (S) and a long (L) fragment. Our results show that SAT and Euro-Asiatic strains have consistent and significant differences in the size of the S fragment of the 5'-UTR. This genomic region, not previously compared between different isolates, has been related to host-specificity [2]. Despite surprising variation in nt sequence, the predicted stem loop secondary structure was conserved in 84% of the isolates. We describe for the first time the presence of cluster-specific indels and oligonucleotide motifs [3] in the L fragment of 5'-UTR present in 100% of the isolates. In a region where only 47% of the bases are invariant, conservation of these motifs suggests biological significance. The 3'-UTR, probably involved in the binding of viral and host proteins, showed for the first time the presence of a putative motif sequence invariant in all isolates. Overall, 46% of the nt were invariant, although the least similar pair of isolates showed 73% nt identity. Surprisingly, coding regions 2B and 3C were the most conserved; while VP1 (but none of the other structural proteins), L^{pro}, 3A and 3B exhibited similar values of variability and undergo diversifying selection. We confirmed and delimited for the first time the range of sequences recognized as cleavage sites between the viral proteins, and novel, highly conserved sequence motifs with probable functional significance were identified within proteins L^{pro}, VP2, VP1, and 3C. The data reveal novel highly conserved genomic regions, indicating functional constraints for variability as well as novel viral genomic motifs with probable biological significance [3].

Characterization of serotype-specific signatures

Current vaccines have good short-term coverage against homologous strains, but are inferior to the protection levels achieved by animals that have recovered from natural infection. Moreover, the antigenic variation of the virus imposes the formulation of local strain vaccines and the constant revision of the isolates included in the vaccine. Serotypic classification of FMDV into seven serotypes was the result of complement fixation cross-reaction tests, which reasonably correlated with lack of cross-neutralisation in vitro and cross-protection in vivo [1,2,5]. Despite the thousands of VP1 sequences available, the efforts to correlate the very well known neutralisation domains of VP1 with phylogenetic lineage and sequence data have been unsuccessful and the molecular determinants of serotypes remain unknown. The molecular characterization of serotypes has obvious applications in vaccine selection and epidemiological control of the disease. The publishing and GenBank annotation of the first 13 complete sequences of the SAT1 and SAT3 serotypes allowed recognition of serotype-specific sequences in the UTRs, including IRES, L^{pro}/capsid, 1C/1D, 3B/3C and 3C/3D cleavage sequences, 2C, 3C and structural proteins. Most remarkable was that adding new sequences to the only SAT2 sequence available in GenBank revealed unusual features of that sequence, and a new prototype sequence for SAT2 serotype [3].

Identification of a novel SAT virus lineage

Phylogenetic analysis of complete FMDV genomes indicates that sequences are monophyletic relative to serotype classification, close to the shape of cladograms based solely on structural proteins. However, when UTRs and non-structural proteins are run separately, a clear bipolar distribution of Euro-Asiatic and SAT clades results, plus a third lineage, not previously described. This lineage has the longest sequence described, which includes sequences that are specific from other SAT and also those specific of Euro-Asiatic lineages. This mosaic of SAT and Euro-Asiatic sequence elements was observed all along the complete genome and confirmed by the phylogenetic and recombination analyses performed.

Identification of strains circulating around the world for epidemiological and forensic attribution

Complete or partial VP1 sequencing has been the hallmark of molecular epidemiology for the last 30 years. With the phylogenetic reconstruction of isolate relationships, we can define with absolute certainty that the closer the geographical origin of two isolates, the closer the genetic relationships, and vice versa [5,6]. Dissimilar genotypes (topotypes) found in the same geographical region are interpreted as the introduction of the strain from other location. However, the latest analysis enlarging the compared sequences from partial to the whole VP1 coding region resulted in a new topotype distribution [7]. Complete P1 sequences of the same isolates suggested recombination [7]. These data indicate that enlarging the regions compared increases the information gained. Full-length genomic analysis, allowed the establishment of the first map of invariant and highly conserved nucleotides and amino acid residues for FMDV. The probability of a given mutation can now be estimated, proving to be an invaluable forensic tool for identifying unnaturally altered viruses and specific signatures in the field [8]. For instance, it could not be determined whether the complete P1 sequence of the FMDV O-PI/04 isolate was derived from the O Brugge or the O Campos lineage. Based on estimated probability of change for specific nucleotide positions between the three isolates, and the kind of substitution incurred, the O-PI/04 was clearly attributed to a passage of the O Brugge strain. Interestingly, highly variable positions within capsid coding region, were not informative enough, while positions from non-structural proteins (2B, 2C and 3D) showed key differences that proved useful in identifying the parental strain. In another instance, the nature of contamination of two samples was deduced. The V2799 isolate presumed to contain SAT 1 Isrl strain, was actually shown to include two additional strains, one 97% similar SAT1/6 SWA (V2338 3 IBRS), and one 99% similar to A Argentina 2001 (V2754 BHK p55). In other cases, the vial was shown to contain one 100% SAT1-Rhodesia 5/66 (V0331, 1 BTT) and another 99% similar to SAT1/6 SWA (V2338 3 IBRS). The high throughput full-length genome sequence resolved the sources of contamination without having to grow the isolates in vitro for plaque purifying and sequencing a representative number of plaques.

Recombination

Adjoining nucleotide sequence phylogenies were constructed separately for UTRs, structural and non-structural coding regions, P1, P2 and P3 polyproteins and for each of the mature protein-encoding regions. Phylogenies constructed with the structural (P1) and non-structural regions of the genome (P2 and P3) showed significantly incongruent results. The study of collateral relationships between isolates using split decomposition analysis indicated high probability of recombination for a number of isolates. Similarity plots and bootscanning analysis across the complete genome identified multiple sites as potential recombination points, distributed preferably throughout the non-structural protein coding regions L^{pro}, P2 and P3, but not within the capsid coding region. Bootscanning analysis identified peaks in the capsid region of possible recombination sites, but it is difficult to distinguish these from functional conservation, common ancestors or convergent evolution. Since it is probable that nucleotide substitution is the dominant evolutionary mechanism of FMDV, evidences of recombination are lost by the rapid accumulation of nucleotide substitutions. On the other hand, the remarkable conservation of VP4 aa and nt sequences between all serotypes may serve as a recombination site at the 5' end of the capsid polyprotein (P1). The importance of the recombination in FMDV evolution and epidemiology in natural samples has been described through the comparison of a significant number of full length sequences. In some cases identification of a parental virus was possible [3], but the most frequent situation was that individual sequences result in complex mosaics of sequences suggestive of multiple recombination events. It is difficult to reach a conclusion since only a small fraction of all circulating FMDV is sequenced; thus we lack the information about all the intermediates between every two possible recombinants.

Assessment of mutation and recombination processes of FMD viruses during natural host replication

RNA viruses evolve rapidly as populations of genetically distinct but closely related variants, termed viral quasi-species, recognized as highly variable genetic and phenotypic mixtures that allow for the emergence of novel immune escape and pathogenic variants [9]. The high potential for variation of RNA viruses, in particular of FMDV, during its replication in cell culture (in vitro) has been extensively reviewed. However, very little is known regarding the effect of serial passage of virus in the natural host (in vivo) [4,9-13]. We used the rapid full-length sequencing method to analyse 20 serial by-contact transmissions of the highly virulent, pig-adapted, O/Taiwan/97 isolate of FMD using experimental groups of two non-immune pigs each, carried out in isolation units. We examined the continuing changes in FMDV master sequences during replication in a total of 37 full-length FMDV genomes from the serially infected pigs. These were compared with 18 genomes from the same O/Taiwan/97, along 23 passages in BHK-21 cells. Results show that intradermal injection of high titre viral suspension ($2.6x10^6$ TCID50) lead to conservation of the master sequence. Genomic variants were detected starting after the first by-contact transmission. Values of in vivo nucleotide substitution rates were similar to in vitro rates ($1.2x10^{-4}$ to $2.6x10^{-3}$ substitutions per nucleotide). However, given the theoretical differences in the number of replicative cycles in vivo and in vitro, up to two log differences in the mutation rate are found between the animal and the tissue culture, probably due to a higher purifying selection in the natural host [14]. The most remarkable differences between in vivo and in vitro samples were the distribution of mutations along the genome and the transition/transversion and synonymous/non-synonymous rate ratios, indicating the activity of different selective pressures. Of particular significance was the complete absence of changes in the region encoding the hyper-variable capsid protein 1D (VP1) in the pig samples. Interestingly, a significant reduction in susceptibility to infection of the pigs was observed along with the serial passages, and interruption of by-contact chain transmission occurred after passages 14, 18 and 20. However, contact transmission capability was transitorily recovered when vesicular fluid of the last infected pig was injected into naïve pigs. This is the first report of an experimental analysis of the mechanisms of transmission animal-to-animal and its evolution. The results revealed that contact transmission of O/Tw97/FMDV in pigs resembles the bottleneck effect and fitness loss observed during serial virus passages from plaque to plaque in vitro [15], described as Muller's ratchet genetic effect. We propose that other genetic mechanisms, such as recombination and phenotypic masking, might be more important than previously thought in the perpetuation of foot-and-mouth virus and disease in nature. In fact, recombination may eliminate multiple accumulated deleterious mutations in a single step, restoring fitness and continued transmission. Thus, perpetuation, escape, and variability of FMDV might be the result of a functional cell receptor, framed in a robust capsid structure genome, randomly combined with the best fitness adapted non-structural sequences in place. This mechanism of recombination would restore the balance of fitness loss that the Muller's ratchet effect imposes as a consequence of the high error rate of the RNA-pol, which, in turn, would be the critical mechanism for the generation of wide-spectrum mutant viral populations, capable of overcoming the struggles and bottlenecks that replication in nature imposes. New in vivo experiments are needed to investigate the rate of recombination further during natural replication in the host and also during circulation of FMDV within a susceptible animal population. Moreover, future analyses based on the invariant residues and motifs described will help improve our knowledge of the genetic basis of virulence, tropism and host-range determinants.

Conclusions

FMDV full length sequencing and comparative genomics have made possible the following conclusions:

(i) a number of novel highly conserved genomic regions, indicating functional constraints for variability as well as novel viral genomic motifs with probable biological relevance;

(ii) molecular characterization of signatures for serotype identification has obvious application in vaccine selection and epidemiological control of the disease;

(iii) a "genotype" classification may be done based on non-structural coding sequences that defines a new lineage of FMDV not previously described;

(iv) attribution and forensic identification of isolates will be possible only if the database of full- length sequences is sensibly increased;

(v) in vivo described genetic variation, selective force for evolution, and mechanisms of transmission need to be better explored;

(vi) genetic recombination of FMDV deserves to be taken into consideration for the complete understanding of FMD epidemiology;

(vii) genomic variants are generated and rapidly overgrow the master population of naturally infected animals despite it fitness;

(viii) transmission events with O/TW/97/FMDV with in-contact pigs resemble the genetic effect described as Muller's ratchet, indicating transmission of very low numbers of infectious particles from one animal to another;

(ix) the rapid accumulation of deleterious mutations during serial transmission by contact in pigs lead the porcine-adapted hyper-virulent isolate O/TW/97/FMDV to extinction;

(x) while the in vitro Muller's ratchet effect led to aberrant indels to avoid extinction, no indels or any other remarkable genetic defects were detected in the sequences from in vivo samples. We propose that genetic constraints in vivo are significantly influencing the quasi-species composition.

Acknowledgements

We thank Dr. J. Lubroth and the Animal Plant Health Inspection Service (APHIS) of the United States Department of Agriculture for providing FMDV isolates and M. Prarat for providing editing assistance.

References

1 Grubman MJ, Baxt B: Foot-and-Mouth Disease. Clin Microbiol Rev 2004;17:465-493.

2 Mason PW, Grubman MJ, Baxt B:Molecular basis of pathogenesis of FMDV. Virus Res 2003;9:9-32.

3 Carrillo C, Tulman ER, Delhon G, Lu Z, Vagnozzi A, Cutis GF, Rock DL: Comparative genomics of foot-and-mouth disease virus. J Virol 2005;79:6487-6504.

4 Carrillo C, Lu Z, Borca MV, Delhon G, Tulman E, Vagnozzi A, Kutish G, Rock D: Genetic and Phenotypic variation of FMDV during serial passages in pigs. Sent to publication.

5 Knowles NJ, Samuel AR: Molecular epidemiology of foot-and-mouth disease virus. Virus Res 2003;91:65-80.

6 Martinez MA, Carrillo C, Plana J, Mascarella R, Bergada J, Palma EL, Domingo E, Sobrino F: Genetic and immunogenic variations among closely related isolates of foot-and-mouth disease virus. Gene 1988;62:75-84.

7 Knowles NJ, Davies P, Midgley R, Valarcher JF: Identification of a ninth foot-and-mouth disease virus type O topotype and evidence for a recombination event in its evolution. European Commission for the control of foot-and-mouth disease (EUFMD). Crete, Greece. 2004; Book of abstracts: 4.

8 Carrillo C, Rock DL: Molecular Epidemiology and Forensics of RNA Viruses, in Breeze RG, Budowle B, Schutzer SE (eds): Microbial Forensics. Elsevier Academic Press, 2005, pp 174-180.

9 Domingo E, Biebricher C, Holland JJ, Eigen M: Quasispecies and RNA virus evolution: principles and consequences. Landes Bioscience, Austin, Tex. 2000.

10 Carrillo C, Plana J, Mascarella R, Bergada J, Sobrino F: Genetic and phenotypic variability during replication of foot-and-mouth disease virus in swine. Virology 1990;179:890-892.

11 Carrillo C, Borca MV, Moore DM, Morgan DO, Sobrino F: In vivo analysis of the stability and fitness of variants recovered from foot-and-mouth disease virus quasispecies J Gen Virol 1998;79:1699-1706.

12 Martinez MA, Carrillo C, Gonzalez-Candelas F, Moya A, Domingo, Sobrino F: Fitness alteration of foot-and-mouth disease virus mutants: measurement of adaptability of viral quasispecies. J Virol 1991;65:3954-3957.

13 Hughes GJ, Mioulet V, Haydon DT, Kitching RP, Donaldson AI, Woolhouse ME: Serial passage of foot-and-mouth disease virus in sheep reveals declining levels of viraemia over time. J Gen Virol 2002;83:1907-1914.

14 Haydon DT, Samuel AR, Knowles NJ: The generation and persistence of genetic variation in foot-and-mouth disease virus. Prev Vet Med 2001;51:111-124.

15 Domingo E, Escarmis C, Lazaro E, Manrubia SC: Quasispecies dynamics and RNA virus extinction. Virus Res 2005;107:129-139.

Dr. C. Carrillo, Plum Island Animal Disease Center (ARS-USDA), PO Box 848, Greenport, N.Y. 11944, USA.
E-mail: ccarrillo@piadc.ars.usda.gov

Vannier P, Espeseth D (eds): New Diagnostic Technology: Applications in Animal Health and Biologics Controls. Dev Biol (Basel). Basel, Karger, 2006, vol 126, pp 31-39.

Detection of Viruses: Atomic Force Microscopy and Surface Enhanced Raman Spectroscopy

M.D. Porter[1], J.D. Driskell[1], K.M. Kwarta[1], R.J. Lipert[1], J.D. Neill[2], J.F. Ridpath[2]

[1] Iowa State University, Institute for Combinatorial Discovery, Departments of Chemistry and of Chemical and Biological Engineering, Ames Laboratory-U.S. DOE, Ames, IA, USA

[2] Virus and Prion Diseases of Livestock Unit, National Animal Disease Center, United States Department of Agriculture, Ames, IA, USA

Abstract: This paper demonstrates the capability of atomic force microscopy (AFM) and surface enhanced Raman spectroscopy (SERS) to function effectively as ultra-sensitive readout tools for chip-scale platforms designed for pathogen detection in complex biological media. AFM allows direct (i.e., label-free) visualization and quantification of nanometer-sized viruses captured on a smooth, selective surface. AFM readout led to optimization of a capture substrate for feline calicivirus (FCV), and yielded a limit of detection of 3 x 10^6 FCV/mL. SERS-based detection of FCV, carried out in a sandwich-type assay, requires labelling of the substrate-bound FCV with a selective extrinsic Raman label (ERL). These studies yielded a limit of detection of 1 x 10^6 FCV/mL. The prospects of these two readout methods as additions to the arsenal of tools in bioterrorism prevention are briefly discussed.

INTRODUCTION

Methods for low-level pathogen detection are vital to human healthcare, veterinary medicine, and bioterrorism prevention [1]. With respect to viral pathogen detection, there is a range of candidate techniques (e.g. electron microscopy, fluorescent antibody labelling of frozen tissue sections (FATS), ELISA, polymerase chain reaction (PCR), DNA hybridization, virus isolation, and serological testing). These methodologies, however, lack the sensitivity, specificity, speed, versatility, portability and/or throughput sought for facile deployment. For example, virus isolation and serological testing require lengthy sample preparation and PCR requires skilled technicians and is affected by trace levels of contamination. To address these needs, efforts to couple novel readout technologies to immunoassays, including atomic force microscopy (AFM) [2], scanning electron microscopy (SEM) [3], fluorescence-based labeling [4] and antibody-modified micro-cantilevers [5], are under development and performance testing.

The work herein extends our efforts [6,7] and that of others [8-10] to the direct, low-level detection of intact pathogenic viruses using two protein-based readout techniques [AFM and surface-enhanced Raman spectroscopy (SERS)] that could potentially increase sample throughput, enhance portability, and lower the limit for virus detection. These techniques each have intriguingly different attributes, all of which make compelling arguments as promising readout strategies. AFM, with its ability to image surfaces at sub-nanometer length scales, offers the unique advantage of label-free detection by direct visualization of viral pathogens by drawing on size and shape for identification and enumeration. The capability to distinguish pathogens based on size and shape also points to opportunities for multiplexed detection of different viruses on a single platform. SERS is a sensitive readout technique with detection limits in the femtomolar range. The key advantages of SERS are its unparalleled ability for multiplexed analysis and portability. Raman spectral features are 10-100 times narrower than fluorescence bands allowing the use of multiple unique Raman labels. Additionally, SERS instrumentation has no moving parts and involves measurements requiring only a single excitation source for multiple Raman reporters. These attributes facilitate field deployment. To demonstrate the capabilities of these two readout methods, feline calicivirus (FCV) was chosen as a model target due to its rigid spherical structure and nanometer dimensions (36-nm diameter), and spent cell culture media was chosen as a mimic of a biological sample matrix.

Scheme 1 illustrates the steps in the two types of immunoassays. First, a monoclonal antibody specific for FCV (anti-FCV mAb) is immobilized via succinimidyl ester chemistry to a gold-bound thiolate adlayer formed from dithiobis(succinimidyl propionate) (DSP). The immobilized layer of anti-FCV mAb specifically extracts virus from a complex biological matrix. The substrate is either imaged with AFM or incubated with ERLs that bind to captured FCV. The ERLs consist of 60-nm gold particles modified by exposure to 5,5'-dithiobis(succinimidyl-2-nitrobenzoate) (DSNB) and anti-FCV mAb. The nitro group of the DSNB-derived adlayer provides an intense SERS signal, while the succinimide group acts as a coupling agent for tethering anti-FCV mAb to the nanoparticle. The quantity of FCV is determined by the number of FCV counted with AFM or the spectral intensity of the symmetric nitro stretch ($\nu_s(NO_2)$) of the DSNB-derived coating using SERS readout. The following sections describe the findings from these investigations.

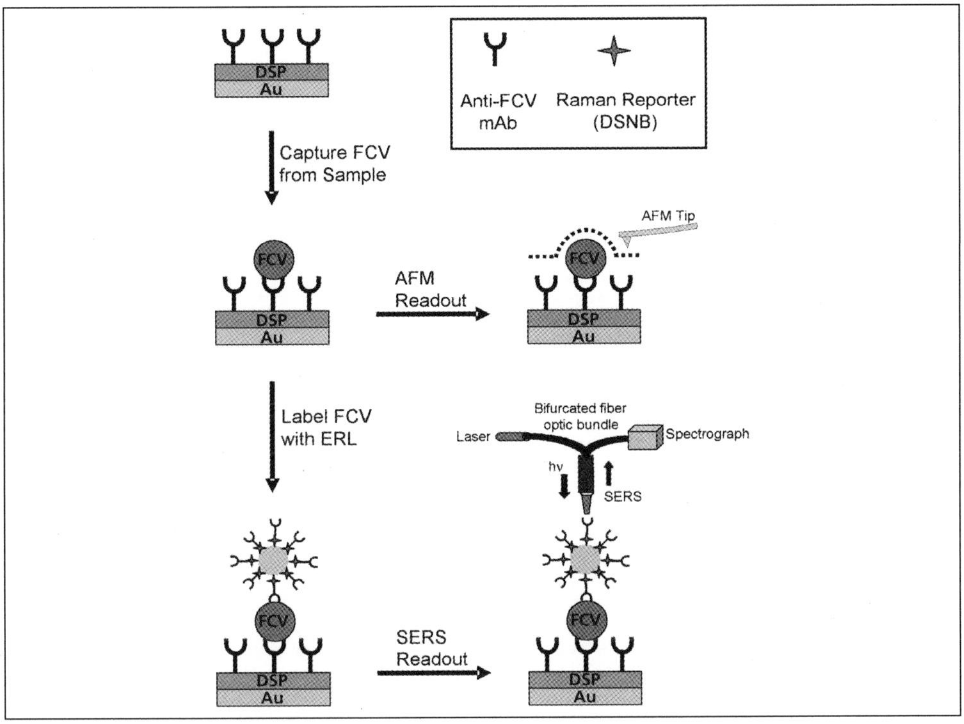

Scheme 1

Materials and Methods

Reagents

The Raman scatterer, DSNB, was synthesized according to a recent literature procedure [7]. Anti-FCV monoclonal antibodies (protein G purified, 1 mg/mL) and FCV, suspended in spent cell culture media (MEM media, CRFK cells), were supplied by the National Animal Disease Center. The FCV concentration of the stock solution was 5.0 x 10^8 viruses/mL, and all dilutions of this stock were made with spent cell culture media.

Capture substrate preparation

Template stripped gold (TSG) [11] served as the capture substrate because its low level of surface roughness facilitated AFM imaging of captured viruses, tens of nanometers in size. The TSG chips were addressed with a polydimethylsiloxane (PDMS) stamping procedure [12]. The substrate was then submerged in a 0.1 mM ethanolic DSP solution for ~12 h, rinsed with ethanol, and dried under a stream of nitrogen.

Anti-FCV mAb (20 μL, 100 μg/mL), diluted in 50 mM borate buffer (pH 8.5), was applied to the sample area for 8 h in a humidity chamber. This step forms a layer of capture antibodies by coupling through the terminal succinimidyl ester of the DSP-derived monolayer [13]. The substrate was then gently rinsed with 10 mM PBS.

AFM immunoassay protocol

The virus solution (20 μL), obtained directly from a freeze/thaw lysate of cultured cells, was pipetted on to the substrates and incubated at room temperature for 16 h. The substrate was gently rinsed with 10 mM PBS (pH 7.4), and then with a 3-s stream of deionized water to remove residual salt for AFM imaging. Finally, the substrate was dried with a stream of nitrogen and imaged with AFM.

Preparation of ERLs

The pH of the gold nanoparticle solution (Ted Pella, 1.0 mL) was adjusted to 8.5 by adding 40 μL of 50 mM borate buffer. Then, 10 μL of 1 mM DSNB in acetonitrile were pipetted into the colloidal solution and mixed for 8 h. Next, 20 μg of anti-FCV mAb were added to the colloidal solution and allowed to react for 12 h. Flocculation tests indicated that this step effectively coated the suspended colloid [12,14].

To stabilize the ERLs and to block unreacted succinimidyl esters, the colloidal solution was modified by the addition of 10% BSA in 2 mM borate buffer (100 μL). After 6 h, the suspension was centrifuged at 2000g for 10 min to remove any residual materials. After decanting the clear supernatant, the loose red sediment was resuspended in 1.0 mL of 2 mM borate buffer containing 1% BSA. The centrifugation/resuspension cycle was repeated twice. The appropriate volume of 1.5 M NaCl was added to yield a final NaCl concentration of 150 mM. As a final step, the labelled nanoparticles were passed through a 0.22-mm syringe filter (Costar) to remove large aggregates.

SERS immunoassay protocol

SuperBlock blocking buffer (20 μL) was pipetted on to the sample area and then rinsed with 10 mM PBS after a 12-h exposure. The capture substrate was exposed to 20 μL aliquots of FCV, diluted in spent cell culture media, for 8 h at room temperature in a humidity chamber and then rinsed with 2 mM borate buffer (1% BSA and 150 mM NaCl). The captured viruses were then exposed to the ERLs for ~12 h at room temperature in a humidity chamber. This step was performed with 1.0 mL of ERLs (1.3 x 10^{10} particles/mL) in a 24-well plate on a shaker (New Brunswick) with agitation (90 rpm). The substrates were rinsed with the noted borate buffer before drying with a stream of nitrogen gas and measuring the SERS signal. These conditions, along with those for the construction of ERLs, reflect the results of a series of experiments that optimized performance with respect to buffer composition and pH, blocking agent, and ionic strength [12].

Instrumentation

(i) AFM - Substrates were imaged using a MultiMode NanoScope IIIa AFM (Digital Instruments) equipped with a 150-μm scanner. The AFM was operated in TappingMode, imaging 5 x 5 mm areas at a scan rate of 1.5 Hz. The setpoint oscillation amplitude was set to 80% of the free oscillation amplitude. The cantilevers were n(+)-silicon TESP probes (Nanosensors).

(ii) SERS - The Raman spectra were collected with a NanoRaman I fibre-optic-based Raman system (Concurrent Analytical), a portable, field-deployable instrument. The light source was a 30 mW, 632.8 nm He-Ne laser. The incident laser light was focused to a 25-μm spot size on the substrate. All spectra were acquired with a 1-s integration.

Results and Discussion

AFM immunoassay

AFM images were collected from substrates exposed to varying concentrations of FCV to evaluate the effectiveness of the method. These results are presented in Figure 1. A scan size of 25 μm² was chosen as a compromise between image resolution to determine the sizes of the imaged objects and image area to sample accurately the capture surface representatively. Captured FCV appear as spherical objects with a height of about 22 nm (Fig. 1). Irregularly shaped objects are also evident in the images and are attributed to cell debris and other components in the spent cell culture media. The observed height of the FCV is somewhat smaller than the expected 36-nm diameter because of dehydration. Moreover, the images show the FCV having varying sizes in the plane of the substrate (i.e., *x-y* plane). This observation is attributed to tip convolution effects which can have a pronounced impact in the lateral resolution of an image [15]. However, AFM can distinguish objects based upon their height (*z*-axis), regardless of tip convolution effects in the *x-y* plane. Importantly, all the captured FCV have the same height (22 nm), which allows for quantification based on virus enumeration.

The evolution of the images follows the expected trend: the number of captured FCV increases as FCV concentration increases. To determine the performance of the assay, five capture substrates were exposed to FCV at each concentration, with five images collected at different locations on each sample for representative sampling. Test solutions covered the range of 3 x 10^8 viruses/mL to 5 x 10^6 viruses/mL and were diluted in spent cell culture media. The number of FCV observed in a 25 μm^2 image was then averaged and standard deviations were calculated for each FCV concentration. The resulting dose-response curve is shown in Figure 2. This plot was used to calculate a limit of detection which was defined as the concentration of virus that yields a surface coverage equal to the average blank signal plus three times the standard deviation of the blank samples. The blank signal, typically 1-2 counts per image, is attributed to particles in solution that are comparable to FCV in size and shape. Collectively, these data translate to a limit of detection of 3 x 10^6 FCV/mL, which is competitive with other methods, offering limits of detection of 10^5 viruses/mL for fluorescence [4] and microcantilever [5] detection. Thus, AFM is competitive with conventional techniques while offering this level of performance without requiring the use of secondary labels and associated processing steps.

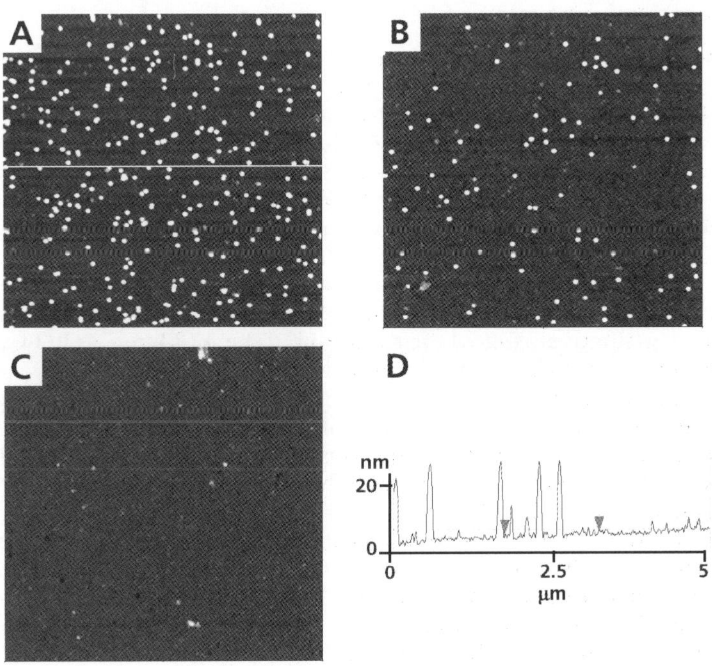

Fig. 1: AFM micrographs (5 x 5 µm) of FCV bound to capture substrates at three different concentrations: (A) 3 x 10^8, (B) 5 x 10^7, and (C) 0 viruses/mL (blank cell culture media). (D) cross-sectional plot of line scan shown in (A).

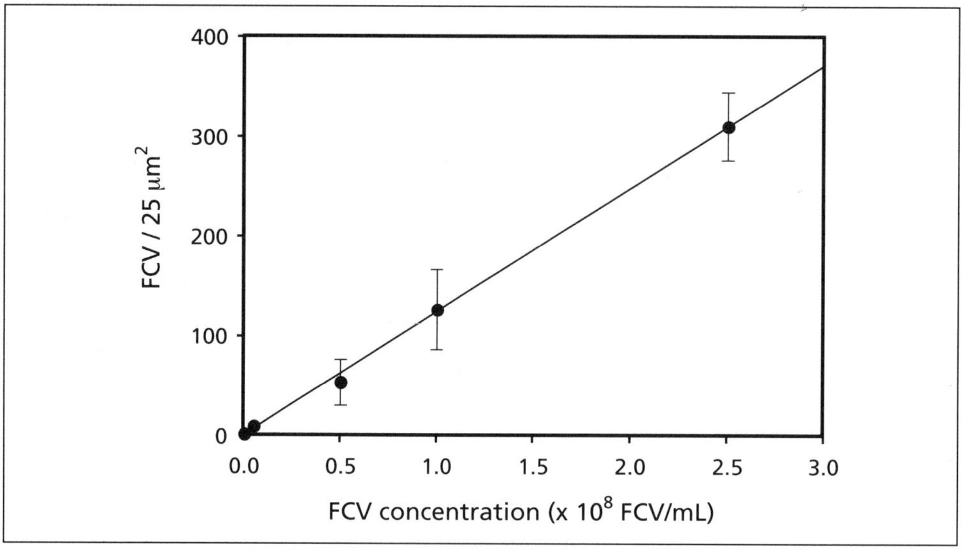

Fig. 2: Dose response curve for substrates exposed to FCV for ~16 h.

SERS-based immunoassay of FCV

The results from the optimization studies, noted earlier, were used to design an effective procedure for a SERS-based immunoassay for FCV. Capture substrates, coated with anti-FCV mAbs, were first incubated with sample solutions (5.0×10^5 to 2.5×10^8 FCV/mL) and then with ERL labelling solution. Representative SERS spectra for each FCV concentration, collected with a 1-s integration time, are given in Figure 3. The SERS spectra confirm the presence of the ERLs, based on features (e.g. the symmetric nitro stretch, $\nu_s(NO_2)$, at 1336 cm^{-1} and an aromatic ring mode at 1558 cm^{-1}) consistent with the DSNB-derived adlayer. The results also show an increase in the intensity of these diagnostic features as FCV levels increase. The spectrum for the blank indicates the presence of a small but readily detectable level of non-specific ERL binding. These responses also qualitatively follow the evolution of the AFM images in Figure 1 in that the number of captured viruses increases as the concentration of FCV increases.

The dose-response curve, created by plotting the $\nu_s(NO_2)$ intensity versus FCV concentration, is shown in Figure 4. In this plot, each data point is the average signal of five measurements from different locations on the same substrate and the standard deviations are represented by the error bars. The limit of detection, as defined previously, was determined to be 1×10^6 FCV/mL.

As a basis for validation of the SERS data, the AFM calibration curve was extrapolated from 25 μm^2 images to 500 μm^2 to approximate the area sampled in the SERS experiments. The correlation between the two data sets show that for every captured virus, the SERS response is ~3 counts/s. Non-specific binding resulted in 70 counts/s for the control; therefore a signal of 210 counts/s confirms specific virus binding. This signal of 210 counts/s, based on the correlation with AFM data, arises from the presence of ~70 captured viruses. In other words, we have the capability of detecting fewer than one hundred binding events, which is of immense value to the ultra-low level detection of viral pathogens and other immuno-diagnostic areas.

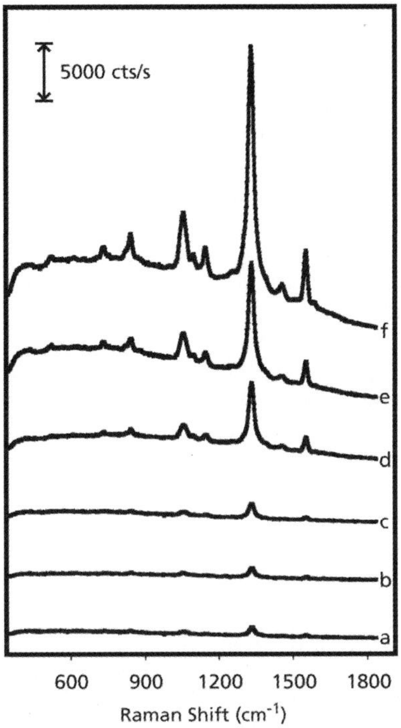

Fig. 3: SERS spectra (1-s integration) measured for various FCV concentrations: a) blank, b) 5.0 x 10^5, c) 5.0 x 10^6, d) 5.0 x 10^7, e) 1.0 x 10^8, f) 2.5 x 10^8 FCV/mL. The spectra are offset for visualization.

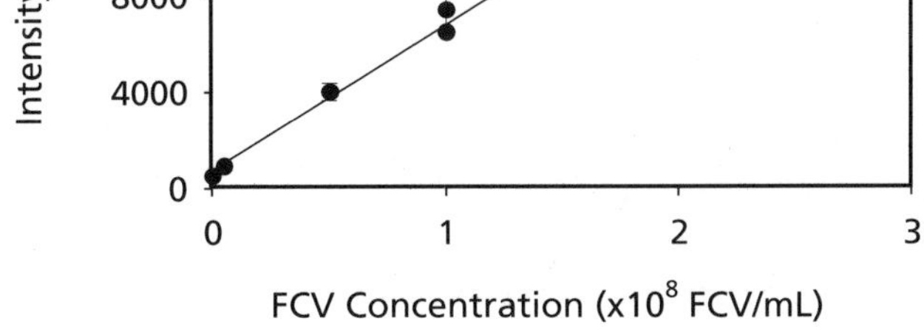

Fig. 4: Dose-response curve for SERS-based detection of FCV in cell culture media.

Conclusions

In summary, we have shown that AFM and SERS-based immuno-assays are competitive with other readout methods yielding limits of detection of 3 x 10^6 FCV/mL and 1 x 10^6 FCV/mL respectively. These techniques offer the added advantages of effortless multiplexing capabilities and portability. Current efforts are focused on lowering the detection limits and increasing mass transport to reduce the assay time. Lower detection limits could be realized by means of increasing AFM imaging rate and area, enhancing the scattered Raman signal, and pre-concentrating the analyte. Decreased assay times could be achieved through electro-kinetic migration, elevated temperatures, magnetic focusing and convection. Preliminary studies have resulted in a 100–fold improvement in the detection limit by sampling a larger portion of the capture substrate, and a stronger Raman scatterer yielded a 200-fold increase in signal. The use of elevated temperatures and convection both led to incubation times as short as 10 min. These combined studies suggest the potential to reach detectable limits of tens of viruses per mL in less than 10 min.

Acknowledgement

This work was supported through a grant from USDA-NADC and by the Institute for Combinatorial Discovery of Iowa State University. J.D. Driskell would like to acknowledge the support of the Mary K. Fassel and Velmer A. Fassel Fellowship. K.M. Kwarta would like to acknowledge the support of the Dow Chemical Fellowship. The Ames Laboratory is operated for the U.S. Department of Energy by Iowa State University under contract W-7405-eng-82.

References

1 Chomel BB: Control and prevention of emerging zoonoses. J Vet Med Educ 2003;30:145-147.

2 Nettikadan SR, Johnson JC, Mosher C, Henderson E: Virus particle detction by solid phase immunocapture and atomic force microscopy. Biochem Biophys Res Commun 2003;311:540-545.

3 Kjeldsberg E: Demonstration of calicivirus in human faeces by immunosorbent and immunogold-labelling electron microscopy methods. J Virol Methods 1986;14:321-333.

4 Donaldson KA, Kramer MF, Lim DV: A rapid detection method for Vaccinia virus, the surrogate for smallpox virus. Biosens Bioelectron 2004;20:322-327.

5 Llic B, Yang Y, Craighead HG: Virus detection using nanoelectromechanical devices. Appl Phys Lett 2004;85:2604-2606.

6 Pris AD, Porter MD: Nanoparticle Coding: Size-Based Assays Using Atomic Force Microscopy. Langmuir 2004;20:6969-6973.

7 Grubisha DS, Lipert RJ, Park H-Y, Driskell J, Porter MD: Femtomolar Detection of Prostate-Specific Antigen: An Immunoassay Based on Surface-Enhanced Raman Scattering and Immunogold Labels. Anal Chem 2003;75:5936-5943.

8 Kuznetsov YG, Malkin AJ, Lucas RW, Plomp M, McPherson A: Imaging of viruses by atomic force microscopy. J Gen Virol 2001;82:2025-2034.

9 Nettikadan SR, Johnson JC, Mosher C, Henderson E: Virus particle detection by solid phase immunocapture and atomic force microscopy. Biochem Biophys Res Commun 2003;311:540-545.

10 Xu S, Ji X, Xu W, Li X, Wang L, Bai Y, Zhao B, Ozaki Y: Immunoassay using probe-labelling immunogold nanoparticles with silver staining enhancement via surface-enhanced Raman scattering. Analyst 2004;129:63-68.

11 Stamou D, Gourdon D, Liley M, Burnham NA, Kuik A, Vogel H, Duschl C: Uniformly Flat Gold Surfaces: Imaging the Domain Structure of Organic Monolayers Using Scanning Force Microscopy. Langmuir 1997;13:2425-2428.

12 Driskell JD, Kwarta KM, Lipert RJ, Porter MD, Neill JD, Ridpath JF: Low-Level Detection of Viral Pathogens by a Surface-Enhanced Raman Scattering Based Immunoassay. Anal Chem 2005.

13 Wagner P, Hegner M, Kernen P, Zaugg F, Semenza G: Covalent Immobilization of native Biomolecules on to Au(111) via N-Hydroxysuccinimide Ester Functionalized Self-Assembled Monolayers for Scanning Probe Microscopy. Biophys J 1996;70:2052-2066.

14 Geoghegan WD, Ackerman GA: Adsorption of Horseradish Peroxidase, Ovomucoid and Anti-Immunoglobulin to Colloidal Gold for the Indirect Detection of Concanavalin A, Wheat Germ Agglutinin and Goat Anti-Human Immunoglobulin G on Cell Surfaces at the Elecetron Microscopic Level: A New Method, Theory and Application. J Histochem Cytochem 1977;25:1187-1200.

15 Montelius L, Tegenfeldt JO: Direct observation of the tip shape in scanning probe microscopy. Appl Phys Lett 1993;62:2628-2630.

Dr. M.D. Porter, Iowa State University, 42 Spedding Hall, Ames, IA 50011, USA.
E-mail: mporter@porter1.ameslab.gov

Vannier P, Espeseth D (eds): New Diagnostic Technology: Applications in Animal Health and Biologics Controls. Dev Biol (Basel). Basel, Karger, 2006, vol 126, p 41.

SESSION II

Standardisation, Validation and Overcoming the Problems of Molecular Diagnostic Methods

Chairman: *Alejandro Schudel,* OIE

Vannier P, Espeseth D (eds): New Diagnostic Technology: Applications in Animal Health and Biologics Controls. Dev Biol (Basel). Basel, Karger, 2006, vol 126, pp 43-51.

Development of a Framework for International Certification by OIE of Diagnostic Tests Validated as Fit for Purpose

P. Wright[1], S. Edwards[2], A. Diallo[3], R. Jacobson[4]

[1] Canadian Food Inspection Agency (NCFAD), Winnipeg, Manitoba, Canada
[2] Veterinary Laboratories Agency (Weybridge), Addlestone, Surrey, U.K.
[3] International Atomic Energy Agency, Vienna, Austria
[4] Eugene, OR, USA

Key words: Infectious diseases, diagnostic tests, validation, certification, OIE Registry.

Abstract: Historically, the OIE has focused on test methods applicable to trade and the international movement of animals and animal products. With its expanding role as the World Organisation for Animal Health, the OIE has recognised the need to evaluate test methods relative to specific diagnostic applications other than trade. In collaboration with its international partners, the OIE solicited input from experts through consultants' meetings on the development of guidelines for validation and certification of diagnostic assays for infectious animal diseases. Recommendations from the first meeting were formally adopted and have subsequently been acted upon by the OIE. A validation template has been developed that specifically requires a test to be fit or suited for its intended purpose (e.g. as a screening or a confirmatory test). This is a key criterion for validation. The template incorporates four distinct stages of validation, each of which has bearing on the evaluation of fitness for purpose. The OIE has just recently created a registry for diagnostic tests that fulfil these validation requirements. Assay developers are invited to submit validation dossiers to the OIE for evaluation by a panel of experts. Recognising that validation is an incremental process, tests methods achieving at least the first stages of validation may be provisionally accepted. To provide additional confidence in assay performance, the OIE, through its network of Reference Laboratories, has embarked on the development of evaluation panels. These panels would contain specially selected test samples that would assist in verifying fitness for purpose.

Background

The OIE's *Terrestrial Animal Health Code* [1] describes international trade standards for terrestrial animals and their products. Within these standards, reference is made to 'prescribed' tests methods. These are the assays required by the *Code* for the testing of animals before they are moved internationally. Prescribed tests may be found in the OIE's *Manual of Diagnostic Tests and Vaccines for Terrestrial Animals* [2]. The *Manual* is the international standard for diagnostic assays and is recognised as such in the Sanitary and Phytosanitary Agreement of the World Trade Organisation. Therefore, the *Code* and the *Manual* are companion standards, so much so, that the disease chapters in the *Manual* are organised in the same fashion as in the *Code* for ease of cross-reference. In addition to prescribed tests, the *Manual* also describes 'alternative' tests. These assays are suitable for the diagnosis of disease within a local setting, and can be used in the import/export of animals but only after bilateral agreement. A compendium of both prescribed and alternative tests, organised by disease, may be found in the *Manual* under the heading 'List of Tests for International Trade'.

The OIE's Biological Standards Commission, which is an elected body, oversees the production of the *Manual* and is responsible for establishing or approving test methods for the diagnosis of disease in mammals, birds and bees. As indicated above, application to international trade has been at the centre of the approval process. In addition to the many disease-specific chapters, the *Manual* begins with ten introductory chapters that cover a broad range of subject areas of general interest to infectious disease diagnostic laboratories. Two of these introductory chapters deal specifically with test method development and validation. The original chapter covers immunological assays [3] and a more recent addition targets genomic assays [4]. Both chapters summarise the essential principles of assay development and validation that have been described in more detail by Jacobson [5]. These introductory chapters are intended to provide the scientific basis for validating assays as well as background information and guidance towards achieving a validated assay.

The OIE *Standard for Management and Technical Requirements for Laboratories Conducting Tests for Infectious Animal Diseases* [6] is a specific interpretation of the more generally stated requirements of the ISO/IEC 17025:1999 international quality standard for testing laboratories. The OIE Quality Standard clearly states that test methods and related procedures must be appropriate for specific diagnostic applications for the test results to be of any relevance. In other words, the assay must be 'fit for purpose'. The Quality Standard further states that for a test method to be considered appropriate, it must be properly validated and that this validation must respect the principles outlined in the validation chapters of the *Manual*.

The test methods currently described in the *Manual* are considered to be standard methods. Many of these test methods are 'classic' in that they have been used in diagnostic laboratories for many decades and their performance characteristics, including limitations, are well known. These assays, while still useful, will remain status quo in terms of reagents and format until replaced with newer technologies. The introduction of new test methods into the *Manual* has for the most part been initiated by test developers in national research establishments involved in regulatory diagnostic services. Before the addition of any new test method to a chapter, the developers are requested by the Biological Standards Commission to submit validation data for review. However, both the submission requirements and the peer review process have been rather informal. Based on reviewers' comments, the Commission would then consider the addition of the method to the chapter and, if appropriate, its recommendation to the International Committee of the OIE for adoption as a prescribed or alternative test for trade. The above process has precluded submissions directly from the commercial sector.

In its role as the World Organisation for Animal Health, the OIE's scope of activities has broadened considerably over the past decade. In great part, this has been driven by the global impact of diseases such as bovine spongiform encephalopathy, foot-and-mouth disease and highly pathogenic avian influenza. So too, the need for appropriate test methods has expanded beyond trade applications alone. Although the *Manual* sets out the general principles of test validation, it was felt by the OIE and its international partners that there was a need for a more prescriptive framework for test developers to follow in the validation process. It was also recognised that the design of the validation studies should reflect the intended application of the assay.

First Consultants' Meeting

The Joint Division of the Food and Agriculture Organization of the United Nations (FAO) and of the International Atomic Energy Agency (IAEA) in Vienna, Austria manages a large programme on 'Nuclear and Related Techniques in Food and Agriculture'. Animal health is a key component of the Animal Production and Health Sub-programme and the IAEA's Seibersdorf Laboratory is an OIE Collaborating Centre for ELISA and Molecular Techniques in Animal Disease Diagnosis. The Joint Division hosted two expert consultants' meetings on the subject of 'OIE Guidelines for Validation and Certification of Diagnostic Assays for Infectious Animal Diseases'.

The first consultants' meeting was convened in Vienna in November, 2002. In attendance were programme experts from international organisations, including FAO, IAEA, OIE, World Health Organisation WHO and Pan-American Health Organization. In addition, diagnostic experts from the U.K., Canada, the U.S.A., Australia, Denmark, and Sweden were invited. Three principal subject areas were addressed in detail:

(i) diagnostic test applications in animal health;

(ii) validation and performance characteristics;

(iii) assay recognition or certification.

Given that animal disease management is required for economic, public health and environmental reasons, that risk assessment is a key component of this management and that risk assessment is highly dependent on the appropriate evaluation of animals and their products, it was agreed that fitness for purpose was integral to assay development and validation. These purposes or applications could be classified into six broad categories:

(i) Demonstration of population 'freedom' from infection (prevalence apparently zero)*:
  - a. 'free' with vaccination;
  - b. historical 'freedom';
  - c. re-establishment of 'freedom' post-outbreak;
  - * i.e. apparent 'freedom', recognising that absolute proof of freedom from infection in populations is not possible;

(ii) certification of freedom from infection or agent in individual animals or products for trade purposes;

(iii) eradication of infection from defined populations;

(iv) confirmatory diagnosis of clinical cases;

(v) estimation of prevalence of infection to facilitate risk analysis (surveys, classification of herd health status, implementation of disease control measures);

(vi) determination of immune status in individual animals or populations (post-vaccination).

It was agreed that the key data required to evaluate an assay comprised the following parameters:

(i) test type (detection of antibodies, infectious agents or their components);

(ii) analytical sensitivity and specificity;

(iii) diagnostic sensitivity and specificity;

(iv) repeatability (within laboratory);

(v) reproducibility (between laboratories);

(vi) quality assurance capability;

(vii) throughput capacity in the laboratory;

(viii) turn-around-time of test;

(ix) technical sophistication (skills and equipment needed);

(x) interpretation skill required.

In addition to these assay performance characteristics, factors such as the epidemiological parameters of the disease and the host population, sampling strategy and the structure and capability of the field veterinary services, all have an impact on the deployment and application of a given test and must be taken into consideration when designing a disease control or testing programme.

With respect to a process for the recognition or certification of assays, the participants proposed the following guidelines:

(i) assay developers should apply standard template requirements towards validation of a new test;

(ii) total validation package should be evaluated by other laboratories (they should not have been involved in the original validation);

(iii) evaluating laboratories must have established records in working with assays for the disease in question (at least one OIE Reference Laboratory if possible);

(iv) the template, with supporting documents, should be submitted to the OIE for evaluation;

(v) the OIE should accept the assay after a positive and independent peer review of results. The OIE should provide an independent opinion on the purpose(s) for which the assay is deemed to be fit at the time of the OIE evaluation. Any subsequent changes need re-evaluation and demonstration of equivalency or improvement.

Based on the above, the consultants' group drafted the following recommendations to the OIE:

(i) the OIE should adopt a process for evaluation of diagnostic tests for specific purposes (based on the list above);

(ii) a standard validation template should be created to provide guidance to assay developers and to facilitate third party evaluation;

(iii) the template should support incremental stages of validation;

(iv) a registration process and a registry of certified test methods should be created and managed by the OIE;

(v) OIE Reference Laboratories should establish characterised serum/sample collections to provide analytical references, evaluation panels and proficiency panels in support of the validation process.

A complete report of the first consultants' meeting was submitted to the OIE and was reviewed by the Biological Standards Commission at their biannual meeting in Paris in January, 2003.

OIE Resolution

At the 71st General Session of the OIE in Paris in May, 2003, the International Committee adopted Resolution XXIX which states that:

(i) fitness for purpose should be used as a criterion for validation;

(ii) the Director General should make provisions to establish a registry for assays with levels of validation specified;

(iii) OIE Reference Laboratories should be intimately involved with validation efforts;

(iv) OIE Reference Laboratories should establish serum/sample reference collections to be used for validation in line with their mandates;

(v) the Director General be given the mandate to review, in close consultation with the Standards Commission, the procedures involved in the timely approval of assays. This may entail the creation of a specific Ad hoc group comprising relevant experts to evaluate any submissions received;

(vi) the Director General be authorised to recover if necessary, any costs incurred in the process of validating such assays.

The above resolution gave the Director General of the OIE a clear mandate to proceed with the establishment of a new process for the evaluation and certification of diagnostic tests.

Second Consultants' Meeting

The second consultants' meeting was convened by both the Joint FAO/IAEA Division and the OIE in Vienna in December, 2003. The same experts were invited who participated in the first meeting. In addition, experts in the validation of assays in the absence of gold standards were invited from Denmark and Belgium. As the OIE resolution no longer places restrictions on who may submit validation dossiers, major stakeholders from the commercial sector were also invited. Three principal tasks were assigned to working groups selected from among the participants:

(i) finalise a prototype validation template;

(ii) propose an operating procedure for dossier submission;

(iii) develop guidelines for evaluation panels.

A detailed prototype template was presented by the working group responsible for this assignment. The prototype was agreed by all participants. The key components of the prototype are listed below:

(i) background information:
- a. test method (functional description);
- b. intended purpose(s) of test;
- c. applicant;
- d. scientific contact;
- e. accreditation status of developing laboratory;
- f. intellectual property considerations;

(ii) test method:
- a. protocol (detailed description of method and reagents);
- b. kit configuration (if commercial);

(iii) validation – Stage I (Analytical):
- a. calibration data;
- b. repeatability;
- c. analytical specificity;
- d. analytical sensitivity;

(iv) validation – Stage II (Diagnostic):
- a. reference animals (description);
- b. threshold (cut-off) determination;
- c. diagnostic performance estimates;
  - 1) conventional estimates of diagnostic specificity and sensitivity using defined reference animals;
  - 2) mathematical estimates of diagnostic specificity and sensitivity using 'no gold standard' models;
  - 3) comparison between currently used test(s) and proposed new test;

(v) validation – Stage III (Reproducibility):
- a. laboratory selection criteria;
- b. evaluation panel (description);
- c. reproducibility analysis;

(vi) validation – Stage IV (Current use):
   a. laboratories currently using test method;
   b. assay application (purpose);
   c. international reference reagents (availability of calibration reagents);
   d. inter-laboratory testing programmes (involving assay);
   e. international recognition (list of agencies/Organisation s endorsing as an 'official' test method).

The second working group developed proposed guidelines for an operating procedure for submission of dossiers for potential inclusion in the OIE registry. The principle essential features of this operating protocol included:

(i) purpose and aims of registry;
(ii) eligibility of submitters;
(iii) review panel selection and composition;
(iv) administrative responsibilities (OIE);
(v) guiding principles for review procedure;
(vi) conflict resolution;
(vii) registration as fit for purpose;
(viii) change control (affecting assay performance once registered);
(ix) registration maintenance and renewal;
(x) registration of current 'standard' methods;
(xi) registration fee structure;
(xii) glossary of terms and definitions (to assist developers and reviewers).

The third working group developed proposed recommendations for the creation and use of evaluation panels. This latter task is somewhat multipurpose as evaluation panels potentially have more than one application in the assessment of both the test method and the laboratories using them. This working group made the following recommendations related to the following:

(i) purpose and aims of evaluation panels;
(ii) confirmation of fitness for purpose assertions (as an adjunct to validation dossier);
(iii) other applications (proficiency testing, serial release testing, etc,);
(iv) composition and characterisation of evaluation panels (disease specific);
(v) responsibility for preparation and application (OIE Reference Laboratories);
(vi) cost recovery of development and application of panels;
(vii) reporting and consequence of results.

A complete report of the second consultants' meeting was submitted to the OIE and was reviewed by the Biological Standards Commission at their meeting in Paris in January, 2004.

Current Status

Over the period from 2004 to the present, a web version of the validation template as described above and application instructions have been developed and published on the OIE website (www.oie.int) and the registry is now open for submissions. A secretariat has been appointed within the OIE and a set of operating procedures established for the processing of applications. The process will be overseen by the Biological Standards Commission and it will appoint expert reviewers to assess validation dossiers. It is recognised that the template is designed around the validation of serological assays; however the same principles apply to other methods such as antigen detection and molecular methods. The template and procedures will be kept under review and will be updated and improved as necessary as in light of comments received from external sources.

Successful applicants will be allowed to use the OIE recognition as evidence of the validity of their test method for the stated purpose(s), and the register itself will be maintained on the OIE website. Tests on the register will be subject to periodic review and re-approval of their status.

Work has begun on the development of guidelines for the creation and application of evaluation panels. These guidelines will be distributed to OIE Reference Laboratories as part of their expanded role in the validation process.

Additional work is planned on the development of a template and guidelines for reviewers. This will ensure a uniform quality of review for all submissions. At the same time, the glossary of terms that currently accompanies the OIE Quality Standard will be reviewed and expanded to assist both test developers and reviewers. The glossary will be expanded to include more comprehensive epidemiological terms, new terms unique to molecular methods, as well as descriptive terms related to fitness for purpose.

The Biological Standards Commission is also developing links with statisticians and epidemiologists with special expertise in the application of latent class and extension of Bayesian theory to new models. This will add a valuable dimension to the design of test evaluation studies.

The OIE Registry and the supporting templates and guidelines have been developed to provide a formal mechanism for the certification and recognition of test methods as fit for purpose. However, all laboratories involved test development or the application of diagnostic test methods are encouraged to use these templates and guidelines whether or not they intend to seek formal OIE recognition. In doing so, the quality of infectious disease diagnosis will be enhanced worldwide.

References

1 Office International des Epizooties (OIE): Terrestrial Animal Health Code, 14th ed OIE, Paris, 2005, 634 pp.

2 Office International des Epizooties (OIE): Manual of Diagnostic Tests and Vaccines for Terrestrial Animals, 5th ed OIE, Paris, 2004, 1178 pp.

3 Office International des Epizooties (OIE): Principles of validation of diagnostic assays for infectious diseases. In Manual of Diagnostic Tests and Vaccines for Terrestrial Animals, 5th ed OIE, Paris, 2004, 21-29.

4 Office International des Epizooties (OIE): Validation and quality control of polymerase chain reaction methods used for the diagnosis of infectious diseases. In Manual of Diagnostic Tests and Vaccines for Terrestrial Animals, 5th ed OIE, Paris, 2004, pp 30-36.

5 Jacobson RH: Validation of serological assays for diagnosis of infectious diseases. Rev Sci Tech Off Int Epiz 1998;17(2):469-486.

6 Office International des Epizooties (OIE): OIE Quality Standard and Guidelines for Veterinary Laboratories, OIE, Paris, 2002, 63 pp.

Dr. P. Wright, Canadian Food Inspection Agency, National Centre for Foreign Animal Disease, 1015 Arlington Street, Winnipeg, Manitoba, R3E 3M4, Canada.
E-mail: pwright@inspection.gc.ca

Vannier P, Espeseth D (eds): New Diagnostic Technology: Applications in Animal Health and Biologics Controls. Dev Biol (Basel). Basel, Karger, 2006, vol 126, pp 53-54.

Standardisation and Validation of Multiplex Real-Time PCR Systems

B. Hoffmann, K. Depner, M. Beer

Friedrich-Loeffler-Institut, Institute of Diagnostic Virology, Greifswald – Insel Riems, Germany

Real-time polymerase chain reaction (qPCR) combines many advantages compared to the conventional gel-based PCR. qPCR is highly sensitive since very small fragments are amplified. Furthermore, qPCR assays are also highly specific, using additional probes for the "real-time" detection of PCR products. Amplicon detection with qPCR systems is time-saving and the risk of cross-contamination due to post-PCR manipulation is reduced. Quantification of template concentrations and the possibility of co-detecting several fragments of the same size in one tube using different labelled probes ("multiplexing") are additional advantages of this relatively new method. The use of probes labelled with different fluorophores allows the integration of internal controls into a cost- saving one tube diagnostic strategy. In addition to the FAM channel, which can be used for the detection of the target sequence, up to three additional fluorescence channels are available, e.g. for the detection of different controls or target sequences. In order to balance non-PCR-based well-to-well differences, the fluorophore ROX can be included as a passive reference dye.

Fluorophore	Functions
FAM	Target sequence
HEX	Internal control
ROX	Passive reference dye
Cy5	Positive control identification

In our qPCR systems for the detection of epizootic diseases, for example Food-and-Mouth Disease (FMD), the FAM channel was used to detect an amplicon of the 3D gene of FMD virus. For the amplification a modified protocol was designed based on a publication of our Danish colleagues [1]. The method described was modified in regard to the primers and probe as well as changed to a one-step RT-PCR.

The HEX channel was used for the detection of a heterologous internal control (IC). The IC is added to each sample during the RNA extraction and is checking the efficacy of the nucleic acid extraction and the occurrence of inhibition factors during the PCR for each sample or sample pool. This heterologous IC is based on the gene of the enhanced green fluorescence protein (EGFP). The EGFP gene is used because no occurrence in animal specimens is expected. Using different forward and reverse primers, amplicon sizes from 132 bp to 712 bp can be produced and chosen for a sensitive multiplex RT-PCR. Our aim was to establish a multiplex assay for FMD with the same sensitivity of the single assay. Therefore we used the following set screws:

-Amplification of a relative small number of the IC (approx. 10 000 copies/well),

-Use of limited primers for the IC amplification and

-Amplification of a larger IC fragment with an sub-optimal PCR efficiency compared to the optimal PCR efficiency for the FMD detection assay.

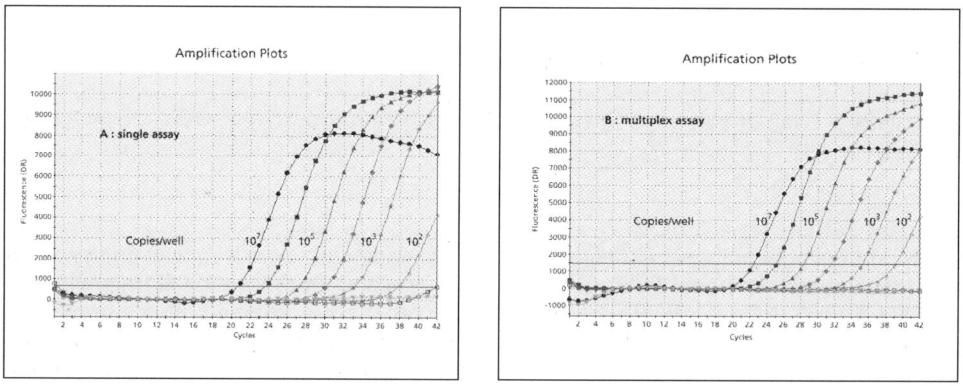

Comparison of the analytical sensitivity of the FMD single assay (A) and the multiplex assay (B) based on a serial dilution of an in vitro transcribed positive control

Finally, the Cy5 channel was used to differentiate the positive control of the target assay from field samples.

In conclusion we were able to develop a multiplex qPCR tests showing identical sensitivities as the single assays. In the four-colour qPCR assay, different controls for sensitive and robust PCR diagnostics were included.

Reference

1 Rasmussen TB, Uttenthal A, de Stricker K, Belak S, Storgaard T: Development of a novel quantitative real-time RT-PCR assay for the simultaneous detection of all serotypes of foot-and-mouth disease virus. Arch Virol 2003;148(10):2005-2021.

Dr. B. Hoffmann, Friedrich-Loeffler-Institut, Institute of Diagnostic Virology, Boddenblick 5a, DE-17493 Greifswald – Insel Riems, Germany.
E-mail: bernd.hoffmann@rie.bfav.de

Vannier P, Espeseth D (eds): New Diagnostic Technology: Applications in Animal Health and Biologics Controls. Dev Biol (Basel). Basel, Karger, 2006, vol 126, p 55.

Detection of Genetically Modified Plants: Limits and Possibilities

Y. Bertheau[1], *A. Kobilinsky*[2], *J. Davison*[3]

[1] INRA PMDV, route de Saint Cyr, 78026 Versailles cedex, France
[2] INRA MIA, Domaine de Vilvert, 78352 Jouy en Josas cedex, France
[3] INRA, Unit of Cellular Biology, route de Saint Cyr, 78026 Versailles cedex, France

Recent European Union regulations, e.g. 1829/2003/EC, restrict the permissible levels of fortuitous presence of genetically modified (GM) plants in non-GM food and feedstuffs. This, in turn, has required the development of several detection methods, and these are still evolving in accuracy, specificity, reproducibility but also in cost-effectiveness. In field situations, protein-based methods are preferred, while, in the laboratory, the more sensitive DNA-based real-time quantitative PCR methods are used, which might induce discrepancies in results. As with all analytical methods, reducing measurement uncertainty is a challenge with economic impact, necessitating rigorous sampling plans and statistical analyses. Several other issues currently remain unsolved. For example, using present techniques, it is difficult to differentiate between contamination by two different GM-seed types, each transgenic for individual transgenes, or a single seed type, containing two different transgenes commonly referred to as gene stacking. An example of the latter would be a seed type simultaneously containing genes for herbicide tolerance and insect resistance after pollination. Another unsolved problem is the detection of unknown transgenes, where the ingredient derivative is unauthorized in the European Union. These new methods, while designed for the detection of GM-plants and their processed products, have with minor modifications, applications to e.g. micro-organisms involved in human, plant and animal diseases, including bio-terrorism. This presentation will discuss the progress and difficulties in solving GMO detection related issues.

Dr. Y. Bertheau, INRA PMDV, Route de Saint Cyr, FR-78026 Versailles cedex, France.
E-mail : bertheau@versailles.inra.fr

Vannier P, Espeseth D (eds): New Diagnostic Technology: Applications in Animal Health and Biologics Controls. Dev Biol (Basel). Basel, Karger, 2006, vol 126, p 57.

Limits and Advantages of New Technologies in Criminology

A. Schlenck, O. Pascal

Centre Hospitalier Régional et Universitaire de Nantes, France

What's up Doc in criminology ?

It is very difficult today to speak about all the new technologies which have appeared recently in the criminal world: fingerprinting (and the associated data bases), toxicology (research of toxics in hairs), virology and bacteriology, smell identification (by "expert" dogs...) ...

The most popular of all is probably DNA fingerprinting, the revolution of the past ten years for the criminal justice authorities. Used for paternity testing as well as for blood stain identification, DNA seems to provide absolute and undeniable evidence. Reality or Dream?

First we have to consider the technical aspect and its progress: Sir Alec Jeffreys in 1985 described identification using restriction enzymes, e.g. Southern blot and radiolabelled probes. At least one week was necessary to obtain the first result: today a complete profile (with more than 10 loci) could be sent to the National Data Base in fewer than six hours, requiring a thousand-fold less DNA than in 1985.

PCR associated with capillary electrophoresis are the main actors of this technological revolution. Unfortunately the contaminations are closely associated. Prevention (separated area for pre- and post-PCR, clothes dedicated for each step of the analysis, quality management systems) and research on these contaminations in the laboratory (after wiping the laboratory equipment for example) or in the samples (using a negative sample) are essential for the biologist.

A large number of forensic laboratories have standardization references such as ISO or 17025.

Secondly we have to consider the use of such evidence in trials. In France, contrary to other countries such as the USA, we were fortunate that DNA profiling was poorly attacked by the defence lawyers. Unfortunately, for a relatively short time, some defence lawyers have been very interested by the weight of such evidence: a cigarette bud with the DNA fingerprint of Mr. X is not proof that Mr. X is a killer. DNA typing is only one item of evidence among others and could not be used alone to send somebody to jail.

Other aspects of this technology could be evoked, such as databases, and the risk of it being used for purposes other than identification, or the future of DNA (determination of age, phenotype, etc.).

Dr. A. Schlenck, CHU Nantes, Quai Moncousu, FR-44093 Nantes Cedex, France.
E-mail : alexandra.schlenck@chu-nantes.fr

Vannier P, Espeseth D (eds): New Diagnostic Technology: Applications in Animal Health and Biologics Controls. Dev Biol (Basel). Basel, Karger, 2006, vol 126, pp 59-60.

Regulations in the United States Having an Impact on the Movement of Biological Materials

L.A. Thomas

United States Department of Agriculure, Animal and Plant Health Inspection Service, Veterinary Services

There are a number of U.S. government agencies that have the authority to regulate the movement of biological materials such as the Centers for Disease Control (CDC), U.S. Fish and Wildlife Services, and Animal and Plant Health Inspection Service (APHIS). However, the focus of this discussion will be based on two critieria (i) those APHIS regulations that have an impact on the movement of biological agents such as bacteria, viruses, and prions; diagnostic reagents; and diagnostic samples and (ii) those that have an impact on animal health.

The Agricultural Bioterrorism Protection Act of 2002 which is a sub-part of the Public Health Security and Bioterrorism Preparedness Response Act of 2002, requires that entities, such as private, State, and Federal research laboratories, universities, and vaccine companies that possess, use, or transfer agents or toxins deemed to be a threat to public health and safety, or to animal or plant health or products, register with either APHIS or CDC. The resulting regulations are known as the select agent regulations (9 CFR Part 121, 7 CFR Part 331, and 42 CFR Part 73). There are 20 "overlap" agents (viruses, bacteria, and toxins) that have been listed that pose a significant risk to both animal and human health. Commonly known "overlap" agents include *Bacillus anthracis, Brucella abortus*, Rift Valley Fever virus, and botulinum neurotoxins. There are 23 "veterinary only" agents. Commonly known "veterinary only" agents include foot-and-mouth disease virus, highly pathogenic avian influenza virus, and bovine spongiform encepthalopathy. Materials that are shipped for clinical diagnosis are exempt from these regulations, as are non-viable agents. Additionally, agents may be excluded from these regulations if it has been determined that they are attenuated and pose no risk to public and/or animal health. *Brucella abortus*, Strain 19, is an example of a strain that has been excluded from these regulations.

Registration of an entity requires that an "Application for Laboratory Registration for Possession, Use, and Transfer of Select Agents and Toxins" be completed and submitted to either CDC or APHIS. Registration also requires that the U.S. Department of Justice complete a security risk assessment for the facility, its owners, the designated responsible official, and all individuals possessing, using, or transferring the agents or toxins. Before registration is granted, the facility must also meet biosafety requirements that are commensurate with the risk that the agent or toxin poses and must establish security measures that provide graded protection in accordance with the threat that the agent or toxin poses. An on-site inspection is also required where the government verifies the biosafety and security measures the entity has in place to safeguard the select agent and/or toxin.

There are further exemptions to the select agent regulations. Products that contain any of these agents or toxins and that are cleared, approved, licensed, or registered under any of the following acts are not subject to select agent regulations: Federal Food, Drug, and Cosmetic Act; Section 351 of the Public Health Service; Virus–Serum–Toxin Act; and the Federal Insecticide, Fungicide and Rodenticide Act. APHIS can determine on a case–by–case basis whether certain investigational products that contain select agents pose no threat to animal health or animal products and may be exempted.

Diagnostic facilities for other animals are exempt if they report the identification of any select agent or toxin and either destroy or transfer the agent or toxin within seven days.

Another regulation having an impact on the movement of biological materials is 9 CFR Part 122 "Organisms and Vectors". Under this regulation, APHIS controls the importation and interstate movement of (i) any culture or collections of organisms or their derivatives, which may introduce or disseminate any contagious or infectious disease of animals (including poultry) and (ii) all animals (including poultry) such as mice, pigeons, guinea pigs, rats, ferrets, rabbits, chickens, dogs, and similar, which have been treated or inoculated with organisms, or which are diseased or infected with any contagious, infectious, or communicable disease of animals or poultry or which may have been exposed to any such disease.

The requirements for the issue of a permit generally include submission of a permit application, laboratory inspection to verify the appropriate biosafety level of the facilities, and a signed agreement by the permittee that he/she will abide by the conditions/restrictions as listed on the permit. Conditions/restriction include such requirements as limiting work to in vitro diagnostics only; preventing further movement of the agent to other laboratories; limiting the studies to laboratory animals (such as mice, rats, or rabbits); or limiting work to certain laboratories in the facility. Conditions/restrictions are determined by the agent being used; the type of research that will be conducted; and the biosafety level of the facility.

Materials from animals that (i) are not considered to be livestock or poutry; (ii) have not been inoculated with or exposed to any infectious or contagious agent of agricultural concern; and (iii) do not contain any other animal, do not require an import permit.

Dr. L.A. Thomas, USDA/APHIS/VS, 4700 River Rd, Rivedale, MD 20737, USA.
E-mail : leeann.thomas@aphis.usda.gov

Vannier P, Espeseth D (eds): New Diagnostic Technology: Applications in Animal Health and Biologics Controls. Dev Biol (Basel). Basel, Karger, 2006, vol 126, pp 61-70.

Transportation of Reagents, Reference Materials and Samples: The International Perspective

J.E Pearson[1], *S. Edwards*[2]

[1] World Organisation for Animal Health (OIE), Paris, France
[2] Veterinary Laboratories Agency, U.K.

Key words: Infectious substances, dangerous goods, United Nations model regulations.

Abstract: The International Regulations for the transport of infectious substances, which could include reagents, reference material and samples, are based on the 13th revision of the United Nations Model Regulations and are the standard for transport of infectious substances by all means of transportation. The 13th revision, effective January 2005 and further amended in March and July 2005, made major improvements in these shipping regulations. They specifically exempt certain substances, including those that have been neutralized or inactivated to destroy any pathogens and samples from "normal" animals. Infectious substances are divided into Category A, which includes primarily cultures of the more pathogenic agents and Category B, which includes all other substances that do not meet the Category A criteria. Tissue specimens, submitted for diagnosis, are included in Category B. Category A shipments must have a Dangerous Goods Certificate and meet other requirements; Category B shipments do not. The National requirements, such as import permits, and certain airline restrictions must also be met.

Introduction

These Model Regulations for the transport of Dangerous Goods were developed by the United Nations Sub-Committee of Experts on the Transport of Dangerous Goods (UN SCETDG) and are applicable to all modes of transportation. These regulations outline the requirements for the international transport of biological materials, which include reagents, reference material and samples. The 13th revision of the Model Regulations made major changes in the shipping requirements for the transport of dangerous goods, including infectious substances [1]. The reference to risk groups was eliminated and less stringent requirements for shipment of diagnostic specimens were included. The 13th revision of the Model Regulations became effect in January 2005. However, at the July and December 2004 meeting of the UN SCETDG, there were major changes to the Model Regulations proposed by the World Organization for

Animal Health (OIE) to facilitate movement of diagnostic specimens from animals. The OIE's proposals, as well as changes proposed by other organisations, were accepted [2]. The International Civil Aviation Organization (ICAO) develops the technical instructions for the safe transport of dangerous goods by air based on the Model Regulations. The International Air Transport Association (IATA) Dangerous Goods Regulations, which are published each year, are the interpretation of ICAO instructions applied to shipments by air. The January 2005 IATA publication included the 13th revision of the Model Regulations [3]; IATA published amendments in March and July 2005 that included the changes to the Model Regulations [4].

Regulations for Shipment of Biological Material

These Regulations, as published by IATA in 2005 [3,4[addresses the shipment of infectious substances, which are defined as: *"Substances which are known or are reasonably expected to contain pathogens. Pathogens are defined as micro-organisms (including bacteria, viruses, rickettsiae, parasites, fungi) and other agents such as prions, which can cause disease in humans or animals"*. Infectious substances are divided into Category A and B. Category A is defined as: *"An infectious substance which is transported in a form that, when exposure to it occurs, is capable of causing permanent disability, life-threatening or fatal disease in otherwise healthy humans or animals. (UN 2814 human or zoonotic, 2900 animal)"*. Category B is defined as: *"An infectious substance which does not meet the criteria for inclusion in Category A (UN3373)."* Category A and B shipments will be discussed in more detail later.

Exempted Substances

The 13th Model Regulations and the IATA Regulations have a specific list of substances that are exempted from the Dangerous Goods Regulations. These exemptions allow the shipment of some reference reagents and samples without restriction. The following is the list of exemptions:

- *"Substances in a form that any present pathogens have been neutralised or inactivated such that they no longer pose a health risk"*. The shipper is responsible for ensuring that pathogens are neutralised or inactivated; if this cannot be ensured the shipment will have to meet the Category A or B requirements

  Note: The guidelines for OIE Reference Sera states [5]: *"The reference standards should be prepared so that they are free of infectious material. To facilitate shipment between countries it is recommended that the standards be irradiated at 25–30 kilogray (2.5–3.0 Mrad). Bovine sera should be from a BSE-free source"*. Consequently, if Reference Serum is inactivated as recommended, it is excluded from the Dangerous Goods Requirements.

- *"Substances which do not contain infectious substances or substances which are unlikely to cause disease in humans or animals".*
- *"Substances containing micro-organisms which are non-pathogenic to humans or animals".*
- *"Environmental samples (including food and water samples), which are not considered to pose a significant risk of infection".*
- *"Dried blood spots, collected by applying a drop of blood on to absorbent material, or faecal occult blood screening tests and blood or blood components that have been collected for the purposes of transfusion or for the preparation of blood products to be used for transfusion or transplantation and any tissues or organs intended for use in transplantation".*
- *"Patient specimens for which there is minimal likelihood that pathogens are present are not subject to these Regulations if the specimen is transported in a packaging which will prevent any leakage and which is marked with the words* ***"EXEMPT HUMAN SPECIMEN"*** *or* ***"EXEMPT ANIMAL SPECIMEN"*** *as appropriate. The packaging should meet the following conditions:*

  (a) The packaging should consist of three components:
  - *(1) a leak-proof primary receptacle(s);*
  - *(2) a leak-proof secondary packaging; and*
  - *(3) an outer packaging of adequate strength for its capacity, mass and intended use, and with at least one surface having minimum dimensions of 100 mm* x *100 mm;*

  (b) For liquids, absorbent material in sufficient quantity to absorb the entire contents must be placed between the primary receptacle(s) and the secondary packaging so that, during transport, any release or leak of a liquid substance will not reach the outer packaging and will not compromise the integrity of the cushioning material;

  (c) When multiple fragile primary receptacles are placed in a single secondary packaging, they should be either individually wrapped or separated to prevent contact between them.

Note: *In determining whether a patient specimen has a minimal likelihood that pathogens are present, an element of professional judgment is required to determine if a substance is exempt under this paragraph. That judgment should be based on the known medical history, symptoms and individual circumstances of the source, human or animal, and endemic local conditions".* This exemption is applicable to samples collected from normal animals for disease surveillance purposes.

There are also exceptions for some biological products; however, not all products are exempt and the shipper must review the regulations carefully. The following is the DGR definition of Biological Products [1,3]: *"Biological products are those products derived from living organisms which are manufactured and distributed in accordance with the requirements of appropriate national authorities, which may have special licensing requirements, and are used either for prevention, treatment, or diagnosis of disease in humans or animals, or for development, experimental or investigational purposes related thereto. They include, but are not limited to, finished or unfinished products such as vaccines".*

Category A Shipments

The pathogens listed in Tables 1 and 2 shall be designated as Category A. In addition, all new and emerging pathogens shall also be designated as Category A. As outlined in Tables 1 and 2, shipments containing cultures must be designated as Category A. Cultures are defined as: *"Cultures are the result of a process by which pathogens are intentionally propagated. This definition does not include patient specimens"*. The only zoonotic or animal pathogens for which both cultures and tissue samples must be included as Category A are Crimean-Congo haemorrhagic fever virus, Monkeypox virus, and Nipah virus. Shipment of any material that might contain one of these pathogens, including diagnostic samples, must be designated Category A. Shipment of tissue samples for the diagnosis of all other listed zoonotic and animal pathogens are designated as Category B.

A Shipper's Declaration of Dangerous Goods must be completed for all Category A shipments and the shipper must have received training on IATA shipping regulations. The package must be labelled as "Infectious Substance Affecting Humans (UN 2814) or Infectious Substance Affecting Animals (UN 2900) and packed as outlined in the IATA regulations (Fig. 1).

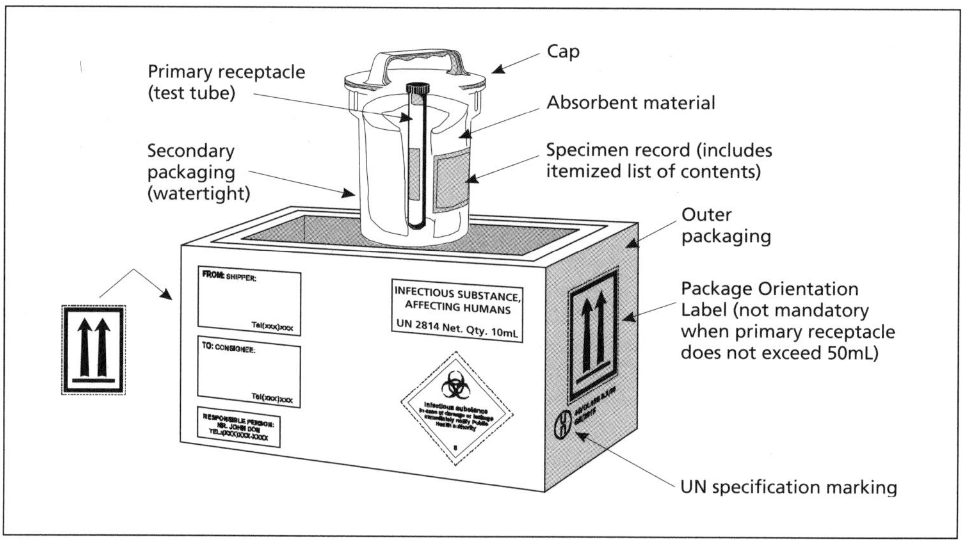

Fig. 1: Example of triple packaging system (UN P620) for the packaging and labelling of Category A infectious substances (figure kindly provided by IATA, Montreal, Canada).

Table 1: Indicative examples of human pathogens forbidden as diagnostic specimens that must be shipped as infectious substances affecting humans (UN 2814).

Bacillus anthracis (cultures only)	Japanese encephalitis virus (cultures only)
Brucella abortus (cultures only)	Junin virus
Brucella melitensis (cultures only)	Kyasanur Forest disease virus
Brucella suis (cultures only)	Lassa virus
Burkholderia mallei – Pseudomonas mallei – Glanders (cultures only)	Machupo virus
Burkholderia pseudomallei – Pseudomonas pseudomallei (cultures only)	Marburg virus
Chlamydia psittaci – avian strains (cultures only)	Mycobacterium tuberculosis (cultures only)
Clostridium botulinum (cultures only)	Monkeypox virus
Coccidioides immitis (cultures only)	Nipah virus
Coxiella burnetii (cultures only)	Omsk hemorrhagic fever virus
Crimean-Congo hemorrhagic fever virus	Poliovirus (cultures only)
Dengue virus (cultures only)	Rabies virus (cultures only)
Eastern equine encephalitis virus (cultures only)	*Rickettsia prowazekii* (cultures only)
Escherichia coli, verotoxigenic (cultures only)	*Rickettsia rickettsii* (cultures only)
Ebola virus	Rift Valley fever virus (cultures only)
Flexal virus (cultures only)	Russian spring summer encephalitis virus
Francisella tularensis (cultures only)	Sabia virus
Guanarito virus	*Shigella dysenteriae* type 1 (cultures only)
Hantaan virus	Tick-borne encephalitis virus (cultures only)
Hantavirus causing haemorrhagic fever with renal syndrome	Variola virus
Hendra virus	Venezuelan equine encephalitis virus (cultures only)
Hepatitis B virus (cultures only)	West Nile virus (cultures only)
Herpes B virus (cultures only)	Yellow fever virus (cultures only)
Human immunodeficiency virus (cultures only)	Yersinia pestis (cultures only)
Highly pathogenic avian influenza virus (cultures only)	

Table 2: Indicative examples of animal pathogens forbidden as diagnostic specimens that must be shipped as infectious substances affecting animals (UN 2900).

African swine fever virus (cultures only)	Peste des petits ruminants virus (cultures only)
Avian paramyxovirus Type 1 – Velogenic Newcastle disease virus (cultures only)	Rinderpest virus (cultures only)
Classical swine fever virus (cultures only)	Sheep-pox virus (cultures only)
Foot and mouth disease virus (cultures only)	Goatpox virus (cultures only)
Lumpy skin disease virus (cultures only)	Swine vesicular disease virus (cultures only)
Mycoplasma mycoides – Contagious bovine pleuropneumonia (cultures only)	Vesicular stomatitis virus (cultures only)

As stated previously, the 13th Model Regulations, without the 2004 revisions, were published [1] and, in some cases, may still be used as the Standard until 2007 for shipments other than by air. These Model Regulations, without the 2004 revisions, state that tissues as well as cultures of all the animal pathogens (UN2900) and some of the zoonotic pathogens must be shipped as Category A and that all cultures must be shipped as Category A. In addition, the list of animal pathogens includes African horse sickness virus and bluetongue virus, even though the agents are non-contagious. The March and July amendments to the IATA regulations include all the 2004 changes proposed by the OIE and others that were accepted by the UN SCETDG [4].

Category B Shipments

As stated previously, those substances that do not meet the Category A requirement can be shipped as Category B (UN 3373). This designation can be used for shipment of all specimens for diagnostic purposes from suspected infectious diseases of animals other than the three zoonotic diseases. Cultures of organisms that do not fit into the definition of Category A infectious substance can be transported as Category B. A Declaration of Dangerous Goods and training on the IATA regulations are not required. The packaging requirements (P650) are slightly less restrictive than for Category A (Fig. 2). Category B shipments shall be labelled ***"DIAGNOSTIC SPECIMENS"*** or ***"CLINICAL SPECIMENS"*** or ***"BIOLOGICAL SUBSTANCE, CATEGORY B"*** and contain the UN3373 label.

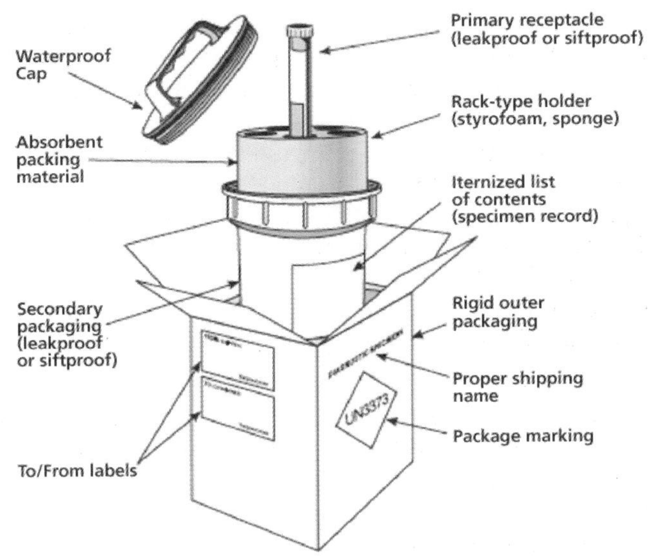

Fig. 2: Example of the triple packaging system (UN650) for the packing and labelling of Category B infectious substances (figure kindly provided by IATA, Montreal, Canada).

Packaging of Category B Shipments

As most of the diagnostic samples are shipped as Category B shipments, the following is an outline of these P650 shipping requirements (Fig. 2). The reader is referred to the IATA regulations for the complete details of the shipping requirements.

i) Infectious substances assigned to UN 3373 "DIAGNOSTIC SPECIMENS" or "CLINICAL SPECIMENS" or "BIOLOGICAL SUBSTANCE, CATEGORY B" must be packed in good quality packaging, which must be strong enough to withstand the shocks and loadings normally encountered during transport. Packaging must be constructed and closed so as to prevent any loss of contents which might be caused under normal conditions of transport.

ii) The packaging must consist of three components:
- a primary receptacle;
- a secondary form of packaging;
- a rigid outer packaging.

iii) For liquid substances:

- the primary receptacle(s) must be leak-proof and must not contain more than one litre; the secondary packaging must also be leak-proof;
- adequate adsorbent material must be packed around the primary receptacle(s) to absorb all the fluid in the primary receptacle(s);
- if multiple primary receptacles are used they should be individually wrapped or separated so as to prevent contact;
- the primary receptacle or the secondary packaging must be capable of withstanding without leakage an internal pressure of 95 kPa in the range of –40°C to 55°C (–40°F to 130°F);
- the outer packaging must not contain more than four litres. This quantity excludes ice, dry ice, or liquid nitrogen when used to keep specimens cold.

iv) For solid substances:

- the primary receptacle(s) must be sift-proof and must not exceed the outer packaging weight limit; the secondary packaging must be sift-proof;
- adequate adsorbent material must be packed around the primary receptacle(s) to absorb all the fluid in the primary receptacle(s);
- except for packages containing body parts, organs or whole bodies, the outer packaging must not contain more than four kg. This quantity excludes ice, dry ice or liquid nitrogen when used to keep specimens cold;
- if there is any doubt as to whether or not residual liquid may be present in the primary receptacle during transport then packaging suitable for liquids, including absorbent materials, must be used.

v) An itemised list of contents must be enclosed between the secondary packaging and the outer packaging.

vi) If shipped at ambient temperatures or higher, the primary receptacle must have a positive means of ensuring that it is leak-proof, such as a leak-proof seal, heat seal or skirted stopper. If screw caps are used they should be sealed with parafilm or tape.

vii) Prefrozen packs or dry ice can be packed around the secondary receptacle. If dry ice is used, there must be an internal support to secure the secondary receptacle in the original position after the dry ice has been dissipated. The outer packaging must permit the release of carbon dioxide. There are additional requirements if liquid nitrogen is used and these are described in the DGR.

viii) Packages containing diagnostic or clinical specimens are not required to have the net quantity marked on the outside of the package. However, where dry ice is used as a refrigerant, the net quantity of dry ice must be shown.

ix) The primary and secondary receptacles must be put into a shipping container with adequate cushioning material.

x) The packaging must be able to withstand a 1.2 metre drop test. (There are additional strength requirements for packaging used for UN 2900 and UN 2814 specimens.)

xi) At least one surface of the outer packaging must have a minimum dimension of 100 mm x 100 mm.

xii) For transport, the label **3373** must be displayed on the external surface of the outer packaging on a background of a contrasting colour and must be clearly visible and legible. The mark must be in the form of a square with an angle of 45° (diamond-shaped) with each side having a length of at least 50 mm, the width of the line must be at least two mm, and the letters and numbers must be at least six mm high. The proper shipping name "DIAGNOSTIC SPECIMENS" or "CLINICAL SPECIMENS" or "BIOLOGICAL SUBSTANCE, CATEGORY B" in letters at least six mm high must be marked on the outer package adjacent to the diamond-shaped mark.

Other Requirements and Restrictions

The following are other shipping requirements that must be complied with:

- Transport of Category A or B infectious substances in carry-on luggage or transport in diplomatic pouches is strictly prohibited by international air carriers.
- Some airlines will not accept any dangerous goods (Category A or B) or require that all Category B shipments be sent as Category A. The IATA web site should be monitored to determine what airlines have additional restrictions. As an example, US Airways requires that all diagnostic or clinical specimens must meet Category A requirements.
- Any National requirement for importation of biological material must be met. This would include obtaining an import permit and, in some cases, an export permit. The OIE Reference Laboratories have been encouraged to establish and distribute blanket import permits to expedite shipments and to take other action that will expedite receiving samples.
- As stated previously, cultures of the highly pathogenic agents must be sent as Category A. This restriction makes it more difficult and expensive to ship cultures for characterisation and epidemiological investigations. The OIE Reference Laboratories are encouraged to provide technical assistance to laboratories wishing to submit cultures for characterisation. International organisations should attempt to find financial and other assistance to ensure cultures are submitted to reference laboratories in a timely manner.

Conclusion

Reference material, reagents and diagnostic samples can often be shipped with a minimum of restrictions. Shipments of cultures of highly pathogenic agents (Category A) must meet all the requirements specified for Infectious Substances. The exemptions and shipping categories are summarised in Figure 3.

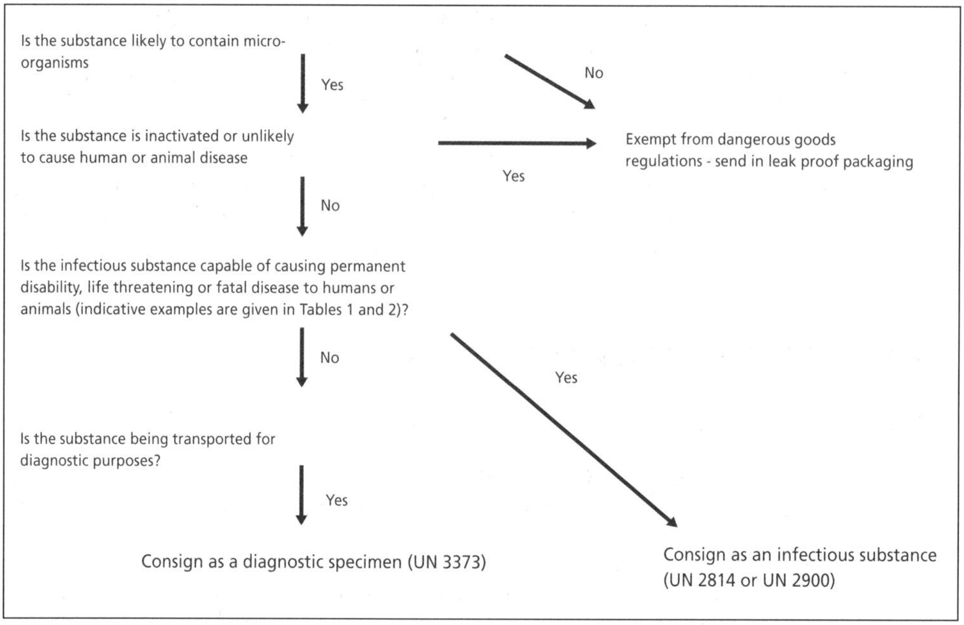

Fig. 3: Summary of shipping categories for biological material.

References

1 United Nations Sub-Committee of Experts on the Transport of Dangerous Goods. 13th Revision of the Model Regulations for the Transport of Dangerous Goods http://www.unece.org/trans/danger/publi/unrec/rev13/13files_e.html.

2 United Nations Sub-Committee of Experts on the Transport of Dangerous Goods. Amendments to the 13th Revision of the Model Regulations for the Transport of Dangerous Goods http://www.unece.org/trans/danger/publi/unrec/rev13/13_amend_e.html.

3 INTERNATIONAL AIR TRANSPORT ASSOCIATION. Dangerous Goods Regulations, 44th Edition. International Air Transport Association, 800 Place Victoria, P.O. Box 113, Montreal, Quebec H4Z 1M1, Canada, 2005, pp 824.

4 INTERNATIONAL AIR TRANSPORT ASSOCIATION. Addendum to Dangerous Goods Regulations, March and July 2005, 46th Edition. http://www.iata.org/whatwedo/dangerous_goods/download.htm.

5 World Organisation for Animal Health (OIE) OIE Quality Standard and Guidelines for Veterinary Laboratories: Infectious Diseases, World Organisation for Animal Health (OIE) 2002, pp 43-51.

Dr. J.E. Pearson, 4016 Phoenix Street, Ames, IA 50014, USA
E-mail: JPearson34@aol.com

Vannier P, Espeseth D (eds): New Diagnostic Technology: Applications in Animal Health and Biologics Controls. Dev Biol (Basel). Basel, Karger, 2006, vol 126, pp 71-76.

Constraints on Transportation of Reagents, Reference Materials and Samples, Difficulties and Possible Solutions: a User's Perspective

P.C. Blanchard

University of California Animal Health and Food Safety Laboratory, Tulare, CA, USA

Key words: Veterinary pathogen importation, transportation, reference materials.

Abstract: The air transportation of infectious materials is regulated by international air transport associations and based on United Nations Model regulations which have become more practical in addressing animal disease agents. However, individual countries' import and interstate requirements determine what materials can be imported and transported, and this approval process can be long, resulting in delays in organism confirmation, use of international OIE and other reference laboratories, and acquisition of reference materials, proficiency test panels, and reagents for performing necessary testing. Delays can be prevented for permits that are required for the routine work performed by a laboratory through the use of comprehensive and annually renewed permits. This process, however, does not address new and exotic agents where time is critical to an effective emergency response. This paper suggests actions by both the OIE and regulatory authorities which can assist in streamlining and expediting the permit process.

INTRODUCTION

The American Association of Veterinary Laboratory Diagnosticians (AAVLD), with a membership in 28 countries of over 1,200 individuals, of whom 85% are based in the United States of America (U.S.A.), has been involved with transportation regulations and their interpretation and impact on laboratories since 1999. AAVLD's initial involvement was related to the interpretation and dissemination of information on the International Air Transport Association (IATA) [1] and United States Department of Transportation (DOT) Dangerous Goods Regulations. Since 2002, the AAVLD efforts have expanded to include the United States Department of Agriculture (USDA) regulations for the importation and interstate transportation of organisms and vectors in the Code of Federal registry, 9CFR122 [2] and transportation of select agents [3]. Agricultural select agents for the U.S. are those listed in 121.3 of 9CFR121, and posted on the USDA website at http://www.aphis.usda.gov/programs/ag_selectagent/ag_bioterr_toxinslist.html. The justification for listing an agent or toxin was based on the effect, virulence or toxicity and method of dissemination among animals or plants and availability and effectiveness of treatment or prevention products or strategies. Both the interstate and select agent regulations apply to international importation of these agents. All U.S. select agent recipients must be a registered facility with USDA. The 9CFR122 interstate and import permits are generally issued to a laboratory for a set period of time (usually one year) whereas select agent permits to transfer an organism or toxin must be pre-approved for every shipment and the recipient must notify USDA within two days of its receipt. The select agent list includes disease agents considered both exotic and endemic in the U.S.

In 2002, the USDA National Veterinary Services Laboratory (NVSL) notified its customers, including veterinary diagnostic laboratories, biologicals companies and university researchers, that they could not ship any live organisms to laboratories which did not provide them with the appropriate interstate shipping permit in compliance with the 9CFR122 regulation originally written in 1958 and last amended in 1994. This regulation pertains to "All cultures or collections of organisms or their derivatives, which may introduce or disseminate any contagious or infectious disease of animals (including poultry)" [2]. The NVSL decision resulted in temporary discontinuation by some laboratories of testing that required live organisms for serology and resulted in the cancellation of the 2002 AAVLD self assessment exercise which was to include five endemic non-select agent organisms isolated from animals and postponed the shipping of laboratory-spiked faecal samples for the annual *Mycobacterium avium ss paratuberculosis* culture test. The same requirements of the 9CFR122 regulation, in 2003-2004, delayed the human health laboratory proficiency testing programme for both select and non-select agents and created delays in dissemination of live reagents to human health laboratories for testing human sera for West Nile Virus and Avian Influenza. When permits require a specific list of agents, each new agent requires a laboratory to obtain an amended permit under 9CFR122. Most of the agents used in the human health laboratory proficiency testing for diseases in humans also cause infection in animals so their transportation is regulated by USDA under 9CFR122. In 2003, the 9CFR122 regulation interpretation began to include some organisms or their derivatives which were classified as "non-viable" and exempt in the select agent rule because they were not capable of causing the disease. Under this interpretation, attenuated live organism vaccine strains such as *Bacillus anthracis* Sterne strain and antigen extracts of organisms such as *Coxiella* require permits under the import and interstate transport regulation (9CFR122).

The preceding is an example of the constraints which can have an impact on laboratories, biologicals companies and researchers. Further examples of transportation constraints which occur when regulations are too broad are seen in the delays which ensue due to: (i) permit requirements to obtain common reference organisms such as American Tissue Culture Collection organisms for antimicrobial testing quality control; (ii) shipment of organisms to biologicals companies for autogenous vaccines; (iii) shipment to laboratories which offer specialized testing such as DNA sequencing to characterize unknown organisms and (iv) the use of proficiency test panels which include common organisms not listed on most laboratory permits such as *Staphylococcus* and *Rhodococcus*. Internationally these constraints can have an impact on multi-national validation studies of new methodologies even when non-viable materials are used. As more countries recognize the need for surge capacity in response to a human zoonotic, food-borne or exotic animal disease outbreak, networks of human, animal and food testing laboratories at the federal, regional and local level are being developed to provide rapid surge capacity to respond to these outbreaks. These laboratories require continuing proficiency testing, may serve as surveillance testing sites and will be involved in testing for a new or exotic disease. During an outbreak these networks of laboratories require standardized reagents, quality control standards and proficiency panels to certify the competence of additional laboratory personnel who are trained to perform testing during the outbreak. Current import regulations may delay acquisition of these materials especially when a number of laboratories are requesting amended permits to provide surge capacity. The U.S. is not unique internationally in the types of permits, time required to obtain such permits which in some cases can be up to three months, and the cost for new and amended permits and laboratory site audits.

Discussion

Impact

The impact of the transportation regulations on laboratories include:

(i) timeliness of confirmation and characterization of organisms;

(ii) cost and delays in obtaining standardized reagents and quality control materials for emerging and exotic diseases;

(iii) cost and delays in performance of proficiency testing for new and emerging agents;

(iv) rapidity of a national laboratory response capability and capacity to a disease outbreak of a new or exotic agent.

The transportation constraints for routine testing and studies performed by a laboratory can be overcome by becoming familiar with the breadth of organisms and derivatives for which permits are required and applying for one permit that includes all the agents or derivatives which are required for quality control, testing reagents, proficiency tests or are likely to be received for further characterization, collaborative studies, biologicals production or antimicrobial monitoring by pharmaceutical companies.

Solutions

Transport authorities (national ground and IATA/ICAO) shipping regulations have become increasingly practical with each revision and are generally based on United Nations Model Regulations. The involvement of OIE as an expert consultant on these regulations has ensured that animal disease agents are appropriately classified. However, specific airlines may have further restrictions and the IATA website should be consulted for amendments and restrictions to the regulations [1].

Laboratories should be familiar with import and other permit requirements for their country and have in place permits for organisms, their derivatives and toxins, and infected tissues, if necessary; to cover the breadth and type of work they perform.

OIE should:

(i) strongly encourage all OIE reference laboratories to obtain broad permits for receipt of suspect infected samples and isolates from other countries for confirmation of the disease for which they are a reference authority;

(ii) continue their vital role as expert consultants for transport regulations which develop from the UN model regulations on dangerous goods;

(iii) endorse scientifically valid methods for the inactivation of organisms and adventitious organisms within sera and for the preparation of non-viable derivatives of organisms for use as proficiency panels, reagents and reference materials;

(iv) encourage the development and validation of non-infectious derivatives or genetic constructs which would be exempt from permit requirements and can be used for quality control reagents and proficiency test panel;

(v) post on the OIE website a check list of common, though not all inclusive, information for importing laboratories to obtain when requesting reagents and reference materials from other countries. This would include types of documents and approvals which may be required by the importing country. Some national laboratories have developed these checklists (NVSL - Barbara Martin personal communication).

Regulatory authorities should:

(i) amend broad regulations and specify the organisms and derivatives of concern so that the laboratories concerned can complete a single permit which will not require continuous amendments;

(ii) specify the exempted organisms or derivatives on a countries import website to expedite the process of shipment when a permit is not required. For example, armoured RNA composed of less than a specified percent of a genome, or stained glass slides of formalin-fixed tissues, would be exempt;

(iii) accept a list of proficiency test organisms and locations that will receive the proficiency test in lieu of requiring permits, when regulations exist which require all organisms to be permitted, and if the organisms do not require special biosafety procedures;

(iv) provide a letter of permit exemption to a company or laboratory for an organism or derivative when review of the inactivation or derivation process indicates a permit is not required.

(v) develop emergency response plans to provide one-day approval of permits in the event of numerous requests from surge capacity laboratories during a disease outbreak;

(vi) develop, with emergency response laboratory network coordinators, a template of required information for amended permits to use in the event of a disease outbreak to include importation and transportation of proficiency test panels, reagents and reference materials;

(vii) recognize the laboratories in an established network as a single entity for the rapid approval of amended permits and for the annual renewal of permits for a standard set of reagents and organisms for which the network is approved to provide testing;

(viii) participate in federal, regional and local emergency exercises by including the actual permit process in the exercise to identify obstacles to the rapid issue of permits and exemptions required by laboratories and biologicals manufacturers;

(ix) include diagnostic kit licensing and importation authorities in the emergency exercise process if the disease is likely to involve the use of a commercially licensed kit from another country to test for a newly introduced exotic disease (i.e. antigen and antibody kits for foot-and-mouth disease, classical swine fever, etc).

Conclusions

The transportation of diagnostic specimens, organisms, their toxin and derivatives between countries is regulated by import restrictions and permits of the recipient's country as well as by international air and national ground transportation requirements. Laboratories, researchers and biologicals companies should become familiar with the regulations for their country and in most routine situations can, through annually renewed and comprehensive permits for the type of work they perform, meet the import and transport requirements with minimal effort after the first year. However, diagnostic laboratories are expected to recognize, diagnose and identify emerging and exotic diseases and to investigate new agents further. The new and exotic agents which may be found each year are not predictable, so approved agent-specific import or interstate transport permits will not include them. Effective and timely responses to this latter group of agents can be seriously affected by regulatory requirements for new permits. A variety of solutions can assist in making the permit process more transparent and user-friendly for routine requests but only a strong commitment by regulatory authorities to expedite the import process in order to implement a rapid emergency response will overcome the delays which can occur in the face of an exotic disease or newly recognized agent outbreak.

References

1 International Air Transportation Association (2005) Dangerous Goods Regulations, 44th Edition. International Air Transportation Association, 800 Place Victoria, P.O. Box 113, Montreal, Quebec H4Z 1M1, Canada, 824pp. Addendum to regulations posted at http://www.iata.org/whatwedo/dangerous_goods/download.htm.

2 Code of Federal Registry, 9 CFR part 122, Organisms and Vectors. at: http://www.access.gpo.gov/nara/cfr/waisidx_00/9cfr122_00.html.

3 Code of Federal Registry, 9 CFR part 121 Possession, Use, and Transfer of Select Agents and Toxins. Federal Register 2005;70(52):13242-13292. Posted at: http://www.aphis.usda.gov/programs/ag_selectagent/ under 9CFR121 and 7CFR331 link

Dr. P.C. Blanchard, University of California Animal Health and Food Safety Laboratory System, Tulare Branch, 18830 Road 112, Tulare, CA 93274, USA.
E-mail: pcblanchard@ucdavis.edu

Vannier P, Espeseth D (eds): New Diagnostic Technology: Applications in Animal Health and Biologics Controls. Dev Biol (Basel). Basel, Karger, 2006, vol 126, p 77.

BSE Diagnostics: What's New

B. Oesch, A. Raeber

Prionics AG, Schlieren, Switzerland

The only reliable molecular marker for prion diseases is PrP^{Sc}, the pathological conformer of the prion protein that accumulates in the central nervous system (CNS) and to a lesser extent in lympho-reticular tissues. For BSE, several commercial diagnostic kits based on the post mortem immunochemical detection of PrP^{Sc} in brain tissue are now available. These rapid screening tests have been used in active surveillance of BSE and have greatly improved the detection of infected cattle before their entry into the human food chain.

Furthermore, we will review the currently approved rapid tests for the post mortem detection of the BSE-associated pathological prion protein as well as emerging technologies for the ante mortem diagnosis of prion diseases. These rapid screening tests have been used in active surveillance of BSE and have greatly improved the detection of infected cattle before their entry into the human food chain. In particular, we will also present the most rapid BSE test (PrioSTRIP) which allows the generation of results within about 1.5 h from sample to result.

At present, no diagnostic test exists for the detection of prion diseases in live animals or humans. New diagnostic techniques aimed at increasing sensitivity and specificity of PrP^{Sc} detection in body fluids and at identifying novel surrogate markers are under development. The transmission of prions from humans to humans has emphasized the need for a diagnostic test able to diagnose prion diseases from blood.

We have been using the prion-specific monoclonal antibody 15B3 to investigate the presence of PrP^{Sc} in brain as well as other tissues/body fluids from cattle and sheep in the light of early detection of infected animals. Results on sensitivity and specificity will be presented.

Dr. B. Oesch, Prionics AG, Wagistr. 27, CH-8952 Schlieren, Switzerland.
E-mail : bruno.oesch@prionics.com

Vannier P, Espeseth D (eds): New Diagnostic Technology: Applications in Animal Health and Biologics Controls. Dev Biol (Basel). Basel, Karger, 2006, vol 126, pp 79-86.

In–House Validation and Quality Control of Real-Time PCR Methods for GMO Detection: A Practical Approach

I. Ciabatti, A. Froiio, F. Gatto, D. Amaddeo, U. Marchesi

Istituto Zooprofilattico Sperimentale Lazio e Toscana, National Reference Center for GMO Analysis, Rome, Italy

Key words: Genetically modified organism, validation, quality control, real-time PCR.

Abstract: GMO detection and quantification methods in the EU are mainly based on real-time PCR. The analytical methods in use must be validated, first on an intra-laboratory scale and through a collaborative trial thereafter. Since a consensual protocol for intra-laboratory validation of real-time PCR methods is lacking, we provide a practical approach for the in-house validation of quantitative real-time PCR methods, establishing acceptability criteria and quality controls for PCR runs. Parameters such as limit of detection, limit of quantification, precision, trueness, linear dynamic range, PCR efficiency, robustness and specificity are considered. The protocol is sufficiently detailed to be directly applicable, increases the reliability of results and their harmonization among different laboratories, and represents a necessary preliminary step before proceeding to a time-consuming and costly full validation study.

Introduction

According to the International Standard EN ISO/IEC 17025, analytical methods must be validated in order to demonstrate that they are fit for the purpose; this concept has also been laid down in the Community legislation on GM food and feed which entered into force by the end of 2003 [1]. GMO detection and quantification methods in the EU are mainly based on real-time PCR (RT-PCR). General guidelines for the validation of methods for the detection of biotechnology-derived products have been established by international organizations, such as CODEX, CEN, ISO, the European Network of GMO laboratories (ENGL) and OIE [2-6]. However, the definition of a practical protocol for intra-laboratory validation of DNA-based methods from these general guidelines is not straightforward and different models have been adopted.

Therefore we propose a practical approach for the in-house validation of quantitative real-time PCR methods with particular reference to GMO analysis, establishing acceptability criteria and quality controls for PCR runs.

Material and Methods

The validation procedure has been developed and tested on an RT-PCR method for the relative quantification of Roundup Ready soy (GTS 40/3/2 line) based on amplification of a construct-specific sequence and a portion of the taxon-specific lectin gene le1 [7].

The validation parameters have been determined on certified reference materials (IRMM CRM) and samples derived from proficiency testing schemes (GEMMA).

For LOD and LOQ estimation, a procedure adapted from [8] was followed: RR soy DNA extracted from Certified Reference Materials was diluted in water or maize DNA, testing each dilution in at least five PCR replicates. The average Ct values obtained from each replicate group were plotted against the log of the copy number of the corresponding dilution level to get a calibration curve:

$$Ct = a \log (\text{copy number}) + b$$

which was used to calculate the genome copy number corresponding to each specific Ct value; for every replicate group the average value, the standard deviation and the relative standard deviation (RSD) of the genome copy number were determined. Then we evaluated the best fitting curve ($R^2 \geq 0.94$) for the relationship between RSD and the genome copy number; the procedure was integrally repeated three times on different sets of measurements, both for the reference and the target amplification. The copy numbers at which the RSD is equal to 33% or 25% were interpolated from each of the three curves and the average value was considered to be the LOD and the LOQ respectively.

Results and Discussion

A protocol for the intra-laboratory validation of RT-PCR methods for GMO detection has been designed, considering the following parameters: limit of detection (LOD), limit of quantification (LOQ), precision, trueness, linear dynamic range, PCR efficiency, robustness and specificity. The validation protocol has been developed and tested on a real-time PCR method which has already been routinely used in our laboratory for more than four years, thus considering a large set of analytical data collected during this period.

The LOD can be determined by diluting target DNA in water or non-target DNA and determining the target copy number at which the relative standard deviation of repeatability (RSD_r) is 33% [4]. Similarly the LOQ can be calculated considering a RSD_r equal to 25% [3,4].

Figure 1a shows a calibration curve for the target amplification, while in an exponential regression curve, describing the relationship between the relative standard deviation and the copy number, is shown. From the exponential regression curve, the copy numbers corresponding to RSD values of 33% and 25% are interpolated. Three calibration curves and the relative regression curves have been used to calculate mean $LOD_{33\%}$ and $LOQ_{25\%}$ both for the lectin (Table 1) and the target sequences (Table 2).

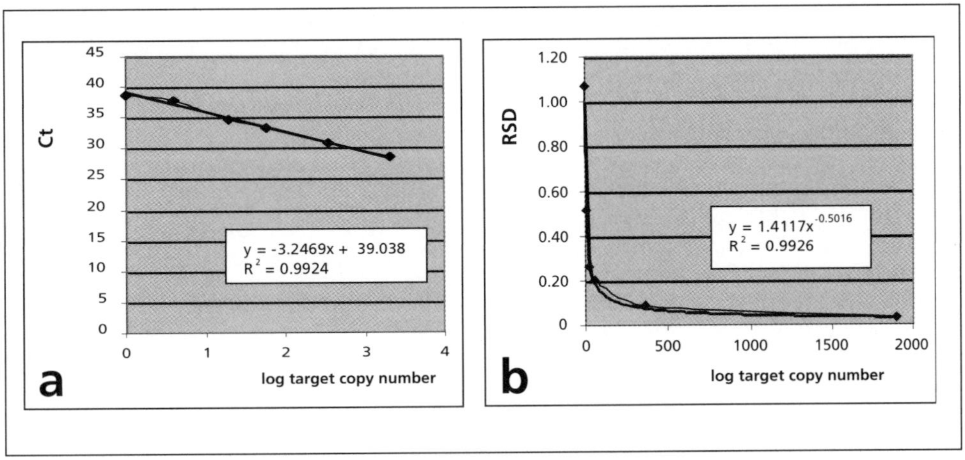

Fig. 1: a) calibration curve for the target sequence, b) exponential regression curve, relative standard deviation versus target copy number.

Table 1: Calibration curves (Ct versus lectin copy number) and exponential regression curves (RSD versus lectin copy number) obtained from three different sets of measurements for the lectin amplification. $LOD_{33\%}$ and $LOQ_{25\%}$ are computed as mean values.

Range of lectin copy number examined	Ct vs lectin copy number	R^2	RSD vs lectin copy number	R^2	LOD $_{RSD\ 33\%}$ (copy n°)	LOQ $_{RSD25\%}$ (copy n°)
41000-3	y = 39.0 - 3.68 x	0.986	$y = 1.45\ x^{-0.550}$	0.974	15	24
41000-2	y = 40.1 - 3.84 x	0.997	$y = 1.02\ x^{-0.368}$	0.940	22	45
10000-1	y = 39.4 - 3.83 x	0.996	$y = 1.12\ x^{-0.391}$	0.975	23	46
				mean	**20**	**38**

Table 2: Calibration curves (Ct versus target copy number) and exponential regression curves (RSD versus target copy number) obtained from three different sets of measurements for the target amplification. $LOD_{33\%}$ and $LOQ_{25\%}$ are computed as mean values.

Range of target copy number examined	Ct vs target copy number	R^2	RSD vs target copy number	R^2	LOD $_{RSD\ 33\%}$ (copy n°)	LOQ $_{RSD\ 25\%}$ (copy n°)
2050-1	y = 39.0 - (3.25) x	0.992	$y = 1.41\ x^{-0.502}$	0.993	18	31
410-1	y = 39.4 - (3.57) x	0.994	$y = 0.97\ x^{-0.402}$	0.952	14	29
2050-4	y = 39.0 - (3.31) x	0.996	$y = 1.28\ x^{-0.506}$	0.986	14	25
				mean	**15**	**28**

The precision can be determined under intermediate repeatability conditions evaluating the RSD of different test results obtained on a number of samples/reference materials at different concentrations within the range of measurement.

The trueness can be calculated on certified reference materials by comparing the average value obtained from a large series of test results and the reference value. In Table 3, estimates of RR soy concentration on certified reference materials, under intermediate precision conditions, were used to calculate mean, standard deviation, RSD and bias. Intermediate precision and trueness can be computed as the highest RSD value and bias respectively. The trueness should be within ± 25% (ENGL, [3]) or ± 30% (CODEX, [2]) of the accepted reference value over the whole dynamic range.

To estimate the trueness of the method on food and feed matrices closer to real samples than CRMs, results obtained from proficiency testing schemes can also be considered.

Table 3: RR soy concentration estimated for each certified reference material in 14 real-time PCR sessions carried out by different operators on six different DNA extractions. RSD and bias were used to assess intermediate precision and trueness.

RR soy concentration estimated for each certified reference material					
RR soy 0.1%	**RR soy 0.5%**	**RR soy 1%**	**RR soy 2%**	**RR soy 5%**	
0.10	0.52	0.93	2.04	5.04	
0.10	0.50	0.98	2.05	4.99	
0.10	0.53	0.94	1.95	5.16	
0.10	0.48	1.38	2.06	4.95	
0.10	0.55	0.94	2.23	4.56	
0.10	0.53	0.91	2.19	4.81	
0.11	0.46	0.81	2.19	5.36	
0.11	0.43	0.88	2.44	4.83	
0.09	0.59	1.01	2.03	4.52	
0.10	0.47	1.02	1.97	5.12	
0.10	0.46	1.05	2.03	4.97	
0.10	0.49	1.01	1.91	5.18	
0.10	0.46	0.95	2.28	4.79	
0.10	0.50	1.02	2.20	4.56	
0.10	0.50	0.99	2.11	4.92	**mean**
0.00	0.04	0.13	0.15	0.25	**standard deviation**
5	9	**13**	7	5	**RSD (%)**
0	0	1	**5.5**	1.6	**bias (%)**

The coefficient R^2 and the slope of regression curves obtained with dilutions of DNA in water or non-target DNA, both for the endogenous (Table 1) and the target amplification (Table 2), provide an indication of the linear dynamic range and the PCR efficiency ($E = 10^{-1/slope} - 1$). In case the ΔC_t method is used for quantification, R^2 and the slope from a number of calibration curves, ΔC_t versus log (% RR soy), should be evaluated (Table 4).

The ENGL document sets the limits for these performance criteria as $R^2 \geq 0.98$ and slope between –3.6 and –3.1 [3].

To measure the robustness of the method, that is its capacity to remain unaffected by small, but deliberate deviations from the experimental conditions described in the procedure [3], the contributions of factors, such as instrumentation, operator, reagent concentration and cycling temperature, cannot be individually considered, but instead they can be collectively evaluated through results obtained under intermediate repeatability conditions. In Table 5 results obtained from two different operators within a one-month period on the same DNA extracted from certified reference materials are presented. The t-test has been used to compare results obtained by the two operators to assess whether they differ significantly from each other within a 95% confidence interval.

Table 4: Equations of calibration curves for the ΔC_t method obtained in a number of amplification runs. Linear dynamic range and PCR efficiency are evaluated through R^2 and slope values.

ΔC_t vs log(%RR soy)	**R^2**
y= -3.3544x+8.469	0.999
y= -3.4096x+8.3914	0.996
y= -3.2382x+8.307	0.994
y= -3.4513x+8.3422	0.998
y= -3.3722x+8.1275	0.991
y= -3.3593x+8.4877	0.998
y= -3.5093x+8.5772	0.992
y= -3.5794x+8.477	0.997
y= -3.4376x+8.4695	0.990
y= -3.3001x+8.9693	0.998
y= -3.4749x+8.5258	1.000
y= -3.5711x+8.575	0.996
y= -3.3763x+8.5892	0.993
y= -3.3632x+8.5915	0.995
y= -3.4686x+8.5137	0.995
y= -3.5535x+8.3106	0.997
y= -3.367x+8.2278	0.991

ΔC_t vs log(%RR soy)	**R^2**
y= -3.167x+8.0913	0.996
y= -3.2487x+7.9719	0.994
y= -3.3572x+8.012	0.992
y= -3.3976x+8.2354	0.996
y= -3.3944x+8.2016	0.998
y= -3.2996x+8.0513	0.997
y= -3.0985x+8.0856	0.996
y= -3.0343x+7.8581	1.000
y= -3.3429x+7.8577	0.992
y= -3.1867x+8.0136	0.999
y= -3.2932x+7.9952	0.999
y= -3.3004x+7.938	0.999
y= -3.1524x+9.0899	0.999
y= -3.346x+8.1106	0.999
y= -3.2872x+8.0771	0.999
y= -3.1976x+7.6848	0.998

Table 5: Results obtained from two different operators within a one-month period on the same DNA extracted from certified reference materials. The t-test has been used to compare results obtained by the two operators.

Operator a

Ref. material	**test 1a**	**test 2a**	**test 3a**	**mean**	**Standard deviation**
RR 0.1%	0.10	0.10	0.10	0.10	0.00
RR 0.5%	0.51	0.48	0.46	0.48	0.03
RR 1%	0.97	1.16	0.95	1.03	0.12
RR 2%	2.47	2.25	2.28	2.33	0.12
RR 5%	4.22	4.17	4.79	4.39	0.34

Operator b

Ref. material	**test 1b**	**test 2b**	**test 3b**	**mean**	**Standard deviation**
RR 0.1%	0.10	0.10	0.10	0.10	0.00
RR 0.5%	0.48	0.48	0.50	0.49	0.01
RR 1%	1.00	1.01	1.02	1.01	0.01
RR 2%	2.40	2.19	2.20	2.26	0.12
RR 5%	4.37	4.71	4.56	4.55	0.17

Ref. material	σ_t	$t_{calculated}$	$t_{tabulated}$ **(95% CI)**
RR 0.1%	0.00	0.00	2.78
RR 0.5%	0.02	0.21	2.78
RR 1%		0.25	4.30
RR 2%	0.12	0.72	2.78
RR 5%	0.27	0.69	2.78

The specificity of the method can be assessed theoretically through nucleotide database search and experimentally against target and non-target sequences from certified reference materials and/or samples from proficiency testing schemes.

The validation process allows evaluation of whether the method is fit for the purpose, at the same time providing the necessary background to establish criteria and requirements to be fulfilled in each PCR run: acceptable values for the slope and the coefficient R^2 of the calibration curves should be settled, as well as the maximum RSD of C_t replicates values. In our experience, analysing a large number of real-time PCR, run within a three-year period, acceptability values for these criteria should be the following:

$-3.6 \leq$ slope ≤ -3.0

R^2 calibration curve ≥ 0.98

RSD_{Ct} of PCR replicates $\leq 1.5\%$

In case the ΔC_t method is used for quantification, Ct values corresponding to LOD and LOQ can be determined for each run and used as cut-off values for the expression of results as non-detectable, detectable but non-quantifiable, or quantifiable.

The minimum number of PCR replicates and calibration points should be carefully established, taking into account the above-mentioned parameters and the influence on measurement uncertainty estimation.

As a preliminary step before the quantitative real-time PCR, a monitor run performed on the endogenous gene at different dilutions allows quality, quantity and amplifiability of the extracted DNA to be estimated, providing a tool to detect inhibition effects.

Finally, to keep the system under statistical control, a control chart should be set up with a CRM or a known positive sample and constantly checked for compliance with acceptability rules (e.g. Westgard rules, [9]).

Conclusion

The practical approach outlined is sufficiently detailed to be directly applicable both to relative and absolute quantification, increases the reliability of results and represents a necessary preliminary step before proceeding to a time-consuming and costly full validation through collaborative trials. Harmonisation of validation procedures is needed.

References

1 Regulation (EC) n° 1829/2003 of the European Parliament and of the Council of 22 September 2003 on genetically modified food and feed. OJ L 268, 18/10/2003, pp 1-23.

2 Consideration of the methods for the detection and identification of foods derived from biotechnology – General approach and criteria for the methods CODEX CX/MAS 05/26/9 (January 2005).

3 Definition of minimum performance requirements for analytical methods of GMO testing – ENGL 25/01/2005.

4 Draft European Standard prEN ISO 24276:2002.

5 Draft European Standard ISO/FDIS 21570:2003(E).

6 OIE Manual of diagnostic tests and vaccines for terrestrial animals, 5th edition, 2004.

7 Hübner P, Waiblinger HU, Pietsch K, Brodmann P : Validation of PCR methods for quantification of genetically modified plants in food. J AOAC Int 2001;84,1855-1864.

8 Schwarz G, Bäumler S, Block A, Felsenstein FG, Wenzel G: Determination of detection and quantification limits for SNP allele frequency estimation in DNA pools using real-time PCR. NAR 2004;32:24.

9 Westgard JO: Internal quality control: planning and implementation strategies. Ann Clin Biochem 2003;40:593-611.

Dr. I. Ciabatti, Istituto Zooprofilattico Sperimentale Lazio e Toscana, via Appia Nuova 1411, IT-00178 Rome, Italy.
E-mail: iciabatti@rm.izs.it

Vannier P, Espeseth D (eds): New Diagnostic Technology: Applications in Animal Health and Biologics Controls. Dev Biol (Basel). Basel, Karger, 2006, vol 126, pp 87-88.

Diagnostic Testing in the Context of International Trade with Animal Products: Challenges for Cross Disciplinary Collaboration and Methodological Research

M. Greiner

International EpiLab, Danish Institute for Food and Veterinary Research, Soeborg, Denmark

Introduction

Diagnostic testing for animal diseases relevant to international trade occurs for various purposes, such as (i) demonstration of freedom from infection in the population; (ii) demonstration of freedom from infection or agent in individual animals or products for trade purpose; (iii) eradication of infection from defined populations; (iv) confirmatory diagnosis of clinical cases; (v) estimation of prevalence of infection to facilitate risk analysis and (vi) determining the immune status in individual animals or populations [1]. It is a commonly accepted notion that the diagnostic performance characteristics of tests should be established and assessed with regard to the specific purpose of the tests ("fitness for purpose" principle).

The Challenge

The underlying principles of the diagnostic process are well established in each of the disciplines such as virology, microbiology, laboratory, statistics and epidemiology. However, since these disciplines approach the issue of diagnostic testing from different angles, they pursue different quality goals and use a different terminology. No overarching theoretical concept exists that could govern the process of test evaluation in a generic fashion. Limited resources and the lack of a "gold standard" are typical constraints of evaluation studies and the possibilities for dealing with these constraints differ from case to case. The evaluation of diagnostic tests is a complex problem, which cannot be solved alone with any laboratory or statistical method used in separation.

Cross Disciplinary Approach

Test validation to the required OIE standards can only be achieved in close cooperation of different professional groups. In order to stimulate such collaboration, a first "International Meeting on the Design and Analyses of Diagnostic Test Evaluation Studies" was conducted in Nairn, Scotland (28-30 March 2005). More than 50 participants from 12 countries with different professional backgrounds (Fig. 1) discussed issues related to the topic. A report of this meeting is available [2].

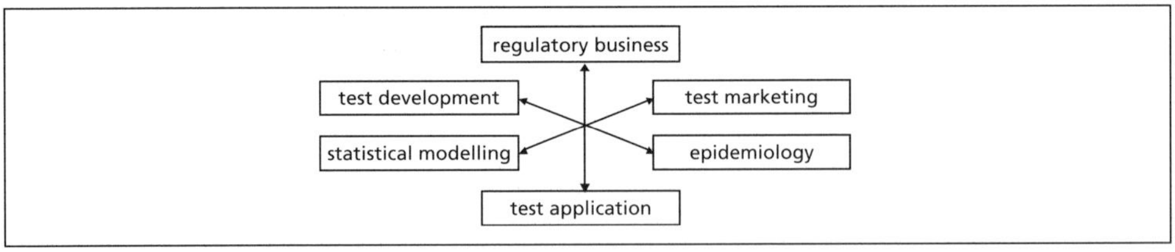

Fig. 1: International Meeting on The Design and Analysis of Diagnostic Test Evaluation Studies. 2005: A place of dialogue among six professional areas of expertise.

The New OIE Approach

The "OIE assay validation template" [3] provides the structure of dossiers submitted to the OIE for certification of diagnostic assays. Herewith, requirements for the process of assay validation are formulated.

Methodological Research Needed

Science-based methods are in principle available to address all aspects of test validation in a manner compliant with OIE standards. However, the practitioner is often confronted with specific situations, which are not considered in the methodological literature. In the following, some examples are given for areas that would benefit from applied methodological research.

Laboratories should be assisted to identify and apply best-suited statistical methods and indices to control the analytical stability of the diagnostic process. Statistical methods for estimation of sensitivity and specificity without a gold standard, such as latent class models (LCM), should be validated for situations encountered in practice, e.g. regarding the quality and amount of prior information, characteristics of tests available and sampling constraints. The value of meta-analysis for the synthesis of empirical knowledge should be re-evaluated in comparison to Bayesian approaches.

In the report of the "International Meeting on the Design and Analyses of Diagnostic Test Evaluation Studies", it was, among others, concluded that

- Guidelines should be written for those are lacking, particularly on the use of LCM.
- Strategic plans for knowledge transfer, particularly on the issue of LCM, should be developed and the dialogue between and within the different expert groups should be continued.
- The feasibility, practicability and efficiency of all procedures should be considered as critical constraints.
- The submission of a validation dossier, following the OIE guidelines, should be financially feasible for the test developer.

Acknowledgements

Marios Georgiadis (Greece), George Gunn (UK), Wesley Johnson (USA), Koen Mintiens (Belgium), Nils Toft (Denmark), Dr. Koen Mintiens (Belgium) and Nils Toft (Denmark) were co-organisers of the cited meeting and Peter Wright (Canada) provided input to the report.

References

1 www.iaea.org/programmes/nafa/d3/mtc/report-cmoie-validation.pdf

2 www.dfvf.dk/Default.asp?ID=9727.

3 www.oie.int/vcda/eng/en_background_VCDA.htm.

Dr. M. Greiner, International Epilab, Danish Institute for Food and Veterinary Research, Morkhoj Bygade 19, DK-2860 Soeborg, Denmark.
E-mail : mgr@dfvf.dk

Vannier P, Espeseth D (eds): New Diagnostic Technology: Applications in Animal Health and Biologics Controls. Dev Biol (Basel). Basel, Karger, 2006, vol 126, pp 89-97.

Practical Experiences of Moving Molecular Diagnostics into Routine Use at the Veterinary Laboratories Agency

J. Sawyer[1], P. Wakeley[1], D. West[1], C. Fearnley[1], S. Anderson[1], M. Flowers[1], K. Webster[1], J. Errington[2], R. Williams[2]

[1] Veterinary Laboratories Agency, Biotechnology Department Weybridge, Addlestone, U.K.

[2] VLA Penrith Laboratory, Penrith U.K.

Abstract: The practical bench application of molecular tests (here defined as nucleic acid-based tests) in a routine testing situation is not without its challenges. This paper outlines the approaches we take at the Veterinary Laboratories Agency (VLA) and highlights some of the practical issues which we have found to be important for the successful introduction and use of molecular tests in a routine testing environment. The potential advantages of molecular tests, and factors which dictate which tests are adopted for routine testing, are discussed before giving some examples of molecular tests in routine use at the VLA. The instrumentation, reagents and assays we use are outlined, followed by sections of how we approach validation and how we manage and resource the transfer of molecular tests into routine use.

Introduction

The practical bench application of molecular tests (here defined as nucleic acid-based tests) in a routine testing situation is not without its challenges. This paper outlines the approach that we take at the Veterinary Laboratories Agency (VLA) and highlights some of the practical issues which we have found to be important for the successful introduction of molecular tests into routine use.

Why Use Molecular Tests?

The driving force for developing new tests (including molecular tests) comes from a need to test for, or deliver something, that existing tests cannot. Molecular tests can be more sensitive, faster, cheaper or allow new levels of discrimination or identification that are not possible using existing approaches. These potential advantages have led to a large number of published assays for animal disease and veterinary testing [1]. We view molecular tests very much as complementary to existing tests and approaches; they are not a panacea to solve all our testing problems, but are a powerful tool in helping to improve the range of tests that the VLA provides.

What Determines which Molecular Tests End up in Routine Use?

Customers are key to determining which tests make the jump from research to routine use. Without a demand and a willingness to pay for tests on a continuing basis then tests will not be taken into routine use, regardless of the availability of a method. Many established tests are fit for the purpose and do not need replacing. Because of the cost of introducing new tests, new testing approaches are likely to be concentrated in new areas of investigation or "politically hot" areas where there are obvious failings in existing tests.

Examples of Molecular Tests Carried out at the VLA

The VLA carries out around 30 molecular tests routinely, some of which are listed in Table 1

For example, VLA is evaluating the use of PCR to detect high temperature rendered animal material in feeds at a lower concentration than is possible by existing protein-based methods. PCR [2] was used in the Classical Swine Fever outbreak in 2000 and provided a rapid result (within a day) as opposed to waiting several days for virus isolation. By the end of the outbreak, clinical in-field decisions were being made on the strength of the PCR results. This is one of the first examples of a PCRbased test being used in a 'frontline' outbreak situation. A recently developed PCR test for *Taylorella equigenitalis* (which causes Contagious Equine Metritis) allows detection in less than a day, in contrast to several days culture. The Bovine Viral Diarrhoea Virus (BVDV) assay is used to confirm positive animals identified using antigen ELISA from blood samples. It can distinguish Type 1 and 2 BVDV virus (PCR reactions are carried out in a single tube), identify persistently infected calves which can be missed by antigen ELISA, and is also used for testing abortion material (which can be problematic with ELISA).

"DNA fingerprinting" is another important application area for molecular diagnostics. The ability to subdivide organisms into different strains or groups much more easily and routinely than when using older methods can help trace outbreaks and allows the spread of organisms to be monitored in the environment. For example, the use of a fingerprinting technique called spoligotyping [3] has allowed the population structure and epidemiology of *Mycobacterium bovis*, the causative agent of Bovine TB, to be monitored. This helps to provide information on the spread of the disease and contributes to the development of control measures [4].

Table 1: Examples of molecular tests run at the VLA.

Target	Test type	Advantage
Bovine Viral Diarrhoea virus (BVDV)	Real-time reverse – transcriptase PCR	Can distinguish Type 1 and 2 virus; can identify PI calves; suitable for use on abortion material
Animal material in feeds	Real-time PCR	More sensitive detection of high temperature rendered animal material in animal feeds
Classical swine fever	Real-time reverse – transcriptase PCR [2]	Faster test than virus isolation
Taylorella equigenitalis	Real-time PCR	Faster than culture; can distinguish *Taylorella asinigenitalis*
Pathogenic leptospires	Real-time PCR	Faster than culture, Leptospira culture takes many months and isolates may be overgrown by other contaminants or not viable
Rabies	Real-time reverse - transcriptase PCR [5]	Rapid test; distinguishes three genotypes
E.coli virulence genes	Real-time PCR	Rapid test, multiplex replaces a variety of tests including Vero cell assays
PrP gene in sheep	DNA sequencing	Identifies individual genotypes which are linked to scrapie susceptibility
Mycoplasma spp.	PCR and DGGE [6]	Allows identification and species/strain identification
Mycobacterium bovis	Spoligotyping [3,4]	Allows strain identification

Tools – Instrumentation, Assays and Reagents

Instrumentation

Improvements in instrumentation, robotics and reagents are making the use of molecular testing in a routine environment more realistic. We view real-time PCR [7] as central to our development of routine molecular diagnostics because of its inherent advantages for routine testing (closed tube format and lack of gel electrophoresis step, simplified result call). Combining robotics (for nucleic acid extraction, reagent and sample dispensing) with real time PCR to allow testing with minimal manual intervention is key to making molecular diagnostics affordable and more efficient. Lower cost of real time PCR instruments is also a major factor in allowing us to make more widespread use of this technology.

We make widespread use of the Roche MagNaPure extraction robot and our "workhorse" real time PCR machine is the Stratagene MX3000P. The MagNaPure instrument is useful as protocols and kits are available "off-the-shelf", requiring minimal user training, and kits make standardisation more straightforward in a routine testing environment. We have found the MagNaPure to be a reliable and useful piece of instrumentation. The alternative is to purchase robotic systems which require programming and development of in-house methods and protocols. This adds flexibility but requires more training and the risk of serious consequences if key staff with programming skills and knowledge leave the organisation.

The limitations of the MagNaPure system include a limited 32-sample capacity and reagents being packaged in aliquots for 32 samples. Although it is possible to process small numbers of samples, the packaging of the reagents means that effectively they must be used in these batch sizes. This means as many reagents are used to process three samples as 32 samples, which leads to a large increase in cost. For some diseases, where rapid turnaround of small numbers of samples is required and batching of samples is not possible, the use of a robot becomes uneconomic. In this case we rely on manual nucleic acid extraction kits. Although Roche have introduced a MagNaPure Compact which is capable of processing a smaller number of samples, the chemistry and reagents used are not identical, which makes it difficult to implement for assays validated on the larger MagNaPure instrument. The range of kits available is also not so comprehensive.

An ideal situation would be one where robotic nucleic acid extraction was completely scalable using the same instrumentation and reagents. In this way, small numbers of routine extractions could be carried out on a day-to-day basis in 'peacetime' using the same reagents and instrumentation as in an outbreak situation. Critically, this means staff are familiar with the technology, and instruments do not have to be re-commissioned after long periods of not being used. The range of instrumentation and companies involved in robotics is increasing all the time and selection of the most appropriate robotic instrumentation should be preceded by a thorough review of what is on the market.

The use of robotics is also useful for the set up of PCR reactions and loading of samples. Robots can be used to set up reactions more accurately and with greater precision than manual set up [8]. We have also found that using robotics to transfer nucleic acids from storage plates to PCR reaction vessels is more accurate and less prone to contamination than even experienced staff can achieve manually.

We try, as far a possible, to encourage common instrument platforms across the organisation to facilitate the transfer of methods between research and routine testing areas and between laboratories. This requires a corporate view to implement and is sometimes hard to achieve. For example, some instruments and reagents may be flexible, suiting research use, but be less suited to routine application. If not taken into account at an early stage, these different requirements might mean that extensive re-design or validation of assays may be required before a research tool can be transferred to routine use.

Reagents

Although ideally we would use a limited range of reagents (for example to simplify ordering, reagent storage and assay set-up), the reality is that we use reagents and enzymes from a wide range of manufacturers. This reflects historical developments, the preferences of research groups and the performance characteristics of different products. Once an assay has been validated it is often not cost effective to revalidate with a new product even if it did offer advantages in terms of cost or convenience. For example, for some older assays we prepare master mixes from individual components (buffers, dNTPs, magnesium chloride) and use non-hot-start Taq rather than commercial master mixes. If these assays were developed today it is likely that we would have used commercial master mixes.

The impact of the availability of commercial 'off-the-shelf' reagents such as hot-start Taq Polymerase, improved enzymes and "ready-to-go" master mixes should not be underestimated. There has been a steady move away from 'home-brew' reagents to "off-the-shelf" commercial reagents and kits that seems likely to continue. Although it is sometimes argued that all assays should be prepared ab initio using individual components ("so you understand how they work") in reality the advantages that master mixes offer (convenience, speed, improved consistency) is difficult to ignore in a routine testing environment, particularly when operating to third party quality accreditations such as ISO17025. Importantly, the validation process will provide data demonstrating the assay (and hence the commercial mastermix) is fit for the purpose.

Real-time PCR chemistry

To a large extent, we rely on TaqMan chemistry for most of our assays as it has proved to be robust and reliable, although we do carry out some hybridisation probe-based assays. Most of our assays are based on standard FAM-TAMRA probes, although we also use other dyes (Yakima Yellow, HEX) as well as dark quenchers for certain applications.

Reagent consistency

Because routine testing can lead to the use of large amounts of reagents, we frequently need to change batches of reagents. In our experience, it is important to check for batch-to-batch consistency by performing parallel testing of old and new stocks on the same test samples (for example batches of primers and probes). This gives reassurance that a test is not 'drifting' over time due to changes in reagents and provides data with which to approach manufacturers if faulty batches are received.

Within the laboratory there is much that can be done to improve consistency by thinking about how assays are used day-to-day and setting up to make routine use as straightforward as possible. Actions as simple as aliquoting reagents into single use aliquots (to prevent freeze-thawing effects and increase convenience) can be very beneficial.

PCR control assays

We have been exploring the use of control assays (internal amplification controls, mimics and housekeeping genes) exploiting the ability of real-time instruments to detect several dyes. We are not yet convinced that the routine use of controls amplified in the same tube (multiplexing) is robust enough for routine use, although we are continuing to explore the possibilities. In assays that require a control gene our general approach would be to run separate single-plex assays for both targets and control genes. Although intuitively less satisfying we find that practically this is generally the best approach.

From an ease of testing point of view, the use of housekeeping genes is our preferred control. These targets have the advantage of being inherently part of the sample being tested and, as such, test for both sample integrity and correct PCR set-up. In our experience, the higher target concentration of typical housekeeping genes often leads to a reduction in sensitivity when testing for the target gene in the same tube, presumably by competition. A housekeeping gene with a low copy number or expression would seem to be a sensible approach and we are currently exploring the possibilities. Advances in reagents and chemistries are also being brought to bear on this problem; for example, a recent product from Qiagen claims to allow multiplexing of targets without the typical reduction in amplification efficiencies of the lower concentration target. This warrants further evaluation in a routine diagnostic setting.

Internal amplification controls or mimics, where artificial nucleic acids are added to the sample, have the advantage that their concentration can be adjusted but they are much more awkward to use routinely (making stocks, aliquoting, possibility of contamination) and do not account for the storage histories of samples in terms of degradation. An advantage of this approach is that it can be adopted in sample types where use of a housekeeping gene might be limited by lack of host nucleic acid (urine, serum).

Multiple outcome PCR

One area where we have exploited the use of several dyes is in the use of what we call "multiple outcome PCR". Here, a single primer set amplifies amplicons (for example from different strains or genotypes) which are distinguished with the use of specific TaqMan probes labelled with different dyes (all present in the same tube). For example, we have used this approach for distinguishing BVDV Types 1, 2 and border disease and also for rabies genotypes [5].

Sequencing

Sequencing is also an essential, and increasingly important, approach in our testing portfolio and we use a capillary sequencer which has simplified routine analysis. In addition to assays where sequencing is the primary testing method (for example, PrP gene analysis for scrapie susceptibility testing in sheep) sequencing PCR products for confirmation is an essential part of validating a PCR-based test. The sequencer is also being used increasingly for methods such as VNTR and MLST typing.

Validation, SOPs and Quality Systems

The key documents relating to the routine use of any test are the validation document and Standard Operating Procedure (SOP). In our experience, the availability of a published research method is often only the starting point in delivering a test that is validated and suitable for day-to-day use on clinical material. Validation of molecular tests by comparing them to a Gold standard test may be complicated by the PCR being more sensitive, resulting in apparent reduced specificity. To an extent this may be resolved by assessing sample provenance, clinical history and sequencing any PCR products to confirm identity. Another major practical issue is access to sufficient samples for the validation process to obtain robust test performance statistics. For many less common diseases, large numbers of known positive samples are difficult (or impossible) to obtain.

It is not always possible to conform to suggested guidelines (e.g. OIE recommendations on test validation [9,10]). Faced with low numbers of samples for test evaluation, or for tests with no gold standard, we have introduced molecular tests as 'partially validated' and then added validation data as we test significant numbers of clinical samples. In this system, positives are confirmed by other means and a sample of negatives are also confirmed as suitable (non-inhibitory) for PCR testing using control genes. This principle of 'continuing' validation allows for the rapid introduction of new tests and reduces the cost of validation. This process is used under defined conditions. It is only applicable when there is sound evidence from testing an appropriate range of known cultures, spiked samples (to provide analytical data) and some clinical samples (to show that the target is available in particular tissues) that a test may be released as partially validated. We find that it is important to have clear guidelines for the validation process in order to maintain a consistent approach across our organisation.

The introduction of a third party accreditation and quality systems (such as IS09001:2000 and ISO17025) have also had a major impact on the practical day-to-day use of molecular tests and improved the quality of testing significantly.

Getting New Tests into Routine Use – Research Versus Routine Use, Management of Projects and Resources

The traditional route to a test in routine use is the development of a research tool which evolves into a routine test over time. At some stage in this process the staff who carry out this test typically change from the researchers who developed the test to routine testing staff.

Although it might be regarded as stating the obvious to say that there is a fundamental difference between research use and routine use of a test, there are real practical implications which are a challenge to overcome.

Within the VLA, the most important difference between research and routine is the amount of validation required before a test can be offered as a routine test. Importantly, the process involves the generation of data to demonstrate that the test is fit for the purpose and performs to a certain standard with the type of samples that will be received for routine testing. In addition, the routine application of tests requires stringent laboratory procedures (for example, ISO17025).

The jump from research to routine use is more difficult and time-consuming than is often appreciated. The generation of validation data can be a mundane and time-consuming process. Researchers can lose interest in a test following initial development (and often publication) and are typically not motivated by validating a test to the standards required for routine application. In addition, it is often difficult to find funding for this kind of work. A typical final milestone in a research proposal is often 'validate the test and introduce the assay as a routine test'. This is easy to write but the reality is that the effort and cost needed to achieve this is often underestimated. This can lead to a transfer gap where tests never make the jump to routine use and are not exploited as fully as they might be.

Recognising this problem, several years ago the VLA set up a technology transfer group to move new tests and testing technology into routine use. Although there has been a focus on the transfer of molecular tests in recent years, the workgroup is also involved in other test development areas. Crucially, the priority of the staff in this unit is to ensure that new tests are in place as part of the official VLA test portfolio (available for routine use with supporting SOPs and validation data). The technology transfer group works closely with research groups and routine testing staff. A typical project would involve all three groups carrying out different roles, which can vary from project to project. However, the goal of each project is the same – moving the test to routine use. It should be emphasised that this effort takes place alongside the more traditional role of research groups in promoting the adoption of new tests and technology (and in some cases they continue to work independently). The extra resources (people and time) and focus of the workgroup just provides an additional means of working towards a modern test portfolio within the organisation.

We have found it essential for the routine testing groups to be involved in any test development project from its inception, even if they do not carry out laboratory work initially. They should be involved in the validation effort and are the best people to finalise an SOP following their experiences of running the test day-to-day.

Human and Other Factors

In our experience, the most important factors influencing the successful use of molecular tests are human factors. Staff must be well trained, there needs to be good communication between different groups of staff and a clear understanding of why the test is required and how it is going to operate at the bench level. Working with customers (e.g. clinicians, veterinarians, epidemiologists, government) to explain the advantages, limitations and interpretation of molecular test results is also essential. There is no substitute for face-to-face meetings and honest relationships between staff.

Changing tests is often more difficult from a political and logistical point of view than a technical one; for example, tests that form part of international trade agreements are more difficult to change.

In reality, the introduction of new tests and technology is challenging. It takes time and is influenced by funding, politics and people being wary of change. Determination, pragmatism and a sense of humour are perhaps the most important thing when working towards the routine use of molecular tests!

Acknowledgements

We would like to thank the numerous VLA colleagues (too many to mention individually) who have contributed to the approach and assays described in this paper. Thanks also to Defra for funding much of the work.

References

1 Belak S, Thoren P: Molecular diagnosis of animal diseases: some experiences over the past decade. Expert Rev Mol Diagn 2001;1(4):434-443.

2 McGoldrick A, Bensaude E, Ibata G, Sharp G, Paton DJ: Closed one-tube reverse transcription nested polymerase chain reaction for the detection of pestiviral RNA with fluorescent probes. J Virol Methods 1999;79(1):85-95.

3 Aranaz A, Liebana E, Mateos A, Dominguez L, Vidal D, Domingo M, Gonzolez O, Rodriguez-Ferri EF, Bunschoten AE, Van Embden JD, et al: Spacer oligonucleotide typing of Mycobacterium bovis strains from cattle and other animals: a tool for studying epidemiology of tuberculosis. J Clin Microbiol 1996;34(11):2734-2740.

4 Durr PA, Hewinson RG, Clifton-Hadley RS: Molecular epidemiology of bovine tuberculosis. I. Mycobacterium bovis genotyping. Rev Sci Tech 2000 Dec;19(3):675-688.

5 Wakeley PR, Johnson N, McElhinney LM, Marston D, Sawyer J, Fooks AR: Development of a real-time, TaqMan reverse transcription-PCR assay for detection and differentiation of lyssavirus genotypes 1, 5, and 6. J Clin Microbiol 2005;43(6):2786-2792.

6 McAuliffe L, Ellis RJ, Lawes JR, Ayling RD, Nicholas RA: 16S rDNA PCR and denaturing gradient gel electrophoresis; a single generic test for detecting and differentiating Mycoplasma species. J Med Microbiol 2005;54(Pt 8):731-739.

7 Mackay IM: Real-time PCR in the microbiology laboratory. Clin Microbiol Infect 2004;10(3):190-212.

8 Bustin SA: Quantification of mRNA using real-time reverse transcription PCR (RT-PCR): trends and problems. J Mol Endocrinol 2002;29(1):23-39. Notes: TTU 263.

9 http://www.oie.int/eng/normes/mmanual/A_00013.htm.

10 http://www.oie.int/eng/normes/mmanual/A_00014.htm.

Dr. J. Sawyer, Veterinary Laboratory Agency, Biotechnology Dept., Woodham Lane, New Haw, Addlestone, KT15 3NB, Surrey, U.K.
E-mail: j.sawyer@vla.defra.gsi.gov.uk

Vannier P, Espeseth D (eds): New Diagnostic Technology: Applications in Animal Health and Biologics Controls. Dev Biol (Basel). Basel, Karger, 2006, vol 126, p 99.

SESSION III

Application in Detection of Pathogens and Screening Programmes

Chairman: *Paul Midtlyng,* Norway

Vannier P, Espeseth D (eds): New Diagnostic Technology: Applications in Animal Health and Biologics Controls. Dev Biol (Basel). Basel, Karger, 2006, vol 126, pp 101-105.

A User's Inter-Laboratory Comparison of Broodfish Screening for Infectious Pancreatic Necrosis Virus Using Molecular and Conventional Diagnostic Methods

A. Storset[1], Ø. Evensen[2], P.J. Midtlyng[3]

[1] Aqua Gen AS, Trondheim, Norway
[2] Norwegian School of Veterinary Science, Oslo, Norway
[3] VESO, Oslo, Norway

Abstract: Individual testing and subsequent removal of eggs from infectious pancreatic necrosis virus (IPNV)-positive parents is required for export of salmonid eggs to some farming countries. Testing by cell culture requires more than three weeks before the eggs can be released from quarantine and incurs significant logistic problems and costs. The feasibility of the RT-PCR testing method as offered by several laboratories was therefore evaluated during the current inter-laboratory comparison study. Frozen kidney sub-samples from 100 motherfish of Atlantic salmon (*Salmo salar* L.) were shipped to three diagnostic laboratories (A, B and C) for testing by RT-PCR and cell culture. Of the 100 examined samples, all proved IPNV-negative by cell culture. Thirty samples were positive by RT-PCR analyses, but only four of these samples were RT-PCR positive in two laboratories and none in any of the three laboratories. From a disease management point of view, the RT-PCR test outcomes gave no reasonable guidance as to which fish were truly infected and which batches of fertilised eggs should be discarded. This is clearly an unacceptable situation and calls for new research to standardise sample conservation, RNA extraction procedures and amplification techniques, to estimate method sensitivity and specificity, and to validate the method's performance and robustness to support disease control measures in salmonid aquaculture.

INTRODUCTION

Infectious pancreatic necrosis (IPN) is a highly contagious viral disease of young salmonid fish held under intensive rearing conditions [1]. The causative agent is a bi-segmented double-stranded RNA virus belonging to the genus *Aquabirnavirus*, family Birnaviridae. The disease has a wide geographical distribution, occurring in most, if not all, major salmonid-farming countries. IPN is considered the most serious viral disease in terms of its impact on Atlantic salmon *(Salmo salar)* production in Europe [2]. IPN virus strains differ considerably in their virulence and persistence [3] and can be found in apparently healthy populations of salmonids belonging to any age group, and in fresh water and seawater environments. Asymptomatic carriers shedding virus are important sources of horizontal transmission of IPNV. In brook trout and rainbow trout it has been shown that carrier broodstock can vertically transmit the virus to the progeny despite surface disinfection of the eggs [4,5] and IPN has therefore long been considered a vertically transmissible disease. For the import of fertilised salmonid eggs, testing and subsequent removal of progeny from IPNV-positive parents is therefore required in some countries, e.g. Chile and Scotland. The only diagnostic method recommended by OIE for IPNV testing of parental fish is cell culture isolation [6], which requires more than three weeks before the eggs can be released from quarantine. For a more rapid and cost-efficient testing alternative, RT-PCR is being offered by various commercial laboratories. As a supplier of salmonid eggs to international customers, AquaGen AS therefore wanted to evaluate the feasibility of the RT-PCR testing method by organising a small and informal inter-laboratory comparison.

Materials and Methods

At spawning, kidney samples were taken from 100 female Atlantic salmon (*Salmo salar* L.) originating from a population that had experienced an IPN outbreak as post-smolts. Immediately before egg collection and sampling, the fish were anaesthetised with benzocain and exsanguinated. Each kidney sample was split into three sub-samples, frozen in liquid nitrogen, packed with dry ice and one sub-sample from each fish was shipped to three diagnostic laboratories (A, B and C) offering IPNV testing services to the aquaculture industry. Each laboratory used its own in-house procedures for the RT-PCR based upon published protocols [7,8]. Laboratories B and C also analysed the samples using cell culture according to OIE recommended procedures [6]. Upon receipt of the results, agreement between the testing outcomes was calculated using the Win Episcope version 2.0 software [9].

Results

Of the 100 examined samples, 70 proved IPNV-negative in all tests. The remaining 30 samples were also cell culture negative in both laboratories performing this method. There was, however, a major and unexpected difference in the outcome of RT-PCR analyses between the three laboratories (Table 1). In laboratory A, 16 of the remaining 30 samples were RT-PCR positive, whereas only one of these 16 samples yielded RT-PCR positive results in laboratory B, and none in laboratory C. Nine out of the 30 samples were RT-PCR positive in laboratory B and laboratory C; however only four samples gave a positive outcome in both. The RT-PCR analyses thus yielded only chance agreement between laboratories B and C, and even poorer than chance agreement between laboratories A and B, and between laboratories A and C (Table 2).

Table 1: Results from 30 out of 100 Atlantic salmon motherfish subjected to tests for the presence of IPN virus in kidney samples. + = IPNV-positive; - = IPNV-negative; (+) = weak IPNV- positive reaction. 70 fish unanimously yielding negative results are not shown.

	Laboratory A	Laboratory B		Laboratory C	
Fish no	**RT-PCR**	**Cell culture**	**RT-PCR**	**Cell culture**	**RT-PCR**
2091	-	-	(+)	-	-
2092	+	-	-	-	-
2098	-	-	(+)	-	(+)
2101	+	-	-	-	-
2107	-	-	-	-	+
2116	+	-	-	-	-
2117	-	-	+	-	-
2118	-	-	-	-	-
2119	+	-	-	-	-
2124	-	-	-	-	+
2125	+	-	-	-	-
2126	-	-	-	-	+
2129	+	-	-	-	-
2130	-	-	+	-	+
2131	-	-	-	-	(+)
2132	+	-	-	-	-
2134	+	-	-	-	-
2135	+	-	-	-	-
2136	+	-	-	-	-
2143	+	-	-	-	-
2151	-	-	+	-	-
2152	+	-	-	-	-
2156	-	-	(+)	-	-
2157	+	-	+	-	-
2160	-	-	-	-	+
2165	+	-	-	-	-
2168	-	-	+	-	+
2177	+	-	-	-	-
2179	-	-	+	-	(+)
2186	+	-	-	-	-
# positive fish	16	0	9	0	9

Table 2: Calculation of agreement between three laboratories testing kidney samples from Atlantic salmon motherfish for the presence of IPN virus by RT-PCR.

	# positive fish in common	**# negative fish in common**	**Observed minus chance proportion of agreement**	**Kappa estimate**
A vs. B	1	76	-0,009	-0,040
A vs. C	0	75	-0,029	-0,130
B vs. C	4	86	0,064	0,389

Discussion and Conclusion

The fact that cell culture unanimously gave negative results suggests that if at all present, the titre of viable virus was very low in all specimens examined. This fact may in part explain the apparent contradictory results among the RT-PCR positive samples, as generally poor test agreement may be expected near any method's limit of detection. In our case, each kidney sample was divided into three parts without prior homogenisation in order to maximise the preservation of virus during transport. The concentration of viable virus or virus genomic material may therefore have varied between the sub-samples.

The outcome of the RT-PCR analyses were quite disappointing and clearly suggests that the RT-PCR protocols used were not standardised, are technically very vulnerable, or both. In 70 of the 100 fish, the RT-PCR analyses gave a unanimous outcome in all three laboratories, suggesting that there was no detectable IPNV genomic material in either sub-sample from these animals. Unfortunately, we do not know how many of the 100 broodfish sampled were indeed carrying IPNV infection in their kidneys and an estimation of sensitivity or specificity of the RT-PCR applications based on the current data is therefore impossible.

From a disease management point of view, the RT-PCR test outcomes gave no reasonable guidance as to which batches of fertilised eggs should be discarded. If we consider all fish testing RT-PCR positive in one or more laboratory being infected, exclusion of breeders based on only one laboratory's test result would allow the release of eight- or 22-egg batches derived from motherfish testing positive in the other laboratories, respectively. If we consider that only cell-culture positive motherfish present a real risk for vertical transmission of IPNV, between 5 and 16% of the egg batches would have been unnecessarily discarded based on the RT-PCR results from one of the laboratories, and 30% based on all three RT-PCR analyses. This is clearly a rather unacceptable situation in an industrial setting.

The control of IPNV in salmonids is still hampered by the lack of knowledge about which level of contamination of sexual products is required to cause vertical transmission, or which virus titre in relevant tissues of parental fish is associated with an increased risk for vertical transfer [10]. The current study shows that thorough sensitivity and specificity assessments using samples with known virus titre, comprehensive standardisation of sample preparation and conservation, RNA extraction and amplification techniques, and comprehensive validation studies are required to make IPNV RT-PCR a reliable and cost-effective tool for disease control measures in salmonid aquaculture.

References

1 Skjelstad B, Brun E, Jensen I, Johansen L-H, Midtlyng PJ, Rimstad E, Stagg R, Evensen Ø: IPN in salmonids – a review. FHL Havbruk, Trondheim, Project report, 2003, p 118. http://www.veso.no/?Trondheim:Prosjektsammendrag_og_rapporter.

2 Aunsmo A, Holten K, Skjelstad B: Erfaringer med IPN i norsk oppdrettsnæring. Project report, FHL Havbruk, Trondheim, 2003, p 52. http://www.veso.no/?Trondheim:Prosjektsammendrag_og_rapporter.

3 Santi N: Infectious pancreatic necrosis virus infection in Atlantic salmon: experimental studies with emphasis on molecular mechanisms of virulence and persistence. Thesis for the degree of Doctor Medicinae Veterinariae, Norwegian School of Veterinary Science, Oslo, 2005, p 95.

4 Bullock GL, Rucker RR, Amend DH, Wolf K, Stuckey HM: Infectious pancreatic necrosis: transmission with iodine-treated and non-treated eggs of brook trout (*Salvelinus fontinalis*). Journal of the Fisheries Research Board of Canada 1976;33:1197-1198.

5 Dorson M, Rault P, Haffray P, Torhy C: Water-hardening rainbow trout eggs in the presence of an iodophor fails to prevent the experimental egg transmission of infectious pancreatic necrosis virus. Bulletin of the European Assocation of Fish Pathologists 1997;17(1):13-16.

6 Anonymous: Manual of diagnostic tests for Aquatic Animals, fourth edition. Office International des Epizooties, Paris, 2003, p 358.

7 Lopez-Lastra M, Gonzales M, Jashes M, Sandino AM: A detection method for infectious pancreatic necrosis virus (IPNV) based upon reverse transcriptase (RT)- polymerase chain reaction (PCR). Journal of Fish Diseases 1994;17:269-282.

8 Taksdal T, Dannevig BH, Rimstad E: Detection of infectious pancreatic necrosis (IPN)-virus in experimentally infected Atlantic salmon parr by RT-PCR and cell culture isolation. Bulletin of the European Association of Fish Pathologists 2001;21:214-219.

9 Thrusfield M, Ortega C, de Blas I, Noordhuizen JP, Frankena K: WIN EPISCOPE 2.0: improved epidemiological software for veterinary medicine, 2001. http://eie.unizar.es/ratio/soft_sp.htm.

10 Olesen NJ, Bovo G, Dannevig B, Munro E, Midtlyng PJ, Håstein T, Hill B: Work package 4 report: Broodfish testing for viral infections. Project report, VESO, Oslo, 2005, p 18. http://www.veso.no/fisheggtrade/literature.html

Dr. A. Storset, AquaGen AS, P.O. Box 1240 Pir-senteret, NO-7462 Trondheim, Norway.
E-mail: arne.storset@aquagen.no

Vannier P, Espeseth D (eds): New Diagnostic Technology: Applications in Animal Health and Biologics Controls. Dev Biol (Basel). Basel, Karger, 2006, vol 126, pp 107-113.

Swine and Circovirus

M. Banks[1], *S. Grierson*[1], *D. Tucker*[2], *M. Bailey*[3], *M. Donadeau*[4], *C. Sargent*[2], *D. King*[5], *M. Mellencamp*[6]

[1] VLA Weybridge, U.K.
[2] University of Cambridge, U.K.
[3] University of Bristol, U.K.
[4] Sygen International, Fyfield, U.K.
[5] IAH Pirbright, U.K.
[6] Sygen International, Kentucky, U.S.A.

Abstract: In the 1990s Post-weaning, multi-systemic wasting syndrome (PMWS) emerged in N. America and Europe as a major disease problem with significant welfare and economic consequences for pig producers. The disease, characterised by wasting, respiratory, enteric and lymphoid system problems in pigs of 4-16 weeks of age, has since spread so that today it has a global distribution. PCV-2 is consistently associated with PMWS, is more abundant in association with PMWS and is considered by many to be the causative agent of the syndrome. However, several lines of evidence indicate that PCV-2 is necessary but not sufficient to cause the full range of clinical signs associated with PMWS, suggesting the involvement of an as yet unidentified factor or factors.
The process of identifying unknown agents and their respective roles in the pathogenesis of complex syndromes now has an ever broadening spectrum of analytical techniques available. Immune phenotyping, cytokine responses, micro-array profiling, and proteomics are just some of the techniques available. This paper describes the philosophy and the application of these and classical techniques in an integrated, holistic manner to the problem of PMWS and circoviruses, by examination of samples collected from a prospective, clinical case-control study, and discusses some of the preliminary findings in relation to the efforts to understand the aetiopathogenesis of PMWS.

INTRODUCTION

Circoviruses are small (17-22 nm) DNA viruses with single-stranded, circular genomes. Porcine circovirus 1 (PCV-1) was first described in the 1980s as a contaminant of certain porcine pig cell lines and was apparently apathogenic [1]. In 1998, Meehan et al [2] described a circovirus isolated from pigs that was similar to PCV-1 and was associated with wasting. This virus had a <80% genetic homology to PCV-1 and was designated porcine circovirus 2 (PCV-2).

The emergence of post-weaning multi-systemic wasting syndrome (PMWS) and porcine dermatitis and nephropathy syndrome (PDNS) poses major financial and welfare problems to pig farmers. Although there is compelling evidence from experimental and field studies that point to an intimate association with PCV-2, the aetiopathogenesis of these syndromes has yet to be fully elucidated. Although PCV-2 is a key requirement in all experimental models that reproduce PMWS, when inoculated alone into conventionally suckled piglets, PCV-2 does not consistently reproduce the full range of disease signs seen in the field. Furthermore, our work using viral DNA extracted from archived, wax-embedded pig tissues, has demonstrated that genetically similar PCV-2 strains have been present in the U.K. pig herd for at least the last 30 years [3,4]. Moreover, a comparative study of the complete PCV-2 genome sequences failed to identify any consistent differences at the nucleotide, amino acid and motif level between strains isolated from pigs on farms free of PMWS and from pigs farms with current PMWS [5].

This demonstration that PCV-2 appears to be a necessary but not sufficient requirement for the expression of PMWS has led to considerable speculation regarding a requirement for additional co-factors such as immunological events, concurrent infection with other agents, husbandry practices, pig genetics or the influence of other environmental factors. In particular, immune stimulation arising from concurrent infection or vaccination is an attractive possibility. Indeed, this is supported by experimental evidence in young animals [6] and in on-farm vaccine trials [7] although a more recent study was unable to reproduce these findings suggesting that the mechanism may be more complex than first imagined [8].

Although many experimental inoculations and co-inoculations had been undertaken and detailed examinations of gross and histopathological changes associated with PMWS end stage disease performed, no study had thus far examined the events occurring in the pig in the prodromal period before the expression of clinical PMWS. Considering the severe immune dysfunction associated with PMWS [9], we hypothesised that there would be discernible changes in the prodromal period that could be associated with the progression to clinical disease. In the absence of clear evidence relating to the timing of these changes in vivo, and making the assumption that they occurred post-natally, it would be necessary to monitor events serially from birth to the age of disease expression. In this respect and in relation to pig numbers required and, inter alia, continuity/conformity of information, blood samples were the only practicably available medium from which to conduct this monitoring.

Because of the uncertainty associated with experimental reproduction of PMWS/PDNS, and because of the high incidence of PMWS in the U.K. during the conception of this study, we opted to conduct a prospective clinical case-control study on a small number of pig units that were experiencing clinical PMWS/PDNS.

This study involved blood sampling ear-tagged piglets at weekly intervals for the first eight weeks of the neonatal period. After the sampling period, animals were monitored until they developed clinical signs of PMWS/PDNS, or reached slaughter weight at approximately 20 weeks of age. To improve the prospects for the detection and characterisation of the changes or "triggers" associated with PMWS/PDNS, the blood samples were to be subjected to a comprehensive range of virological, serological, immunological and genetic and other examinations comparing samples from pigs that went on to develop disease with those that remained free of clinical signs of PMWS/PDNS.

A major benefit of this field approach is that the information obtained and factors having an impact on PMWS/PDNS expression will be relevant to current disease on farms. This is in contrast to experimental models where the environment and pig immune status often do not mimic the exact conditions on farm.

The concept of applying a wide range of analyses upon the samples was designed to provide an information matrix, with the results of the analyses informing each other and facilitating a holistic appraisal of immune and other events occurring in the blood of neonatal pigs up to the expression of clinical PMWS/PDNS. This paper describes these approaches and philosophies in relation to the investigation of a complex disease syndrome that has proved refractory to more conventional approaches.

Materials and Methods

Animals

A total of 125 apparently healthy new-born pigs (both sexes) were selected for inclusion in the study. This sample set comprised five piglets, from five litters, at five different pig farms. In-door, farrow-to-finish herds in the U.K. that were currently experiencing PMWS/PDNS were chosen for this study. Additional farm selection criteria included herds with conventional health status and vaccination schedules as well as good farm record keeping. To ensure that animals were effectively tracked throughout the study they were ear-tagged (All-Flex) before collection of the first blood sample. A clinical observation sheet recording all clinical signs but in particular those associated with PMWS/PDNS (including pallor, wasting, palpable lymph nodes, dyspnoea, arthralgia and dermatitis) was filled in for each pig daily.

Sample collection

The pigs were blood-sampled at weekly intervals from one week to eight weeks of age (outlined in Fig. 1). A parental sample was obtained from the dam of each litter on a single occasion at one week post-partum. 5 ml blood samples were collected into EDTA tubes (S-Monovette, Sarstedt). 2.5 ml of the EDTA-blood was immediately drawn into RNA stabilisation tubes (PAXgene, PreAnalytiX, Qiagen) and stored at 4°C for subsequent mRNA expression analysis. The PAXgene tube and remaining EDTA blood was transported overnight on ice to the laboratory for aliquoting and storage at 4, -20 or -80oC as appropriate for the anticipated analysis (a schematic of the panel of laboratory assays used is outlined in Fig. 2). After the eight-week sampling period, the animals were monitored weekly and where clinical signs indicative of PMWS/PDNS developed, the animal was killed and frozen and fixed tissues (inguinal, mesenteric and tracheal/bronchial lymph nodes and kidney) collected. Surviving pigs (at approximately 20 weeks) were sent to VLA regional laboratories for slaughter, post-mortem examination and tissue collection as for the affected pigs.

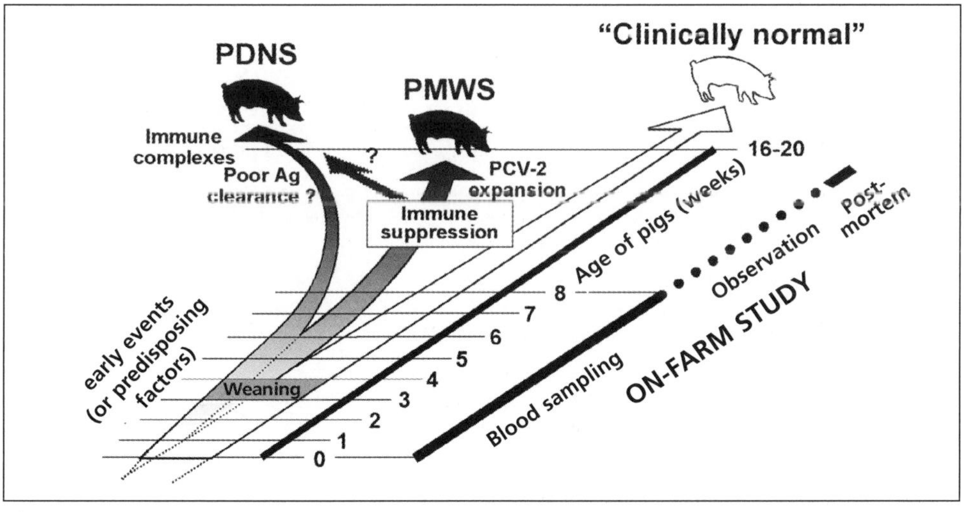

Fig. 1: The aim of the on-farm study is to investigate early events that occur before the expression of the clinical signs of PMWS/PDNS.

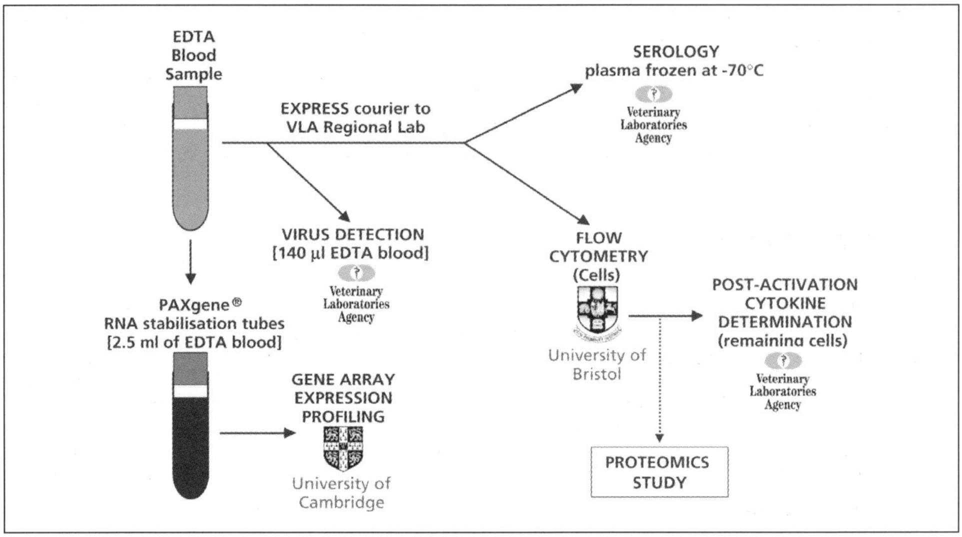

Fig. 2: A schematic overview of the procedure for processing the pig blood samples. 2.5 ml of blood will be immediately drawn into RNA stabilisation (PAXgene) tubes. The remainder of the EDTA blood will be shipped overnight at 4°C by courier for flow cytometric analysis. Other samples (plasma and blood for virological analysis) will be aliquoted and stored at -70°C until processed.

Case definition

The presence of PMWS or PDNS on each farm was confirmed before the start of the sample collection according to the case definition of [10]. This same definition was to be used in the selection of samples for all subsequent analyses.

Virology

In the light of the pivotal events in immunological development that occur in the neonatal period, it is possible that the timing of infection with PCV-2 (or additional pathogens) may play a critical role in the subsequent immune competence of the host, leading to development of PMWS/PDNS. Although many previous studies have detected the presence of PCV-2 (and additional pathogens) in end-stage disease, a temporal study investigating the timing of infection/exposure in individual animals that subsequently go on to develop PMWS or PDNS has not been published. In this study, the PCV-2 viral load will be determined in DNA extracted from the sequential bloods and post-mortem tissues using quantitative TaqMan® PCR. Serology for PCV-2 will also be carried out by indirect-ELISA. Additional implicated agents such as porcine parvovirus and porcine reproductive and respiratory syndrome virus will also be assayed in the samples using PCR and RT-PCR methods.

DNA microarray analysis

The PAXgene tubes containing the stabilised RNA were frozen at –70 °C until the end of the study when the clinical outcome of individuals was known. RNA from affected and un-affected pigs will be prepared using defined clinical and histopathological criteria. These samples will then be subjected to DNA micro-array analysis. Micro-array analysis is a powerful technique that allows the relative abundance of the mRNA transcripts of tens of thousands of genes to be assessed simultaneously (reviewed in [11]). The goal is to attempt to identify unique genes that are differentially regulated between animals that develop PMWS/PDNS and those that remain unaffected in the period leading up to, and overlapping with, the development of disease. In these experiments, the ability of the extracted RNA pools from affected and unaffected animals to hybridise to prepared pig gene libraries will be compared. In subsequent studies with tissues collected from animals sacrificed over the period of the study, these arrays may be supplemented with organ-specific libraries from appropriate tissues.

Table 1: The panels of monoclonal antibodies selected for flow cytometry analysis of EDTA-blood samples.

Panel	Specificity of Mab	Marker present on	Purpose of panel
1	CD45	Leukocytes	Leukocytes
2	1 CD4a	T_h (helper) cells	Activation and education status of T-lymphocytes
	2 CD8a	T_c (cytotoxic) cells	
	3 CD45RC	Naive/Inexperienced	
	4 CD25	Activated/Memory	
3	1 SWC3a	Monocytes+Granulocytes	Monocyte maturation
	2 SWC8	Granulocytes	
	3 CD163	Monocyte maturation	
4	1 sIgG	B cells	Expression of MHC class II on leukocyte populations
	2 CD3	T cells	
	3 MHCII dR	Antigen presenting cells	
	4 CD14	Monocytes	

Immunology

The immunological status of animals was assessed by immune phenotyping of leukocytes to detect the proportion and activation status of T- and B-lymphocytes and monocytes [12]. The four panels of monoclonal antibodies selected for this study are outlined in Table 1. After washing, bound monoclonal antibody was detected using isotype-specific fluorochrome-conjugated reagents, allowing the relative proportions of the cell populations to be quantified using a four-colour flow cytometer (Epics® XL-MCL [Coulter]).

In addition, whole blood cell cultures were set up, with and without the mitogen ConA (5 μg/ml). Cell pellets and cell-free supernatants were harvested and will be assayed by RT-PCR and ELISA for regulatory cytokine activity (IL-4, IL-10 and Interferon-γ).

Humoral immunity will also be assessed in the animals using ELISAs to quantify immunoglobulin classes and isotypes in plasma samples collected at the different times, in addition to detection of antibody to specific agents by ELISA.

Results and Discusssion

A total of 125 pigs were blood-sampled weekly between nought and eight weeks and then observed for PMWS/PDNS until slaughter. Sample collection, delivery and aliquoting have been successfully completed. Flow cytometry immune phenotype data have been obtained and preliminary analysis done. Immunohistochemistry (IHC) has been done on selected samples to confirm the status of pigs with respect to PMWS/PDNS. Limited serology has been done (initially concentrating on sib pairs). Preliminary assessment of clinical observation data has been completed.

In terms of significant clinical signs, relatively little was seen at farm D, but disease was seen to varying degrees in the sampled pigs of the other four herds. One case of PDNS was seen. For the breakdown of litters and herds with pigs that exhibited wasting, see Table 2.

Interestingly, in relation to the limited serology, many of the piglets sero-inverted to PCV-2 following decay of maternal antibody and were PCV-2 antibody negative by eight weeks of age. Almost all of these had sero-converted by the time of the 20 week

Table 2: Clinical outcome. Numbers of pigs in each litter that exhibited wasting at one or more time points.

Farm code	Litter number					Total
	1	2	3	4	5	
A	2	0	3	1	0	6
B	3	0	2	0	2	7
C	0	3	3	2	3	11
D	0	0	0	1	1	2
E	3	0	2	0	2	8

samples. Quantitative PCR will be used to correlate the PCV-2 level with the passive and active humoral immunity to this virus.

All the leukocyte phenotype data have been collected and transformed. There are some patterns but clearly, all neonatal conventional pigs will be experiencing rapid clonal expansion of leukocyte populations whether affected by the agents of PMWS or not. So, currently, in collaboration with Bristol University Veterinary School, sophisticated statistical packages are being applied to try and tease out specific markers for disease progression. In particular we will try and answer the question of which comes first, immune insult or PCV-2 expansion.

The micro-array data on selected samples will be assessed alongside the immune phenotype, serology, virology, cytokine (and potentially proteomics) data from the samples.

Conclusions

Following the failure of more conventional approaches in unravelling the aetiopathogenesis of PMWS/PDNS, this multidisciplinary study brings together a range of techniques, new and old, in order to understand the events at the molecular and whole animal level that precede the clinical expression of PMWS. By using naturally infected conventional pigs we avoided many of the potential pitfalls of experimental infections using CD/CD pigs. This combination of techniques and approaches, in particular the integrated study of host transcriptional/translational events preceding and during disease, represents a modern paradigm for the study of complex diseases of unknown or uncertain aetiology, facilitating the development of intervention and other therapeutic strategies for disease control.

Acknowledgements

This work is a collaborative project funded by DEFRA (ROAME project SE0517 and VLA seedcorn projects SC0119), the Veterinary Laboratories Agency, University of Bristol, University of Cambridge, Sygen International, PLC and PIC Europe. We acknowledge the help of Andrew Gresham and Graham Jackson of VLA Bury St Edmunds for assistance with the recruitment of pig veterinarians/herds and Paul Townsend of the VLA animal services unit for arranging the requisite Home Office licence courses for several of the pig veterinarians. This study would not have been possible without the co-operation of the participating pig farmers and practitioners.

References

1 Tischer I, Geldglom H, Vettermann W, Koch MA: A very small porcine virus with a circular single-stranded DNA. Nature 1982;295:64-66.

2 Meehan BM, McNeilly F, Todd D, Kennedy S, Jewhurst VA, Ellis JA, Hassard LE, Clark EG, Haines DM, Allan GM: Characterization of novel circovirus DNAs associated with wasting syndromes in pigs. J Gen Virol 1998;79:2171-2179.

3 Sandvik T, Grierson S, King DP, Spencer Y, Banks M, Drew TW: Detection and genetic typing of porcine circovirus DNA isolated from archived paraffin embedded pig tissues. ssDNA viruses of plants, birds, pigs and primates symposium, St-Malo, France 2001;111.

4 Grierson SS, King DP, Hicks D, Spencer Y, Drew TW, Sandvik T, Banks M: Pigs in the UK have been infected with PCV-2 since the 1970's. Arch Virol 2004a;149:1171-1183.

5 Grierson SS, King DP, Wellenberg GJ, Banks M: Genome sequence analysis of 10 Dutch porcine circovirus type 2 (PCV-2) isolates from a case-control study. Res Vet Sci 2004b;77:265-268.

6 Krakowka S, Ellis JA, McNeilly F, Ringler S, Rings D M, Allan G: Activation of the immune system is the pivotal event in the production of wasting disease in pigs infected with porcine circovirus-2. Vet Pathol 2001;38:31-42.

7 Kyriakis S C, Saoulidis K, Lekkas S, Miliotis CC, Papoutsis P A, Kennedy S: The effects of immuno-modulation on the clinical and pathological expression of postweaning multi-systemic wasting syndrome. J Comp Path 2002;126:38-46.

8 Ladekjær-Mikkelsen A–S, Nielsen J, Stadejek T, Storgaard T, Krakowka S, Ellis J, McNeilly F, Allan G, Bøtner A: Reproduction of postweaning multi-systemic syndrome in immunostimulated and non-immunostimulated 3-week-old piglets experimentally infected with porcine circovirus type 2. Vet Micro 2002;97-114.

9 Rosell C, Segalés J, Plana-Durán J, Balasch M, Rodríguez-Arrioja GM, Kennedy S, Allan GM, McNeilly F, Latimer KS, Domingo M: Pathological, immunohistochemical, and in-situ hybridisation studies of natural cases of Postweaning Multi-systemic Wasting Syndrome (PMWS) in pigs. J Comp Pathol 1999;120:59-78.

10 Sorden SD: Update on porcine circovirus and post-weaning multi-systemic wasting syndrome (PMWS). Swine Health Prod 2000;8:133–136.

11 Lockhart DJ, Winzeler EA: Genomics, gene expression and DNA arrays. Nature 2000;405:827-836.

12 Haverson K, Bailey M, Stokes CR, Simon A, LeFlufy L, Banfield G, Chen Z, Hollemwequer E, Ledbetter JA: Summary of the first round analyses of the third international workshop on swine leukocyte differentiation antigens. Vet Immunol Immunopathol 2001:80:25-34.

Dr. M. Banks, Veterinary Laboratories Agency, New Haw, Addlestone, Surrey, KT15 3NB. U.K.
E-mail: m.banks@vla.defra.gsi.gov.uk

Vannier P, Espeseth D (eds): New Diagnostic Technology: Applications in Animal Health and Biologics Controls. Dev Biol (Basel). Basel, Karger, 2006, vol 126, pp 115-116.

Does Real-Time RT-PCR for CSF Mark the Beginning of a Paradigm Shift in the Control of CSF?

K. Depner, B. Hoffmann, M. Beer

Friedrich-Loeffler-Institut, National Reference Laboratory for Classical Swine Fever, Greifswald – Insel Riems, Germany

Introduction

When "one conceptual world view is replaced by another" we are talking about a paradigm shift [1]. Based on new scientific data, Kuhn's concept of paradigm shifting will be used in this presentation to propagate the necessity of changes in the policy to control and eradicate Classical Swine Fever (CSF).

The present CSF control strategy is more than 20 years old. Two fundamental changes have occurred which justify a new policy. In the first place there is no longer a rationale for the strict non-vaccination strategy. Culling of healthy pigs has become an ethical and animal welfare issue and is no longer tolerated. Secondly, significant advances in diagnosis and modern vaccines have been made (real-time RT-PCR, DIVA strategy). The foundations for new concepts to control CSF based on novel diagnostic tools and vaccines have been laid.

The Current Policy to Control and Eradicate CSF (*Old Paradigm*)

According to EU Directive 2001/89/EC [2] the general measures for controlling CSF are:

- culling of all pigs within the infected premises,
- installation of protection and surveillance zones,
- restrictions for trade and transport.
- extensive epidemiological, clinical and laboratory examinations.

Additional special measures, such as culling of all pigs within a 1000m. radius and emergency vaccination can be used. However, vaccination has not been applied so far.

The logic behind the actual CSF policy is that only sero-negative pig populations are free from risk, while pigs with CSF antibodies are a risk factor. Vaccination becomes impossible since it would be a "proof" for the presence of CSF. Consequently, freedom from CSF is defined as freedom from CSF antibodies. Therefore, there is a general request for sero-negative pig populations to permit the international trade of pigs and pig products. In this strategic concept the serological diagnosis, which is an indirect diagnostic approach, is of paramount significance (the basis for the old paradigm).

The Future Policy for Controlling and Eradicating CSF (*New Paradigm*)

For avoiding mass culling of healthy uninfected animals a diagnostic strategy must be used which guarantees that only pig herds with virus-positive animals are removed and CSF virus no longer circulates within the pig population. The diagnostic approach has to focus on the presence of virus and no longer on the presence of antibodies. This means that the focus of disease control has to be shifted from the indirect approach (serological testing) to the direct approach (direct detection of virus). The logic behind the future CSF policy would be that only CSF virus-free pig populations are free from risk. Pigs with CSF antibodies are no risk factor. Thus, vaccination becomes a useful tool for CSF control. However, this shift can only be done if appropriate diagnostic tools are available.

The real time RT-PCR has the potential to replace the conventional diagnostic methods used at present [3]. With this tool we are for the first time in a position to change the strategy and to fulfil the above-mentioned requirement (control of virus freedom). The real time RT-PCR-based strategy has a higher reliability, since it is a direct search for CSF virus. Antibodies or cross reactions with other pestiviruses do not disturb the diagnosis. Vaccination does not interfere with the diagnosis. In the event of a CSF outbreak emergency vaccination can be used. Culling of healthy pigs within a 1000m. radius will be avoided since freedom from CSF virus will be checked by rapid PCR testing. The risk will not be higher than with the present strategy based on antibody control.

During the presentation, different models of the new CSF control strategy will be presented and discussed. The advantages of the PCR-based system compared to the present diagnostic approach are shown.

References

1 Kuhn T: The strategy of Scientific Revolutions, The University of Chicago Press, Chicago, 1962.

2 Anonymous: Council Directive 2001/89/EC of 23 October 2001 on Community measures for the control of classical swine fever.

3 Hoffmann B, Depner K, Beer M: Standardisation and validation of multiplex real-time PCR systems, in Vannier P, Espeseth D (eds): New Diagnostic Technology: Applications in Animal Health and Biologics Controls. Dev Biol (Basel). Basel, Karger, 2006, vol 126, pp 53-54

Dr. K. Depner, Friedrich-Loeffler-Institut, National Reference Laboratory for Classical Swine Fever, Boddenblick 5a, DE-17493 Greifswald – Insel Riems, Germany.
E-mail: robert.depner@rie.bfav.de

Vannier P, Espeseth D (eds): New Diagnostic Technology: Applications in Animal Health and Biologics Controls. Dev Biol (Basel). Basel, Karger, 2006, vol 126, pp 117-122.

Application of Molecular Diagnostic Methods to Penaeid Shrimp Diseases: Advances of the past 10 Years for Control of Viral Diseases in Farmed Shrimp

D.V. Lightner, B.T. Poulos, K.F.J Tang-Nelson, C.R. Pantoja, L.M. Nunan, S.A. Navarro, R.M. Redman, L.L. Mohney

OIE Reference Laboratory for Crustacean Diseases, Dept. Veterinary Science & Microbiology, University of Arizona, Tucson, AZ, USA

Key words: PCR, RT-PCR, *Penaeus*, shrimp viruses

Abstract: The most important diseases of farmed penaeid shrimp have infectious aetiologies. Among these are diseases with viral, rickettsial, bacterial, fungal and parasitic aetiologies. Diagnostic methods for these pathogens include the traditional methods of gross pathology, histopathology, classical microbiology, animal bioassay, antibody-based methods, and molecular methods using DNA probes and DNA amplification. While methods using clinical chemistry and tissue culture are standard methods in veterinary and human diagnostic laboratories, the former has not been routinely applied to the diagnosis of penaeid shrimp diseases and the latter has yet to be developed, despite considerable research and development efforts that have spanned the past 40 years. No continuous shrimp cell lines, or lines from other crustaceans, have been developed. Hence, when molecular methods began to be routinely applied to the diagnosis of infectious diseases in humans and domestic animals in the mid- to late 1980s, the technology was applied to the diagnosis of certain important diseases of penaeid shrimp for which only classical diagnostic methods were previously available. A DNA hybridization assay for the parvovirus IHHNV was the first molecular test developed for a shrimp disease. This was followed within a year by the first PCR test for MBV, an important baculovirus disease of shrimp. Today, shrimp disease diagnostic laboratories routinely use molecular tests for diagnostic and surveillance purposes for most of the important penaeid shrimp diseases.

Introduction

Half of the world's production of the tropical marine penaeid shrimp now comes from farms. This is a noteworthy achievement for a major food-producing industry that began commercial development only 30 years ago [1,2]. Despite the phenomenal growth of this industry, it has been plagued during the past two decades with diseases that have severely curtailed its development and profitability [2-6]. During this time, at least four virus-caused pandemics (IHHNV, WSSV, YHV and TSV) have occurred and cost the industry hundreds of millions of dollars in lost revenue, millions of jobs in developing nations, and threatened the existence of some segments of the industry. The recovery of the fledgling shrimp farming industries of Asia and the Americas in the wake of the viral pandemics over the past few decades, and its recent phenomenal growth, have been dependent on effective disease prevention and control [2].

Traditional Diagnostic Methods

The most important diseases of farmed penaeid shrimp have infectious aetiologies. Among these are diseases with viral, rickettsial, bacterial, fungal and parasitic aetiologies. Diagnostic methods for these pathogens include the traditional methods of gross pathology, histopathology, classical microbiology, animal bioassay, antibody-based methods, and molecular methods using DNA probes and DNA amplification (Table 1). To accomplish the tasks of disease diagnosis and control in the early years of shrimp farming, diagnosticians adapted traditional methods of bacteriology, parasitology, and histopathology [3-5]. As is the case in veterinary and human disease diagnosis, the traditional diagnostic methods have remained as the cornerstone in shrimp disease diagnosis. In recent years, a few antibody-based diagnostic methods have been developed and some of these are becoming widely used by the industry [2-4,7]. Although commonly used in the diagnosis of many human and veterinary diseases, methods using classical haematology, clinical chemistry and continuous cell lines (tissue culture) have not been developed for use in routine diagnosis of shrimp diseases or for disease research applications [3-5,12-14]. This is especially the case for cell culture methods. Despite the investment of multiple millions of dollars in research funding over the past 30 years for the development of cell lines for the penaeid shrimp (and other crustaceans such as artemia, crabs, crayfish, freshwater prawns and lobsters), no continuous crustacean cell lines have ever been established [4,7,13-14]. Hence, those working with obligate pathogens of shrimp and other crustaceans have not had cell culture diagnostic methods available to them for the demonstration of CPE, plaque assays, neutralization assays, etc., nor have they had access to cell culture as an in vitro method to grow viruses and other cell associated disease agents for research or diagnostic procedure development.

When the viral diseases of shrimp emerged in the early 1990s as the most economically important diseases of farmed shrimp, the absence of cell culture methods that might have been applied for direct diagnostic applications or for use in the development of sufficient viral antigen to develop antibody-based diagnostic methods, posed a major constraint to the development of diagnostic tests for these diseases. This led those working in shrimp disease diagnostic method development to investigate the potential application of molecular methods that were being developed for disease diagnosis in human and veterinary medicine [4,9].

Table 1: Summary of the available diagnostic methods for infectious and non-infectious disease of penaeid shrimp.

Method	Viral	Bacterial & fungal	Parasitic	Abitotic*
History	+++	+++	+++	+++
Gross signs	++	++	++	++
Direct microscopy	+	++	+++	-
Histopathology	+++	+++	+++	+
Electron microscopy	++	+	++	+
Culture and biochemical identification	-	+++	-	-
Antibody-based tests	++	+	-	-
Hematology and clinical chemistry	-	+	-	+
Toxicology and chemical analysis	-	-	-	++
DNA probes	+++	+	+	-
Gene amplification (PCR/RT-PCR)	+++	+/r&d	+/r&d	-
Tissue culture	-/r&d	-	-	-

*Diseases with aetiologies due to toxicants, nutritional imbalances, environmental extremes, or other non-infectious causes.

Molecular Diagnostic Methods

Nearly 15 years ago, the first molecular methods for the diagnosis of the shrimp virus diseases IHHNV and MBV were reported [15,16]. For IHHNV, a parvovirus, a membrane (dot-blot) hybridization method was developed for the detection of infected carriers of the virus. The IHHNV dot-blot method became commercially available to the shrimp farming industry in kit form that same year [4,16]. The PCR test was developed for detection of the shrimp baculovirus MBV [15]. The development and application of additional DNA hybridization and amplification (PCR and RT-PCR) methods for the diagnosis of the major shrimp viral diseases soon followed [2,4,9,11]. At the present time, DNA probes, standard one-step and nested PCR/RT-PCR, and real-time PCR/RT-PCR methods are commonly used by central shrimp disease diagnostic laboratories, as well as by many laboratories located at shrimp hatcheries and farms [2]. By the year 2000, the OIE Diagnostic Manual for Aquatic Animal Diseases listed molecular test methods (dot-blots, ISH and PCR/RT-PCR) for six of the eight listed penaeid shrimp diseases [6]. The dependence of shrimp disease diagnosticians on molecular methods is best illustrated by the 4th Edition of the OIE Diagnostic Manual [7] in which molecular test methods are the recommended for the screening of seven of the nine listed crustacean diseases (Table 2).

The selection of 'gold standards' for the validation of diagnostic methods using DNA hybridization or amplification for shrimp diseases has been problematic. This is because these molecular methods are generally one or more orders of magnitude more sensitive than the most commonly used traditional diagnostic methods available. Hence, because the molecular methods used for disease diagnosis are more sensitive and/or more rapid than the most often used traditional methods (i.e. histopathology),

Table 2: Summary of current OIE recommended methods* for surveillance and diagnostic confirmation for the listed penaeid shrimp diseases.

Disease agent	Surveillance	Confirmatory diagnosis
TSV	RT-PCR	DNA probes, AB, histology, RT-PCR**
WSSV	PCR, AB	DNA probes, AB, histology, bioassay, PCR**
YHV/GAV	RT-PCR	DNA probes, AB, histology, bioassay, RT PCR**
IHHNV	PCR, DNA probes	DNA probes, histology, PCR**
BP/MBV	PCR, DLM, histology	DNA probes, histology, DLM, PCR**

*Definitions of non-standard abbreviations: AB = antibody-based methods; DLM = direct light microscopy.

**Indicated DNA amplification using PCR or RT-PCR, but with a different methods such as amplification of a different region of the genome or real-time PCR/RT-PCT.

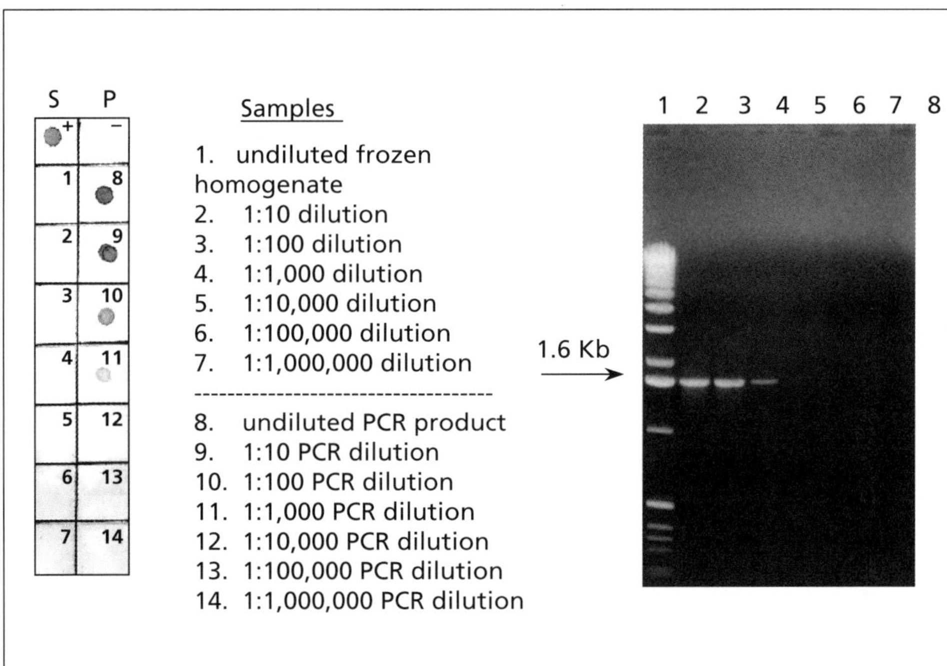

Fig. 1: Sensitivity determination of DNA hybridization and PCR assays for IHHNV by dilution of an infected sample.

the validation of molecular methods has been partly dependent on the use of appropriate positive and negative controls and on the successful application of any given molecular diagnostic method over time [3-7]. Few if any of the molecular methods currently in use by shrimp disease diagnostic laboratories have been formally validated [4,7,9]. Nonetheless, molecular methods readily lend themselves to standardization and validation at least to the level where their sensitivity and specificity have been scientifically documented. In most of the available DNA hybridization (dot-blot) and PCR/RT-PCR methods available for shrimp disease diagnosis, sensitivity of the assay has been demonstrated, usually by testing the method with either a dilution of a known amount of the target pathogen, or by dilution of the clone containing a fragment of the pathogen's target DNA which is a template to which the primers (or probe) bind. Figure 1 illustrates the process used to determine the sensitivity of a DNA hybridization (dot-blot) assay and a PCR method using the same sample sets of a series of dilutions of a sample containing IHHNV.

Likewise, laboratory validation of the specificity of a new molecular test method is typically accomplished by testing the method with samples of either related pathogens or unrelated pathogens likely to be encountered in clinical specimens from disease outbreaks. The application of real-time PCR/RT-PCR methods, especially those that incorporate a labelled DNA probe that is internal to the primer binding sites, to shrimp disease diagnosis has provided another mechanism to validate other molecular methods, and it is increasingly being used as the 'gold standard' to which other traditional and molecular diagnostic tests are compared [2]. Although not all the molecular tests listed in Table 2 have undergone formal standardization and validation, their routine nature and the fact that they have been used widely without dubious results, makes them reliable test methods that are acceptable for both routine surveillance and diagnostic purposes. The shrimp farming industry relies heavily on DNA-based diagnostic and surveillance testing to manage disease. Throughout the shrimp farming industry, even in relatively undeveloped countries, shrimp disease diagnostic laboratories routinely use molecular tests for diagnostic and surveillance purposes for most of the important penaeid shrimp diseases.

Acknowledgements

Much of the work summarized here was supported by the U.S.D.A., C.S.R.E.S., Marine Shrimp Farming Program.

References

1 Tacon AGJ: Aquaculture production trends analysis, in Review of the State of World Aquaculture. FAO Fisheries Circular, 2003, 886(2), pp 5-29.

2 Lightner DV: Biosecurity in shrimp farming: pathogen exclusion through use of SPF stock and routine surveillance. J World Aquaculture Soc 2005;36:229-248.

3 Lightner DV (ed): A Handbook of Shrimp Pathology and Diagnostic Procedures for Diseases of Cultured Penaeid Shrimp. Baton Rouge, Louisiana, USA, World Aquaculture Society, 1996a, 304 pp.

4 Lightner DV, Redman RM: Shrimp diseases and current diagnostic methods. Aquaculture 1998b;164:201-220.

5 Bondad-Reantaso MG, McGladdery SE, East I, Subasinghe RP (eds): Asia Diagnostic Guide to Aquatic Animal Diseases. FAO Fisheries Technical Paper 402, Supplement 2. Rome, FAO, 2001, 240 pp.

6 OIE: Manual of Diagnostic Tests for Aquatic Animal Diseases. 2000, 3th ed.

7 OIE: Manual of Diagnostic Tests for Aquatic Animal Diseases. Paris, France. Office International des Epizooties, 2003, 4th edition, 358 pp.

8 OIE: Diagnostic Manual for Aquatic Animal Diseases. Paris, France. Office International des Epizooties, 2000, 3rd Edition, 237 pp.

9 Walker P, Subasinghe R: DNA-based Molecular Diagnostic Techniques. Research Needs for Standardization and Validation of the Detection of Aquatic Animal Pathogens and Diseases. FOA Fisheries Technical Paper 395. 2000, 93 pp.

10 Flegel TW: Major viral diseases of the black tiger prawn (*Penaeus monodon*) in Thailand. World J Microbiol Biotechnol 1997;13:433-442.

11 Flegel TW, Alday-Sanz V: The crisis in Asian shrimp aquaculture: current status and future needs. J Appl Ichthyol 1998;14:269-273.

12 Alday de Graindorge V, Flegel TW: Diagnosis of Shrimp Diseases with Emphasis on the black tiger shrimp (*Penaeus monodon*). {Multimedia CD}. FAO and Multimedia Asia Co., Ltd., Bangkok, Thailand. ISBN 974-662-093-2, 1999.

13 Jean-Yves Toullec JY: Crustacean primary cell culture: A technical approach. Methods Cell Sci 1999;21:193-198.

14 Crane MStJ: Mutagensis and cell transformation in cell culture. Methods Cell Sci 1999;21:245-253.

15 Vickers JE, Lester RJG, Spradbrow PB, Pemberton JM: Detection of *Penaeus monodon*-type baculovirus (MBV) in digestive glands of postlarval prawns using polymerase chain reaction, in Shariff M, Subasinghe RP, Arthur JR (eds): Diseases in Asian Aquaculture I. Manila, Philippines, Fish Health Section, Asian Fisheries Society. 1992, pp 127-133.

16 Lightner DV, Poulos BT, Bruce L, Redman RM, Mari J, Bonami JR: New developments in penaeid virology: application of biotechnology in research and disease diagnosis for shrimp viruses of concern in the Americas, in Fulks W, Main K (eds): Diseases of Cultured Penaeid Shrimp in Asia and the United States. Makapuu Point, Honolulu, Hawaii, USA, The Oceanic Institute, 1992, pp 233-253.

Dr. D.V. Lightner, University of Arizona, Dept. Veterinary Science & Microbiology, Building #90, Room 106, Tucson, AZ 85721 USA.
E-mail: dvl@u.arizona.edu

Vannier P, Espeseth D (eds): New Diagnostic Technology: Applications in Animal Health and Biologics Controls. Dev Biol (Basel). Basel, Karger, 2006, vol 126, pp 123-132.

A PCR-Based Method for Tuberculosis Detection in Wildlife

S. Hénault, C. Karoui, M.L. Boschiroli

Unité Zoonoses Bactériennes (AFSSA-Lerpaz), Maisons Alfort, France

Abstract: Bovine tuberculosis (TB) is a major worldwide zoonotic disease. Foremost of the drawbacks in the control campaigns is the slow growth of its causative agent, *M. bovis*, as bacteriology remains the "gold standard" for the diagnosis of this disease. Rapid alternative molecular biology methods for TB diagnosis have long been hampered due to mycobacterial-linked difficulties when conventional DNA extraction techniques are applied. Moreover, the correct specificity is difficult to achieve because of the similarities in genetic background between *M. bovis* and other ubiquitous mycobacterial species. Nevertheless, much technological progress has been achieved in recent years, allowing the development of accurate molecular diagnosis.

One of the main problems for bovine TB control is the existence of *M. bovis* wildlife reservoirs, a source of re-contamination in bovine TB-free herds. PCR seems an interesting alternative method for rapidly screening these species in epidemiological enquiries and immediate decision-making to avoid transmission to livestock. We describe here the validation process for a PCR diagnostic method compared to bacteriology in a wildlife TB survey.

INTRODUCTION

Bovine tuberculosis (BTB) is an old-established disease present worldwide as a consequence of cattle husbandry. Its causative agent, *Mycobacterium bovis*, belongs to the *Mycobacterium tuberculosis* complex group, a phylogenetically-defined group to which *Mycobacterium tuberculosis*, the main causative agent of human tuberculosis, also belongs. Although the latter affects mostly human beings and primates, *M. bovis* infects not only cattle but also any mammal species, including man. Bovine tuberculosis is therefore a zoonotic disease, with a different impact on human beings depending on the endemic situation in the cattle population. Indeed, in industrialised countries, eradication control programmes for animal tuberculosis, as well as milk pasteurisation, has considerably reduced the disease incidence due to *M. bovis* from domestic animals to man. However, in developing countries where TB control measures in cattle are only sporadically applied or not at all, and pasteurisation is barely practised, animal tuberculosis is still widely disseminated and can account for up to 5-10% of human tuberculosis cases [1].

Due to the wide spectrum of infection in numerous regions of the world *M. bovis* is one of the main bacterial diseases in wild animals. If generally it originates in a domestic bovine livestock population, it can afterwards be maintained in an autonomous way in certain free wildlife animal populations. In several countries where BTB eradication control campaigns have almost achieved their goal, these external sources have become of particular concern, as they are the main risk of re-introduction of the disease in TB-free cattle farms. The two main examples are the badger in Great Britain [2], and the possum in New Zealand [3].

Public health reasons alone should be enough for applying cattle control programmes. But it must be acknowledged that the productivity losses that tuberculosis brings about, together with trade restrictions imposed to supply "clean" products to the market place, are clearly the primary driving forces for the eradication of tuberculosis nowadays.

These control campaigns are based on animal surveillance by the hyper intra dermo reaction (IDR), or tuberculin test diagnosing. In France, positive IDR animals are culled and on *post mortem* examination, tuberculous-compatible lesions are submitted to bacteriology, which is the "gold standard" diagnostic method for declaration of a tuberculosis outbreak. Total herd elimination for stamping out the disease and subsequent breeder's compensation follow. An essential supplementary means for monitoring the disease comes from general routine abattoir inspection of carcasses: when organs or lymph nodes present tuberculosis-compatible lesions they are submitted to bacteriology for TB confirmation. Because *M. tuberculosis* complex organisms are slow growing, isolation and identification may take up to three months. This waiting period is too long for breeders, unbearable psychologically and economically, as farm production is practically stopped officially while waiting for confirmation of an outbreak.

As for wildlife, there is no common and reliable diagnostic method other than bacteriology for tuberculosis detection. Again the length of diagnostic confirmation can be too long when disease management practices have to be adopted to control its spread.

Rapid TB detection by molecular biology techniques have long been hampered by mycobacterial recalcitrance to physical or chemical DNA extraction but also by the few bacilli present in infected tissues. Nowadays, many technical improvements have been accomplished that can circumvent these drawbacks, which challenged us to set up a PCR diagnostic method based on the wildlife model. In turn this technique may be used for the cattle model.

This paper describes the key steps for achieving maximal *M. bovis* DNA amplification based on an optimal combination of primers and DNA extraction methods; the comparison between this PCR-based diagnostic technique and the bacteriology "gold standard", is performed in a wildlife-TB epidemiological survey.

Materials and Methods

Choice of genetic target and primers

Primers that were used in this study anneal to the insertion sequence *6110* (IS*6110*) (Gene bank X17348) and are described in Table 1. Most of them were chosen from previous work but new ones were designed with Oligo software. Nested PCR reactions were performed with different primer combinations as explained in Table 1. For the first round, amplification programmes were the same as those explained in papers where primers were originally described. For primers designed in this study, the amplification programmes were: 5 cycles at 94°C for one min, 30 sec at 65°C and 2 min at 72°C, followed by 35 cycles at 94°C for one min, 60°C for one min, 72°C for one min and extended to 10 min at 72°C. For the second amplification round, 15 cycles at 94°C for one min, 30 sec at 60°C and 1 min at 72°C were used for every primer combination. Amplification products (1/10 of the reaction mixture) were visualised by electrophoresis in EthBr-stained agarose gel. DNA amplification was performed in a Perkin Elmer Gene Amp 9600 Thermocycler with *Taq* polymerase in bufferB from Promega.

To assess the existence of inhibiting factors in DNA samples, amplification with universal 16S*rDNA* prokaryotic primers BS1/BS2 [4] was performed under the same conditions as described by the authors.

Specimen description, processing and culture

Lymph nodes from four cattle with tuberculous lesions detected at slaughter houses were used to set up assays. Respiratory lymph nodes, whether presenting tuberculous lesions or not from 95 red deer, 36 roe deer, 93 wild boars, 36 badgers and 5 non-identified wild animals (total: 265) were used for validation of the PCR methodology. These animals are from a French forest where a BTB outbreak is taking place.

Tissues were sliced with a scalpel and homogenized by stomaching in the presence of 4% H_2SO_4 as decontaminating solution [10]. After neutralisation with 6% NaOH, samples were inoculated into both Lowenstein-Jensen solid medium slopes and used for DNA extraction and subsequent PCR.

PCR sensitivity and specificity assays

Sensitivity

M. bovis (field strain) pure genomic DNA at a concentration of 1.066 $x10^{-5}$ µg/µl was serially diluted and different primer combinations were used for determining the best amplification rates.

Specificity

Mycobacterial species found in the environment or potentially pathogenic for animals are sometimes co-isolated in *M. bovis* infected tissues. Some of these species, which are genetically linked to *Mycobacterium bovis*, were studied to assess if the chosen IS*6110* primers produced amplification cross-reactions (Table 2).

Table 1: Primer-combinations tested in different nested PCR based on IS*6110* amplification. **R:** reverse primer, **F:** forward primer. Bold text indicates primers designed specifically for this study.

	First round amplification		**Second round amplification**	
	IS*6110* sequence position	**Reference**	**IS*6110* sequence position**	
A	F: 695-724/ R: 885-914	[5]	F: 750-770/ R: 866-885	A1
B	**F: 607-631/R:1072-1096**	This study	F: 750-770/ R: 866-885 F: 695-724/ R: 885-914 **F: 666-680/ R:1072-1096** F: 633-652/ R: 856-875 F: 633-652/ R: 718-738	B1 B2 **B3** B4 B5
C	F: 226-246/ R:866-885	[6]	F: 750-770/ R: 866-885 F: 633-652/ R: 856-875 F: 633-652/ R: 718-738	C1 C2 C3
D	F: 633-652/ R: 856-875	[7]	F: 633-652/ R: 718-738	D1
E	F: 216-236/ R: 858-877	[8]	F: 633-652/ R: 856-875 F: 633-652/ R: 718-738	E1 E2
F	F: 750-770/ R: 866-885	[9]	**F: 816-836/ R: 866-885**	**F1**

Table 2: Mycobacterial strains used for PCR specificity set-up assays.

Mycobacterial species	Reference
M. gordonae	Mycobacterial National Reference Centre Pasteur Institute (human isolate)
M. flavescens	"
M. simiae	"
M. chelonae	"
M. gastri	"
M. vaccae	"
M. fortuitum	"
M. xenopi	"
M. kansasii	"
M. intracellulare	ATCC 13950
M. avium subsp. *avium*	ATCC 25291
M. avium subsp. *silvaticum*	ATCC 49884

DNA extraction method

Tissue homogenates (2 ml) were centrifuged for five minutes at 8000 rpm and the pellets washed several times with a volume of TE until a clear supernatant was obtained. At that moment, tissue pellets were left in a 200 μl volume and used for subsequent extraction methods. These were:

(i) Commercial kit only: the High Pure PCR Template (Roche) was used according to the manufacturer's recommendations. Bacterial lysis was achieved by proteinase K+SDS incubation. 5 μl of the extracted DNA were used for PCR.

(ii) Mechanical disruption + commercial kit: the same procedure as described above was applied, but a mechanical lysis step by shaking the sample in FastPrep Bio 101 at 6 m/s for 45 seconds was performed before adding the kit's lysis solution. 5 ∝l of the extracted DNA were used for PCR.

(iii) Ribolysing + sequence capture: samples were lysed both by mechanical disruption and proteinase K digestion followed by a sequence capture step on the specific *M. tuberculosis* DR region as described before [11]. The biotynilated CapDRa and CapDRb were supplied by MWG. M-280 Strepavidin Dynal Beads (Dynal) were used and separated in a Dynal magnetic bead separator device. 5 μl of extracted DNA were used for PCR.

Analysis of PCR results

Results were analysed using the formulae described in [12]:

Positive agreement (PA): samples positive for both methods
Positive disagreement (PD): PCR-positive/Bacteriology negative case ("false positive")
Negative disagreement (ND): PCR-negative/Bacteriology positive case ("false negative")
Negative agreement (NA): samples negative for both methods
N+: total number of bacteriology positive samples
N-: total number of bacteriology negative samples
Sensitivity (SE): PCR ability to detect the micro-organism detected by bacteriology = **(PA/N+)x100**
Specificity (SP): PCR ability not to detect the micro-organism not detected by bacteriology = **(NA/N-)x100**
Accuracy (AC): degree of correspondence between the two methods = **(PA+NA)/total N**

RESULTS

PCR set up

Choice of primers choice

After extensive bibliographic revision, the insertion sequence *6110* (IS*6110*), which is specific for the *M. tuberculosis* complex species and therefore extensively used for human tuberculosis molecular diagnosis, was chosen as the genetic target. A nested PCR was decided for amplification strategy not only since it increases PCR amplification sensitivity but also because it improves the specificity of the test. Primer combinations used in this study are shown in Table 1.

The first criterion for primer selection was sensitivity on a *M. bovis* DNA dilution range. The second criterion was the presence of a single distinctive amplification band on these DNA dilutions. The third criterion was amplification specificity using DNA from mycobacterial species other than tuberculosis, which can sometimes be isolated from animal tissues (Table 2).

From these set-up assays, two primer combinations, **B3** and **D1**, were chosen. Although D1 gave slightly better sensitivity results on *M. bovis* DNA dilutions, B3 showed a better specificity as hardly any amplification signals were detected on other mycobacterial species (data not shown).

Extraction method

Three DNA extraction methods were tested on well characterised bovine samples which have previously been analysed by bacteriology and proved to be infected by *Mycobacterium bovis*. These extraction procedures were:

(i) commercial kit alone;

(ii) mechanical lysis + commercial kit;

(iii) mechanical lysis + sequence capture [11].

From these three, using either primer combination B3 or D1, only methods (ii) and (iii) gave perfectly accurate results according to bacteriology (data not shown), therefore becoming the methods of choice in the subsequent application study.

APPLICATION IN A WILDLIFE EPIDEMIOLOGICAL SURVEY

Pathological samples from 265 wild animals (95 red deer, 36 roe deer, 93 wild boar, 36 badgers and 5 non-identified wild animals) from a TB outbreak in wildlife, were firstly submitted to homogenisation and decontamination as described in materials and methods. This was the starting material for both bacteriology and DNA extraction methods which were performed in parallel.

All in all, wildlife samples were tested through the four different PCR methods which are described in Table 3.

The four samples which were PCR positive for method 3-B3 but negative for the gold standard method, were in fact true positive samples, as they were also positive in the other three PCR methods applied in this study.

Diagnostic accuracy (AC), sensitivity (SE) and specificity (SP) formulae as explained in materials and methods, were calculated from the above results (Table 4):

From these four methods, 3-B3 was the best as it showed not only better sensitivity but also better specificity results; 3-B3 is our most accurate method compared to bacteriology.

When tested in parallel on the same type of extracted DNA, primer combinations gave the results shown in Table 5.

For combination 3-(B3-D1), although specificity is slightly reduced, sensitivity is sensibly increased compared to when B3, the best of the two primer combinations, is used alone. This is due to the fact that primers B3 and D1 did not detect the same samples with extraction method 3. As for extraction method 2, the use of primer combinations in parallel does not improve the overall results. Indeed, for this extraction method, every sample revealed as positive with primers B3 is also positive with primers D1, therefore not ameliorating the range of detection.

Table 3: Bacteriology (*M. bovis* isolation) and PCR results.

PCR method		Bacteriology	
		+	-
2-B3	+	40	8
	-	14	203
2-D1	+	43	5
	-	11	206
3-B3	+	44	4
	-	10	207
3-D1	+	44	9
	-	10	202
Total bacteriology		**54**	**211**

Table 4: Accuracy (AC), Sensitivity (SE) and Specificity (SP) deduced from the results in Table 3.

PCR method	AC	SE	SP
2-B3	0.92	74	96.2
2-D1	0.94	79.6	97.3
3-B3	0.95	81.5	98.1
3-D1	0.93	81.5	95.7

Table 5: Accuracy (AC), Sensitivity (SE) and Specificity (SP) calculated when both primer combinations were used for each type of extracted DNA.

PCR method	AC	SE	SP
2-(B3-D1)	0.92	79.6	94.8
3-(B3-D1)	0.95	88.9	97.2

Discussion and Conclusions

The aim of the present study was to evaluate different primer systems but also DNA extraction protocols so as to obtain a reliable and sensitive PCR diagnostic method to detect bovine tuberculosis directly from tissue samples. Once an appropriate PCR system was decided, it was evaluated and compared to the bacteriology "gold standard" on a representative wildlife sample collection.

Our report shows that PCR 3-B3, based on *M. bovis* sequence-capture DNA extraction with the primers designed for this study, combination B3, is the best of the four PCR methods assessed with a 81.5% sensitivity and 98.1% specificity compared to bacteriology. However, if two different nested PCR are performed in parallel with primer combinations B3 and D1 on this same DNA, an even better sensitivity result of 88.9% is obtained, although with a slightly reduced specificity of 97.2%. Even though this strategy implies an additional step, it does not delay result acquisition as both nested PCR are performed in parallel. Indeed, the most critical and cumbersome time-consuming step is the DNA extraction itself.

It has been argued that sequence-capture PCR performances are excellent for pauci-bacillary samples [12] as they not only concentrate specific DNA but also eliminate PCR inhibiting factors better. This report emphasises the superior performance of this extraction method compared to other conventional ones even applied in wild-animal tissues.

Moreover, a PCR strategy is ideal when a large number of samples are to be studied for obtaining TB prevalence estimations in a short time delay. Even though its sensitivity values are 10% less important compared to bacteriology, our PCR diagnostic method is capable of detecting additional true positive samples not detected by bacteriology. This adds not only sensitivity but specificity to our method, rendering it more appealing for direct diagnosis. But above all, it allowed us to gain three months in declaring that a TB outbreak was taking place in Brotonne Forest. This shortening of the diagnosis response permits the immediate implementation of control measures to prevent subsequent outbreaks or diminish TB prevalence. Some of the measures to be taken without delay that can contribute to breaking the transmission cycle within wildlife animals is increasing pressure on hunting so as to diminish the wildlife population density. In addition, banning artificial feeding to avoid contact between infected and healthy animals is necessary. To prevent transmission from wildlife reservoirs to livestock, fencing cattle pastures to preserve livestock from free-ranging infected wild animals is one of the immediate measures to be adopted.

Another advantage of PCR over bacteriology is that it is not a laboratory invasive methodology. Indeed, when analysing a large number of samples, solid medium slants may occupy incubators exclusively for long periods, hampering other bacteriology manipulations. This is even more critical for bio-safety 2 or 3 level laboratory facilities, where space for additional incubators is more restricted that in conventional laboratories.

As for extraction method 2, i.e. mechanical lysis followed by a commercial kit extraction, although slightly less sensitive, this is a simple technique that could be easily automated and, combined with a real time PCR, could allow for a standardised routine detection of the pathogen in diagnostic laboratories. This automation is of particular interest on large samplings such as our wildlife survey, for which intensive manipulations can lead to an amplification in cross-contamination.

To our knowledge, this is the first report describing a PCR-TB diagnostic method in wildlife.

Compared to bovine pathological samples, which are taken in proper clean conditions at slaughterhouses, and for which TB PCR diagnosis was reported before, wildlife animal organs are mostly manipulated in the open air by non-experienced hunters, facilitating contamination both with fast-growing ubiquitous bacteria and environmental mycobacteria. PCR is an excellent alternative when contaminations render *M. bovis* isolation impossible. Moreover, wildlife animals might be infected with other mycobacterial species as well [1]. A specific PCR allows detection and identification of *M. tuberculosis* complex species at the same time, unlike bacteriology, for which once a strain is isolated, an identification technique follows to determine the species level.

Due to time-saving resulting from these rapid detection methods, nowadays they are widely used for infectious disease management in livestock. However, culture is still the official diagnostic method for confirmation of an outbreak of TB in cattle. Indeed, total herd elimination is a decision of great magnitude for sanitary services to make, which must be supported by the most reliable existing diagnostic method. Furthermore, molecular typing of *M. bovis* isolated strains permits molecular epidemiology monitoring to understand transmission origins and pathways.

Yet, to ameliorate the time delay for bacteriological diagnosis, new culture systems have proved to be useful in the early detection of *M. tuberculosis* complex organisms, which are all slow growers. Mainly with the Bactec 960 MGIT system, *M. bovis* growth can be detected two to three weeks after inoculation. A specific PCR can then be applied to this liquid culture, therefore leading to identification one month after the bacteriological manipulation has began, compared to the three months that are necessary for conventional mycobacteriology [13].

Nevertheless, on suspicion of a TB outbreak based on a positive tuberculin reaction, a test which has proved to be of limited sensitivity and specificity on low cattle prevalence, a quick PCR positive result might give sounder grounds for applying an official tuberculosis surveillance programme while waiting for bacteriology results.

Our PCR is based on IS*6110* amplification, which is a genetic target present in all *M. tuberculosis* complex organisms. Although this test is therefore not specific to the identification of *M. bovis*, it could be useful for detecting tuberculosis caused by other highly pathogenic members of this complex such as *M. caprae*, which has also been isolated from bovines and wild animal species in our laboratories and elsewhere [14,15].

Acknowledgements

We are grateful to Bruno Garin-Bastuji, Karine Laroucau and Victoria de los Angeles Boschiroli for advice and helpful comments.

References

1 Biet F, Boschiroli ML, Thorel MF, Guilloteau LA: Zoonotic aspects of *Mycobacterium bovis* and *Mycobacterium avium-intracellulare* complex (MAC). Vet Res 2005;36:411-436.

2 Clifton-Hadley RS, Wilesmith JW, Stuart FA: *Mycobacterium bovis* in the European badger (*Meles meles*): Epidemiological findings in tuberculous badgers from naturally infected population. Epidemiol Infect 1993;9-19.

3 Coleman JD, Cooke MM: *Mycobacterium bovis* infection in wildlife in New Zealand. Tuberculosis 2001;81:192-202.

4 Guillou JP, Henault S, Ostyn A, Thorel MF: Use of the in vitro enzymatic amplification method for the detection of *Mycobacterium paratuberculosis* in feces. Rev Sci Tech 1993;12(2):605-615.

5 Scarpellini P, Racca S, Cinque P, Delfanti F, Gianotti N, Terreni MR, Vago L, Lazzarin A: Nested polymerase chain reaction for diagnosis and monitoring treatment response in AIDS patients with tuberculous meningitis. AIDS 1995;9(8):895-900.

6 Liébana E, Aranaz A, Mateos A, Vilafranca M, Gomez-Mampaso E, Tercero J C, Alemany J, Suarez G, Domingo M, Dominguez L: Simple and rapid detection of Mycobacterium tuberculosis complex organisms in bovine tissue samples by PCR. J Clin Microbiol 1995;33(1):33–36.

7 Hermans P W, van Soolingen D, Dale J W, Schuitema A R, McAdam R A, Catty D, van Embden J D: Insertion element IS*986* from *Mycobacterium tuberculosis*: a useful tool for diagnosis and epidemiology of tuberculosis. J Clin Microbiol 1990;28(9):2051–2058.

8 Marchetti G, Gori A, Catozzi L, Vago L, Nebuloni M, Rossi MC, Degli Esposti A, Bandera A, Franzetti F: Evaluation of PCR in Detection of *Mycobacterium tuberculosis* from Formalin-Fixed, Paraffin-Embedded Tissues: Comparison of Four Amplification Assays. J Clin Microbiol 1998;36(6):1512–1517.

9 Eisenach KD, Cave MD, Bates JH, Crawford JT: Polymerase chain reaction amplification of a repetitive DNA sequence specific for *Mycobacterium tuberculosis*. J Infect Dis 1990;161(5):977-981.

10 David H, Lévy-Frébault V, Thorel MF: Méthodes de laboratoire pour mycobactériologie clinique. Paris, Institut Pasteur, 1989, pp 29-53.

11 Taylor MJ, Siobhan Hughes MS, Skuce RA, Neill SD: Detection of *Mycobacterium bovis* in bovine clinical specimens using Real-Time fluorescence and fluorescence resonance energy transfer probe rapid-cycle PCR. J Clin Microbiol 2001;39(4):1272-1278.

12 Hoorfar J, Wolffs P, Rådström P: Diagnostic PCR: validation and sample preparation are two sides of the same coin. APMIS 2004;112:808-814.

13 Chien HP, Yu MC, Wu MH, Lin TP, Luh KT: Comparison of the BACTEC MGIT 960 with Lowenstein-Jensen medium for recovery of mycobacteria from clinical specimens. J Tuberc Lung Dis 2000;4(9):866-870.

14 Gortazar C, Vicente J, Samper S, Garrido JM, Fernández-De-Mera IG, Gavín P, Juste RA, Martín C, Acevedo P, De La Puente M, Höfle U: Molecular characterization of *Mycobacterium tuberculosis* complex isolates from wild ungulates in south-central Spain. Vet Res 2005;36:43-52.

15 Prodinger WG, Brandstätter A, Naumann L, Pacciarini M, Kubica T, Boschiroli ML, Aranaz A, Nagy G, Cvetnic Z, Ocepek M, Erler W, Niemann S, Pavlik I, Irmgard Moser I: Characterization of *Mycobacterium caprae* by mycobacterial interspersed repetitive units (MIRU) for epidemiological and phylogenetic purposes. J Clin Microbiol 2005;43(10):4984-4992.

Dr. M.L. Boschiroli, Bacterial Zoonoses Unit-Animal Pathology and Zoonoses Research Laboratory, French Food Safety Agency (AFSSA), 23 av. du Général de Gaulle, FR-94706 Maisons Alfort, France. E-mail: ml.boschiroli@afssa.fr

Vannier P, Espeseth D (eds): New Diagnostic Technology: Applications in Animal Health and Biologics Controls. Dev Biol (Basel). Basel, Karger, 2006, vol 126, pp 133-145.

Development, Application and Validation of a Taqman® Real-Time RT-PCR Assay for the Detection of Infectious Salmon Anaemia Virus (ISAV) in Atlantic Salmon (*Salmo salar*)

M. Snow[1], *P. McKay*[1], *A.J.A. McBeath*[1], *J. Black*[1], *F. Doig*[1], *R. Kerr*[1], *C.O. Cunningham*[1], *A. Nylund*[2], *M. Devold*[2]

[1] FRS Marine Laboratory, Aberdeen, U.K.
[2] Department of Biology, University of Bergen, Bergen, Norway

Key words: Infectious salmon anaemia virus, ISAV, real-time PCR, qRT-PCR, Taqman.

Abstract: Infectious salmon anaemia (ISA) is a disease of cultured Atlantic salmon (*Salmo salar*) which was successfully eradicated from Scotland following its emergence in 1998. The rapid deployment of sensitive diagnostic methods for the detection of ISA virus (ISAV) was fundamental to the swift eradication of ISA disease in Scotland and continues to be of crucial importance to surveillance of the aquaculture industry. This study reports the development, validation, application and interpretation of two independent, highly sensitive and specific semi-quantitative Taqman® real-time RT-PCR (qRT-PCR) methods for the detection of ISAV. Such technology offers considerable advantages over conventional RT-PCR methods in current routine use for ISAV surveillance. These include an increased sensitivity, enhanced specificity, semi-quantification using endogenous controls, a lack of subjectivity in results interpretation, speed of processing and improved contamination control.

Introduction

Infectious salmon anaemia virus (ISAV) is a pathogen of cultured Atlantic salmon (*Salmo salar*) which has caused significant production losses in Norway, Scotland, Canada, the U.S.A., and the Faroes. Emergence of ISA disease in Scotland in the late 1990s prompted a swift and apparently successful eradication programme [1]. The disease has, however, become established in Norway and Canada where in 1999 alone it was estimated to be responsible for losses totalling \$11 and \$14 million, respectively [2].

The rapid deployment of sensitive diagnostic methods for the detection of ISAV was fundamental to the swift eradication of ISA disease in Scotland and remains equally important in managing the impact and spread of this virus in counties where it is endemic. Three diagnostic assays are in widespread use for ISAV diagnosis, which include virus isolation on permissive cell lines such as SHK-1 [3], indirect fluorescent antibody testing (IFAT) [4] and reverse transcriptase polymerase chain reaction (RT-PCR) [5]. Of these assays, the RT-PCR is generally regarded as the most sensitive [6-8] although a recent study highlighted a lack of reproducibility between laboratories and emphasized a need to standardize inter-laboratory protocols [9]. Much of the variability was attributed to the use of different external controls and methods applied to interpret and confirm results [9].

Recent developments in real-time PCR technology offer the potential to overcome many of the limitations of conventional PCR including an improved sensitivity of ISAV diagnostics. Indeed, a sensitivity increase of 100-fold over conventional RT-PCR was recently reported using a SYBR green-based assay for the detection of ISAV [10]. Further advantages of Taqman® real-time PCR chemistry include improved specificity of amplification, since signal generation is reliant on the annealing of a highly specific secondary probe to the primary amplification product. In addition the development of an endogenous control allows the confirmation of target integrity as well as facilitation of relative quantification of viral load and subsequent comparison between samples. Real-time PCR thus offers considerable advantages over conventional RT-PCR-based protocols and has the potential to reduce problems in the interpretation and reproducibility of results between laboratories.

In this report we describe the development, validation and application of Taqman®-based assays targeting two independent genes of ISAV coupled to an assay targeting the constitutively expressed endogenous control translation elongation factor 1 alpha (ELF1α) [11]. The comparative efficiencies of these assays were determined and a method for the relative quantification of viral target-level developed. Assays were compared with the conventional RT-PCR in widespread use for ISAV diagnostics [5] throughout an experimentally induced clinical outbreak of ISAV and in samples analysed for routine ISAV diagnostics in our laboratory in 2004.

Materials and Methods

Development of Taqman Real-Time RT-PCR method

Primers and probe design

RT-PCR primer pairs and Taqman® MGB probes targeting ISAV segments 7 and 8 were selected from alignments of all available sequences using published sequence data and the Primer Express software (Applied Biosystems), and are detailed in Table 1. Both primers and probes were targeted to conserved gene regions to ensure detection of all documented variants of ISAV. Primers designed for amplification of the salmon ELF1α gene were those recently published by Moore et al. (2005). The ELF probe was designed to target mRNA specifically by spanning two exons such that it would not amplify contaminating salmon genomic DNA.

Table 1: Real-time PCR primers and Taqman® MGB probes.

Organism	Gene target	Upstream primer	Downstream primer	Taqman® trobe	Amplicon size (bp)
ISAV	Segment 7	CAGGGTTGTA TCCATGGTTG AAATG	GTCCAGCCCT AAGCTCAACT C	FAM-CTCTCTCATTG TGATCCC-MGB	155
ISAV	Segment 8	CTACACAGCA GGATGCAGAT GT	CAGGATGCCG GAAGTCGAT	FAM-CATCGTCGCT GCAGTTC-MGB	104
Salmo salar	ELF1α mRNA	CCCCTCCAGG ACGTTTACAA A	CACACGGCCC ACAGGTACA	FAM-ATCGGT#GGTA TTGGAAC	57

RNA extraction

Tissues were stored in 1 ml RNALater (Ambion) and stored at -20°C before RNA extraction. For field samples, 50-100 mg tissue was homogenised in 1ml TRIZOL using a disposable pestle and RNA subsequently extracted according to the manufacturer's protocol (Invitrogen). RNA was finally resuspended in DEPC-treated H_2O, quantified by measuring the OD at 260 and 280 nm on a Genequant Pro spectrophotometer (Amersham Biosciences) and stored at -80°C. In the case of experimental samples, 10mg tissue was homogenised in 600 µl RLT buffer (Qiagen) for 1min using the TissueLyser system (Qiagen). RNA was subsequently extracted using the RNeasy kit according to the manufacturer's protocol for the extraction of RNA from animal tissues (Qiagen). Total RNA was finally resuspended in 200 µl DEPC-treated H_2O and stored at -80°C. For routine diagnostics, negative extraction controls (designated A control) were conducted by performing a blank extraction. Controls were taken through subsequent RT and PCR steps.

Reverse transcription

For field diagnostic samples, reverse transcription reactions were conducted by first preparing dilutions containing 2 µg RNA in 9.5 µl DEPC-treated dH2O. RNA samples were incubated at 55-60°C for 10min and then placed on ice. Reverse transcription was conducted in a total volume of 20 µl which contained 9.5 µl RNA prepared as above, 5 ngµl^{-1} random hexamers (Promega), 1x 1st strand buffer, 1mM each dNTP, 1Uµl^{-1} RNase OUT and 10Uµl^{-1} MMLV reverse transcriptase (all from Invitrogen). Reactions were incubated at 37°C for 1h. In the case of experimental samples, reverse transcription was performed using the Taqman® RT Kit (Applied Biosystems) in a final reaction volume of 25 µl. Firstly, 9.125 µl RNA & 1.25 µl of 50 µM oligo $d(T)_{16}$ were mixed and denatured at 70°C for 10 min. Reactions were placed on ice and mixed with 1x reverse transcriptase buffer, 5.5 mM $MgCl_2$, 2 mM dNTPs, 0.4U RNase Inhibitor and 1.25U Multiscribe reverse transcriptase. Reactions were incubated at 48°C for 90 min followed by 95°C for five min. Negative reverse transcription controls were conducted by performing reactions containing no target (designated B control). Controls were taken through subsequent PCR steps.

Real-time RT-PCR (qRT-PCR) reagents and cycling

Assays were performed on an Applied Biosystems 7000 Sequence Detection System programmed to conduct the following cycling profile: 50°C for two min (AmpErase uracil N-glycosylase incubation), 95°C for 10 min (AmpliTaq Gold polymerase activation) followed by 45 cycles of 95°C for 15s (denaturation) and 60°C for one min (annealing/extension). Reactions were conducted in a final volume of 25 µl containing 1 µl cDNA template, 1x Taqman® Universal PCR mastermix, 900nM each primer and 250 nM Taqman® probe. Controls were conducted for each primer and probe set containing no target (designated C control).

Validation of reaction efficiencies and relative quantification

To determine whether the salmon host cell endogenous ELF1a assay control might permit the relative quantification of ISAV and ensure efficient performance of each assay, a validation experiment was conducted to determine whether the efficiency of all assays was comparable. For each assay, triplicate reactions were conducted on each dilution of a 10-fold serial dilution of cDNA prepared from concentrated standards. Standard curves were generated by plotting the relative dilution of RNA versus the cycle number required to elevate the fluorescence signal above the threshold for the ISAV Segment 7, 8 and ELF assays. Reaction conditions with respect to probe and primer concentrations were independently evaluated as described in the Applied Biosystems literature. The effect of reducing the total reaction volume from 50 μl to 25 μl was also determined.

Sensitivity

The same 10-fold dilution series of ISAV cDNA was used to evaluate the comparative sensitivity of both Taqman® qRT-PCR methods with that of conventional RT-PCR. Triplicate individual reactions were conducted for each test using the methods described above.

Specificity

ISAV is the only orthomyxovirus from fish to be characterised at the molecular level, and as such has been designated within the genus Isavirus. Nucleotide sequence of probes was used to search GenBank using BLAST X [12] to identify any known organisms with which they might cross-react. Segment 7 and 8 probes and primer sets were tested against cDNA derived from a range of other RNA viruses capable of infecting Atlantic salmon including Atlantic salmon paramyxovirus, viral haemorrhagic septicaemia virus (VHSV), infectious haematopoietic necrosis virus (IHNV), infectious pancreatic necrosis virus (IPNV) and pancreas disease virus (PD). To ensure the detection of all forms of ISAV, segment 7 and 8 primers and probes were designed from sequence alignments containing representatives of all known subtypes of ISAV, which are classified according to the sequence of a highly polymorphic region (HPR) of the haemagglutinin esterase (HE) gene. Such alignments included the HPR0 subtype from which all other HPR subtypes of the virus which are associated with clinical disease in aquaculture may have been derived [13,14].

Application of relative quantification to the detection of ISAV in experimentally infected fish

Pathogen-free fish

Atlantic salmon post smolts (mean weight +/- SE, 89.4 +/- 14.69 g) were reared at the FRS Marine Research Unit, Aultbea, Ross-shire, Scotland. Before the experimental challenge, four fish were screened for the presence of ISAV using RT-PCR and qRT-PCR as described above. Fish were allowed to acclimatize for seven days and starved for 24 h before experimental infection. Water temperatures were maintained at 11°C for the duration of the experimental infection. Seawater was extracted from the North Sea and passed through a sand filter and UV disinfection treatment before entering fish tanks.

Experimental infection with ISAV

Atlantic salmon post smolts (n=75) were anaesthetised by immersion in methane tricaine sulphonate (3-aminobenzoic acid ethyl ester of MS-222; Sigma) at a concentration of 100 mg l-1, infected intra-peritoneally with a dose of 2 x 10^6 $TCID_{50}$ $fish^{-1}$ in an inoculum volume of 100 μl, and distributed between two 360 l aquaria (Tanks 1 & 2). Controls consisted of 75 fish similarly anaesthetised and injected with an equivalent volume of culture medium containing no virus, and distributed between a further two tanks (Tanks 3 and 4). Two additional tanks were similarly set up, each of which contained 25 virus infected and non-infected fish respectively. These tanks were used to monitor the efficacy of the viral challenge. All fish were monitored for the duration of the experimental infection period (20d) and all dead fish were recorded and removed twice daily.

Sampling and comparative PCR analysis

From each set of virus-infected (Tanks 1 and 2) and control tanks (Tanks 3 and 4), a total of four fish were sacrificed and sampled on each of days 1,2,3,4,5,6,7,9,11,15 and 19 post infection. Kidney tissue was excised, sampled into RNALater (Ambion) and stored at -80°C for analysis by PCR. RNA extraction and Taqman® qRT-PCR was performed on experimental kidney samples using the methods described above for amplification of ISAV segment 7, 8 and ELF1a. Samples were also processed for conventional RT-PCR detection of ISAV segment 8 which was previously described [5] and is in routine diagnostic use within our laboratory.

Application of qRT-PCR to the detection of ISAV in field samples during routine ISAV surveillance in Scotland

Field samples intended for ISAV diagnosis during the routine ISA surveillance programme conducted by FRS in Scotland were processed in parallel throughout a nine month period in 2004, using both the RT-PCR method and qRT-PCR methods reported in this study. During this period the presence of ISAV was suspected in Scotland, although it was not attributed as the cause of clinical disease [15]. During this investigation a total of 279 kidney and 87 gill tissue samples were processed for ISAV diagnosis by both qRT-PCR and conventional RT-PCR.

RESULTS

Development of Taqman real-time RT-PCR method

Validation of reaction efficiencies and relative quantification

The amplification efficiency graphs obtained from serial dilutions of ISAV-positive fish tissue are indicated in Figure 1. Each of the ISAV assays exhibited comparable efficiencies with that of ELF1α as indicated by them exhibiting a difference in slope (ΔS) of <0.1 [16]. This indicates that the ELF1α endogenous control is a suitable reference for the relative quantification of ISAV over the range of starting concentrations used in this study. For each reaction series, there was also a high correlation between cycle number and dilution factor (R^2 =1.0 in all cases) and slope values approached the optimum theoretical value of -3.32, indicating efficient amplification of all assays. Primer and probe concentrations were optimised at 900 nM and 250 nM respectively and no loss in sensitivity was recorded by reducing the total reaction volume to 25 μl.

Sensitivity

The segment 8 Taqman® qRT-PCR assay consistently detected ISAV at a log higher dilution that the segment 7 Taqman® qRT-PCR assay which itself consistently detected ISAV at a log higher than the conventional RT-PCR (data not shown). In addition, very faint products were observed nearing the endpoint of detection for conventional RT-PCR which required experience in their interpretation.

Specificity

Both segment 7 and 8 assays yielded no detectable product when using cDNA derived from viruses other than ISAV. Virus identified in field samples during 2004 proved to be of the HPR0 type, which demonstrated efficient detection of variants of ISAV [15].

Application of relative quantification to the detection of ISAV in experimentally infected fish

Experimental infection of salmon with ISAV

The mortality obtained in the unsampled tanks used to monitor the efficacy of the experimental infection is shown in Figure 2. Mortality commenced on day 11 post-infection and all fish in the ISAV infected group were dead by day 20. All dead fish showed clinical signs consistent with ISA disease which included pale gills, darkened livers and the presence of ascites. No mortalities were recorded in the control tanks throughout this period.

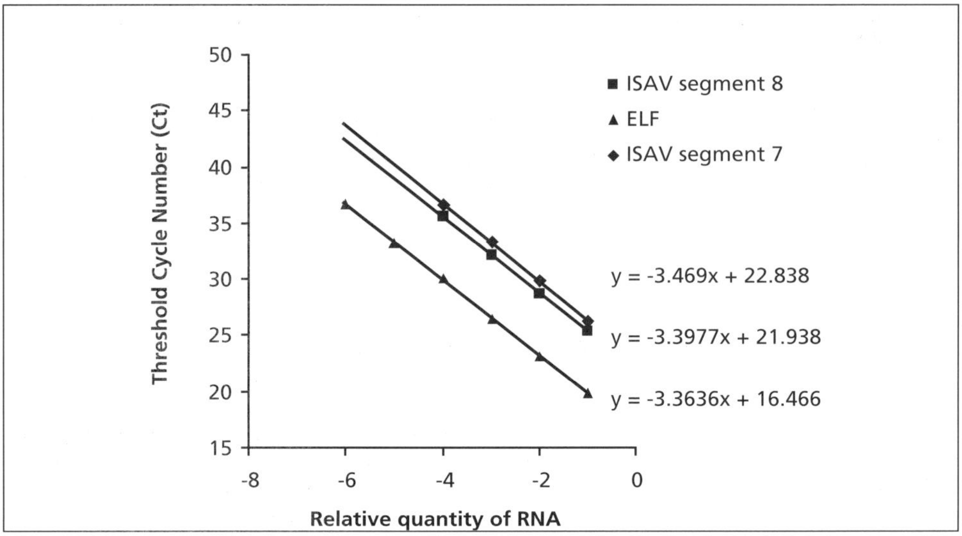

Fig. 1: Standard curves obtained by plotting the relative dilution of RNA versus the cycle number required to elevate the fluorescence signal above the threshold for the ISAV Segment 7, 8 and ELF assays. Equations were derived from linear regression.

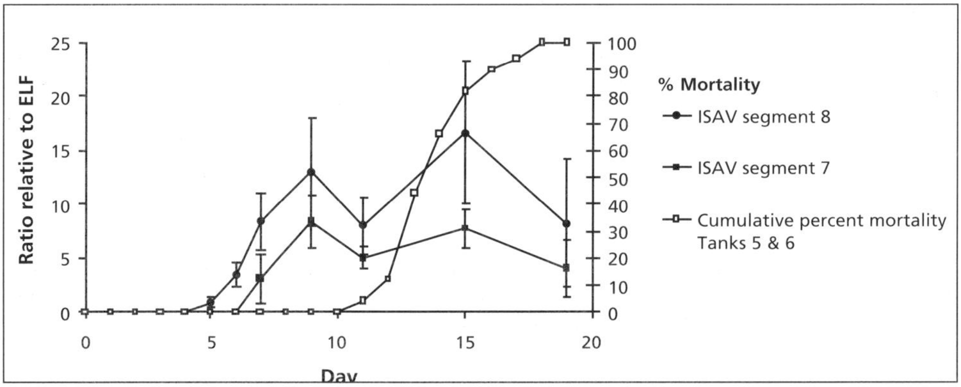

Fig. 2: **Graph indicating the mean level of ISAV segment 7 and 8 detected at different time-points following the experimental infection of Atlantic salmon with ISAV**. Mean levels were derived from four fish sampled at each time-point and represent the level (in arbitrary units) of each ISAV gene expressed as a proportion of the endogenous ELF-1α control signal detected in the same sample. Error bars indicate the standard error of the mean. The mean cumulative percent mortality obtained in un-sampled ISAV challenge reference tanks is indicated.

Sampling and comparative PCR analysis

Results obtained from the qRT-PCR analysis are detailed in Figure 2. The relative signal corresponding to target ISAV RNA was normalized to that obtained with the ELF1α endogenous control in each case. Values were expressed in arbitrary units. This calculation was performed according to the comparative Ct method described in the literature supplied with the ABI 7000 Sequence Detection System. All samples tested positive for ELF1α indicating efficient RNA extraction and reverse transcription in all cases. All fish testing positive for ISAV segment 8 also tested positive for ISAV segment 7. At day 1 post infection, two of the four fish sampled tested positive for ISAV, but at all subsequent time-points all fish tested positive. The mean values obtained for segment 8, however, were consistently higher that those obtained for the segment 7 assay, when expressed as a ratio to all ELF1α (Fig. 2). No contamination was recorded in any of the control fish sampled or in any of the RNA extraction (A), reverse transcription (B), real-time or conventional PCR controls (C) performed as described above.

Results obtained on the same sample set using a conventional RT-PCR targeting ISAV segment 8 corresponded with those obtained with the qRT-PCR methods, with the same two fish testing negative. In many of the positive cases, however, very weak products were observed on conventional agarose gels, which required experienced interpretation. In some cases non-specific products were also observed alongside the expected 155bp product.

Application of relative quantification to the detection of ISAV in field samples during routine ISA surveillance in Scotland

Two hundred and seventy nine kidney samples (each a pool of up to five fish) were screened for the presence of ISAV using Taqman® qRT-PCR assays targeting segments 7 and 8 in addition to conventional RT-PCR which also targets segment 8. Six samples were identified as positive using the conventional RT-PCR. All six samples were also positive using both ISAV segment 7 and 8 Taqman® qRT-PCR assays. qRT-PCR assays for segment 7 and 8 identified a further 11 and 12 positive samples, respectively (corresponding to 17 individual samples), which tested negative by conventional RT-PCR. Of these, six samples were positive using both assays, but five and six samples were positive using only the segment 7 or segment 8 assay respectively. Higher relative quantities of ISAV segment 7 and 8 were identified in samples testing positive for conventional RT-PCR as opposed to those testing negative for RT-PCR and positive by either one or both of the Taqman® assays (Figure 3). Interestingly, some of the values obtained from field samples were of a similar magnitude to those obtained during the course of clinical infection under experimental conditions.

During surveillance for ISAV, gill samples were also taken from some fish. Of a total of 87 fish samples processed, 41 yielded positive results using conventional RT-PCR and both segment 7 and segment 8 Taqman® qRT-PCR assays. No samples were positive using only conventional PCR. Seventeen samples were positive using both segment 7 and segment 8 assay but were negative using the conventional RT-PCR assay. Three samples were positive using the segment 7 Taqman® qRT-PCR assay only and a further 12 samples were positive using only the ISAV segment 8 Taqman® qRT-PCR. Relative quantities of virus detected are indicated in Figure 4. Samples which were positive using all methods yielded high relative quantities of virus. For those which tested negative using conventional RT-PCR, but positive using both Taqman® qRT-PCR assays, low relative values were obtained for both segment 7 and segment 8 assays. Similarly samples testing positive using only one of the qRT-PCR assays also yielded low relative values.

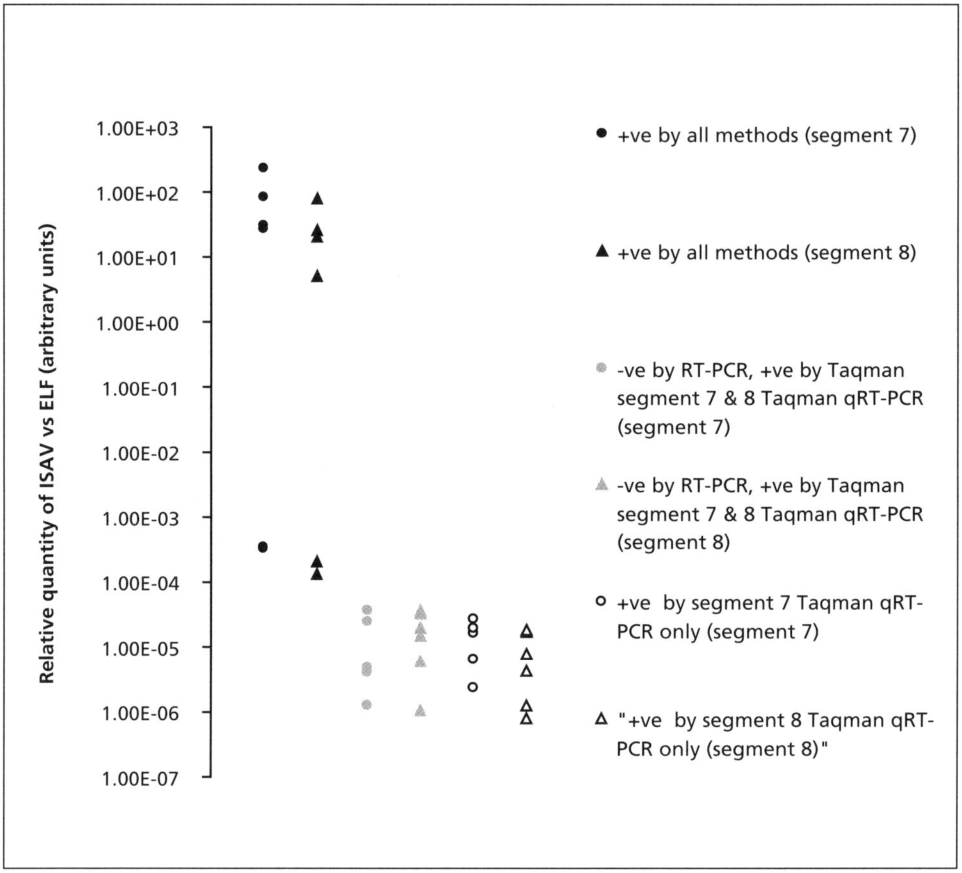

Fig. 3: **Relative levels of ISAV segment 7 and 8 detected in field kidney samples analysed during routine surveillance for ISAV in Scotland during 2004.** ISAV Levels obtained using Taqman® segment 7 (indicated by circles) and segment 8 (indicated by triangles) semi-quantitative qRT-PCR assays and expressed as a proportion of the endogenous ELF-1α control signal detected in each sample are detailed. Results are indicated according to the number of the three diagnostic tests (RT-PCR, Taqman® qRT-PCR segment 7, Taqman® qRT-PCR segment 8) generating positive results in each case.

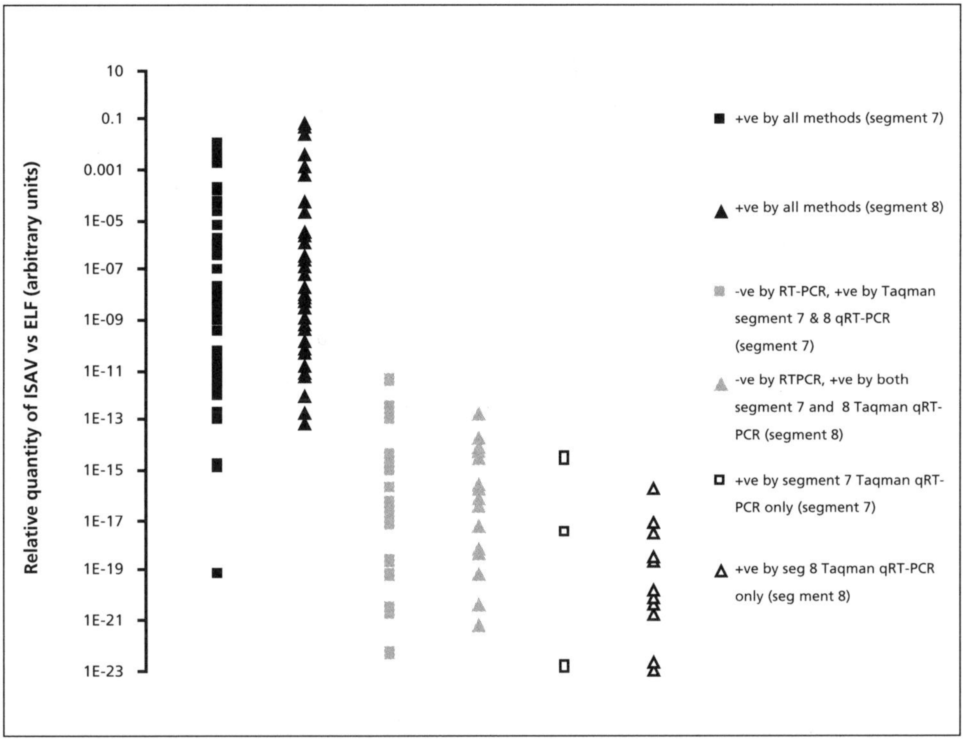

Fig. 4: **Relative levels of ISAV segment 7 and 8 detected in field gill samples analysed during routine surveillance for ISAV in Scotland during 2004.** ISAV Levels obtained using Taqman® segment 7 (indicated by squares) and segment 8 (indicated by triangles) semi-quantitative qRT-PCR assays and expressed as a proportion of the endogenous ELF-1α control signal detected in each sample are detailed. Results are indicated according to the number of the three diagnostic tests (RT-PCR, Taqman® qRT-PCR segment 7, Taqman® qRT-PCR segment 8) generating positive results in each case.

Discussion

The results presented describe two independent highly sensitive and specific two-step Taqman® qRT-PCR methods for the detection of ISAV in Atlantic salmon tissues. Both methods proved to be more sensitive than the conventional RT-PCR method in routine use for ISAV diagnosis [5]. In addition, the application of semi-quantitative Taqman® qRT-PCR to ISAV diagnostics offers a number of advantages associated with this technology including an increased sensitivity, enhanced specificity, quantitative expression against endogenous controls, a lack of subjectivity in the interpretation of results, speed of processing and contamination control.

The segment 8 qRT-PCR assay was shown to be more sensitive than the segment 7 assay during initial experimental dilution series experiments. This was supported by the consistent detection of higher levels of segment 8 in experimentally infected fish during a clinical outbreak of ISA disease and the fact that more field kidney and gill samples were positive by the segment 8 assay only than by the segment 7 assay only. Since the assays displayed similar theoretical efficiencies, such a finding is most likely related to the differential expression of segment 7 and 8 segments throughout the viral infection cycle. Both segments 7 and 8 of ISAV have been shown to encode two proteins, based on removal of an intron from the primary transcript [17] or from overlapping open reading frames, respectively [18]. Segment 8 has been shown to encode a major structural protein [19], while segment 7 has recently been demonstrated to encode at least one protein which mediates the host interferon defence system [20]. An increased abundance of segment 8 mRNA generated during the viral life cycle may thus explain the apparent improved sensitivity of the Taqman® segment 8 assay.

Both Taqman® assays performed at least as well as the RT-PCR segment 8 assay in the analysis of experimentally infected fish. This was perhaps not surprising since the fish had been injected with a relatively high dose of ISAV which, based on the severity of the challenge, was actively replicating. Analysis of serial dilutions, however, suggested an increased sensitivity of the Taqman® assays. This increased sensitivity may be related to the specificity and sensitivity of Taqman® chemistry which permits additional cycles of PCR to be conducted (n=45) when compared to conventional PCR (n=35). Furthermore, the generation of the very small amplicons in both ISAV Taqman® assays may contribute to the improved sensitivity observed. Improved sensitivity of both Taqman® assays over conventional RT-PCR was also demonstrated during evaluation of both kidney and gill samples collected as part routine diagnostic surveillance of Scottish salmon aquaculture during 2004. Indeed, many samples testing negative by conventional PCR tested positive by both segment 7 and segment 8 Taqman® assays. However, the relative quantity of ISAV detected was lower than in those samples which tested positive by conventional RT-PCR. In these cases, the agreement of the independent Taqman® assays, coupled to the detection of signals beyond the apparent detection threshold for conventional RT-PCR strongly supports the fact that these were indeed true positive samples. In some cases, samples were negative by conventional RT-PCR and positive by only one of the segment 7 or 8 Taqman® assays. In all these cases, relative levels of ISAV detectable were also very low. The demonstrated difference in sensitivity of the segment 7 and 8 assays, which is probably related to differences in viral gene expression could account for such differences and supports the view that the results observed are not due to a lack of assay specificity.

The specificity of detection using Taqman® chemistry is significantly improved as compared to either conventional RT-PCR or real-time PCR based on chemistry such as SYBR green. Indeed, detection of Taqman® PCR products is dependent on the annealing of highly specific probes. Experience in our laboratory and in other studies has shown the method to be capable of the absolute discrimination of PCR products based on the presence of a single nucleotide mismatch in the probe binding region. Such specificity raises the prospect of false negatives using a single probe, should mutation occur within this region. To reduce the likelihood of false negatives, probes in this study were designed from absolutely conserved regions based on alignments of all known ISAV sequence types. The parallel use of two independent assays in this study provides further confidence in the characterisation of true negatives and adds significant power to the detection of rapidly mutating organisms such as RNA viruses.

The development of a quantitative method for ISAV diagnosis using the ELF1α endogenous control provides a significant advantage over the conventional RT-PCR method in ensuring quality control of RNA and subsequently generated cDNA quality and quantity. The elongation factor 1 alpha gene has been previously demonstrated to be stably expressed between different individuals, and as such represents a suitable calibrator for semi-quantitative real-time PCR [21]. ISAV load expressed as a function of ELF1α is thus a measure which can be compared across individuals sampled at different times. In a diagnostic context such information can be used to express confidence levels in positive results obtained and as a basis to predict the likelihood of other tests (e.g. conventional RT-PCR or virus culture) also yielding positive results. Results might also be used to predict the clinical status of fish once a more comprehensive understanding of the relationship between viral RNA levels and pathology emerges. Such a method thus offers considerable scope both for improving the understanding and interpretation of PCR results, generating more information than a simple positive/negative result obtained using conventional RT-PCR. Indeed, within the current study of ISAV experimentally infected fish, we have demonstrated the range of signals expected in kidney tissues throughout a clinical disease outbreak. Results from kidney samples obtained in field samples during 2004 [15] were in some cases of a similar magnitude to these values, although in others relative values were significantly lower. Such knowledge might in future assist us in identifying carrier fish populations and those in which clinical disease is expected. In this case, however, further characterization of detected virus identified it as an HPR0 ISAV subtype, the relevance of which to development of ISA disease is not fully understood.

Taqman® qRT-PCR does not require any post-PCR processing and in particular does not require the subjective interpretation of weak products visualised on agarose gels. As such, interpretation of its results is more objective and less variable and suited to implementation in routine viral diagnostic testing. The lack of post-PCR processing also reduces the time taken to complete the method (approximately four to five h) and minimises the potential for obtaining false positives due to cross contamination. The Taqman® method also replaces dTTP with dUTP during PCR amplification. Subsequent PCR mixes are treated with the enzyme uracil N-glycosylase (UNG) which specifically degrades PCR products from previous amplifications in order to reduce the risk of amplicon carry-over resulting in sample contamination [22].

In conclusion, two highly sensitive and specific Taqman® qRT-PCR assays targeting ISAV segment 7 and segment 8 have been developed and evaluated. Use of an internal endogenous control ELF1α allows semi-quantitative expression of viral transcript levels. The application of Taqman® technology to ISAV diagnosis offers a number of significant advantages over conventional RT-PCR and lends itself well to implementation within a routine viral diagnostic environment.

References

1 Stagg RM, Bruno DW, Cunningham CO, Raynard RS, Munro PD, Murray AG, et al: Epizootiological investigations into an outbreak of infectious salmon anaemia (ISA) in Scotland. Aberdeen: FRS Marine Laboratory; 2001 April 2001, Report No 13/01.

2 Cipriano RC, Miller O: Infectious Salmon Anaemia: The Current State of Our Knowledge, in Miller O, Cipriano RC (eds): International response to Infectious salmon anemia: prevention, control and eradication. New Orleans, LA, U.S. Department of Agriculture, Animal and Plant Health Inspection Service, U.S. Department of the Interior, U.S. Geological Survey; U.S. Department of Commerce, National Marine Fisheries Service, 2003, pp 1-11.

3 Dannevig BH, Falk K, Namork E: Isolation of the causal virus of infectious salmon anaemia (ISA) in a long-term cell line from Atlantic salmon head kidney. J Gen Virol 1995;76(6):1353-1359.

4 Falk K, Namork E, Dannevig BH: Characterization and applications of a monoclonal antibody against infectious salmon anaemia virus. Dis Aquat Organ 1998;34(2):77-85.

5 Mjaaland S, Rimstad E, Falk K, Dannevig BH: Genomic characterization of the virus causing infectious salmon anemia in Atlantic salmon (*Salmo salar* L.): an orthomyxo-like virus in a teleost. J Virol 1997;71(10):7681-7686.

6 Devold M, Krossoy B, Aspehaug V, Nylund A: Use of RT-PCR for diagnosis of infectious salmon anaemia virus (ISAV) in carrier sea trout Salmo trutta after experimental infection. Dis Aquat Organ 2000;40(1):9-18.

7 Opitz HM, Bouchard D, Anderson E, Blake S, Nicholson B, Keleher W: A comparison of methods for the detection of experimentally induced subclinical infectious salmon anaemia in Atlantic salmon. Bulletin of the European Association of Fish Pathologists 2000;20(1):12-22.

8 Snow M, Raynard RS, Murray AG, Bruno DW, King JA, Grant R, et al: An evaluation of current diagnostic tests for the detection of infectious salmon anaemia virus (ISAV) following experimental water-borne infection of Atlantic salmon, *Salmo salar* L. J Fish Dis 2003;26(3):135-145.

9 Nerette P, Dohoo I, Hammell L, Gagne N, Barbash P, Maclean S, et al: Estimation of the repeatability and reproducibility of three diagnostic tests for infectious salmon anaemia virus. J Fish Dis 2005;28(2):101-110.

10 Munir K, Kibenge FS: Detection of infectious salmon anaemia virus by real-time RT-PCR. J Virol Methods 2004;117(1):37-47.

11 Moore LJ, Somamoto T, Lie KK, Dijkstra JM, Hordvik I: Characterisation of salmon and trout CD8alpha and CD8beta. Mol Immunol 2005;42(10):1225-1234.

12 Altschul SF, Gish W, Miller W, Myers EW, Lipman DJ: Basic local alignment search tool. J Mol Biol 1990;215:403-410.

13 Cunningham CO, Gregory A, Black J, Simpson I, Raynard RS: A novel variant of the infectious salmon anaemia virus (ISAV) haemagglutinin gene suggests mechanisms for virus diversity. Bull Eur Ass Fish Pathol 2002;22:366-374.

14 Nylund A, Devold M, Plarre H, Isdal E, Aarseth M: Emergence and maintenance of infectious salmon anaemia virus (ISAV) in Europe: a new hypothesis. Dis Aquat Organ 2003;56(1):11-24.

15 Anon: Epizootiological investigation into a case of suspicion of infectious salmon anaemia (ISA) in Scotland in 2004. Aberdeen: FRS Marine Laboratory; 2005 June 05, Report No 11/05.

16 Gut M, Leutenegger CM, Huder JB, Pedersen NC, Lutz H: One-tube fluorogenic reverse transcription-polymerase chain reaction for the quantitation of feline coronaviruses. J Virol Methods 1999;77(1):37-46.

17 Ritchie RJ, Bardiot A, Melville K, Griffiths S, Cunningham CO, Snow M: Identification and characterisation of the genomic segment 7 of the infectious salmon anaemia virus genome. Virus Res 2002;84(1-2):161-170.

18 Clouthier SC, Rector T, Brown NE, Anderson ED: Genomic organization of infectious salmon anaemia virus. J Gen Virol 2002;83(2):421-428.

19 Biering E, Falk K, Hoel E, Thevarajan J, Joerink M, Nylund A, et al: Segment 8 encodes a structural protein of infectious salmon anaemia virus (ISAV); the co-linear transcript from Segment 7 probably encodes a non-structural or minor structural protein. Dis Aquat Organ 2002;49(2):117-122.

20 McBeath AJA, Collet B, Paley R, Duraffour S, Aspehaug V, Biering E, et al: Identification of an interferon antagonist protein encoded by segment 7 of infectious salmon anaemia virus. Virus Research 2006;115(2):176-184.

21 Ingerslev HC, Pettersen EF, Jakobsen RA, Petersen CB, Wergeland HI: Expression profiliing and validation of reference gene candidates in immune relevant tissues and cells from Atlantic salmon (*Salmo salar* L.). Molecular Immunology 2006;43(8):1194-1201.

22 Longo N, Berninger NS, Hartley JL: Use of uracil DNA glycosylase to control carry-over contamination in polymerase chain reactions. Gene 1990;93:125-128.

Dr. M. Snow, FRS Marine Laboratory, 375 Victoria Rd, Aberdeen, AB11 9DB, U.K.
E-mail: M.Snow@marlab.ac.uk

Vannier P, Espeseth D (eds): New Diagnostic Technology: Applications in Animal Health and Biologics Controls. Dev Biol (Basel). Basel, Karger, 2006, vol 126, pp 147-148.

Development of a Comprehensive Set of Real-Time RT-PCR Assays for the Detection of Porcine Respiratory and Reproductive Syndrome Virus in Semen, Serum, and Tissue.

J. Callahan[2], J. Christopher-Hennings[1], W.M. Nelson[2]

[1] Dept. of Veterinary Science, South Dakota State University, Brookings, SD, USA
[2] Tetracore, Inc., Gaithersburg, MD, USA

Introduction

There are many challenges to the development of a Real-Time PCR assay for the detection of all the isolates of this rapidly evolving highly geographically diverse RNA virus, Porcine Respiratory and Reproductive Syndrome Virus (PRRSV), requiring a broad international partnership involving academia, industry, and government. Porcine Reproductive and Respiratory Syndrome Virus has been identified in boars and known to be transmitted via semen. Since large numbers of semen doses are distributed world-wide, with the potential to introduce new isolates into naïve populations, a rapid, reliable, commercially-available and user-friendly assay is needed to identify PRRSV-contaminated semen. We have developed two highly multiplexed assays, one covering the "US-like" isolates, and one covering the "Euro-like" isolates. These highly sensitive assays are very successful in detecting the virus in a very difficult sample matrix such as semen. In validation studies they have equalled or exceeded both culture and nested PCR. The application of a real-time, quantitative assay has broad potential for commercial use in boar studs, diagnostic laboratories and other facilities for monitoring PRRSV biosecurity.

Material and Methods

A real-time, single-tube, RT-PCR assay for the detection of U.S. or LV/European-like PRRSV was used to detect PRRSV RNA. A minor groove binding (MGB) 5' nuclease probe and primers were designed from the 3'UTR PRRSV genomic region by alignment of GenBank isolates and based on conserved areas of the 3'UTR primer and probe region. The PRRSV RNA was transcribed in a single tube using a 25 µl reaction volume consisting of Tetracore U.S. PRRSV Master Mix [18.9 µl Master mix, 2 µl Enzyme mix 1, 0.1 µl Enzyme mix 2] and 4 µl of extracted RNA. The thermal cycler programme for the U.S. PRRSV real-time assay consisted of 52°C for 1800 s, 95°C for 900 s, and 45 cycles at 94°C for 30 s, 61°C for 60 s and 72°C for 60 s. For the LV/European-like PRRSV assay, thermal cycling times consisted of 60°C for 1200 s, 95°C for 15 s and 45 cycles at 95°C for 3 s, 60°C for 30 s.

Results

RNA viruses often exist as "quasi-species", making the design of a comprehensive assay difficult [1, 2]. The first two PRRS viruses that were completely sequenced, North American (NA) isolate 16244B and the European (Eu) Lelystad strain, demonstrated that their genomic organization was the same, but they only had a 63.4% overall nucleotide identity [3]. The 3" UTR had an overall identity of only 76%, but a number of short specific areas were conserved at a much higher degree. Two assays were designed with a single probe and four primers using these highly conserved regions to cover both strains. As more isolates were sequenced, these two original assays appeared by in-silico testing to detect all strains (36 PRRSV isolates were aligned). In vitro testing at the South Dakota State University (SDSU) used 134 clinical isolates obtained through the diagnostic laboratory from various farms within a wide geographical area in the U.S. Fifty-two of these samples were PRRSV positive by both real-time and nPCR, and had been selected with diverse predicted RFLP cut patterns by genomic sequencing of ORF 5. In addition, 82 isolates were PRRSV positive samples submitted to the diagnostic laboratory which had been tested in parallel with real-time PCR and VI with corresponding results [4].

Using serial cell culture dilutions for the U.S. (SD-23983) isolate, the real-time PCR level of detection matched nPCR, which has been previously reported to detect 10 $TCID_{50}$/ml (14). The minimum level of in-vitro transcript that was consistently detected by the RT-PCR reaction was 33 copies/ml. This assay did not react with other common swine viruses. An inter and intra-assay variability of 2.7-8.8% and 0.1-5.7%, respectively was observed. Using serum and semen samples from boars experimentally infected with PRRSV, a 95% correlation between nPCR and real-time PCR was observed with a kappa score of 0.821, which is considered a very good agreement.

Discussion

Porcine reproductive and respiratory syndrome is a widespread and economically devastating disease affecting the swine industry. Recent developments in real-time PCR technology have allowed for fast and accurate quantification of virus levels. In this study we developed a commercially available real-time PCR assay for the detection and quantification of PRRSV in serum and semen samples. Our goal was to develop a real-time PCR assay that would have equal sensitivity to nPCR, which had previously corresponded to the "swine bioassay" gold standard and had been used extensively in previous experimental studies and for diagnostic testing.

References

1 van Vugt JJ, Storgaard T, Oleksiewicz MB, Botner A : High frequency RNA recombination in porcine reproductive and respiratory syndrome virus occurs preferentially between parental sequences with high similarity. J Gen Virol 2001 ;82:2615-2620.

2 Goldberg TL, Lowe JF, Milburn SM, Firkins LD: Quasispecies variation of porcine reproductive and respiratory syndrome virus during natural infection. Virology 2003;317:197-207.

3 Allende R, Lewis TL, Lu Z, Rock DL, Kutish GF, Ali A, Doster AR, Osorio FA: North American and European porcine reproductive and respiratory syndrome viruses differ in non-structural protein coding regions. J Gen Virol 1999 ;80:307-315.

4 Christopher-Hennings J, Nelson EA, Nelson JK, Hines RJ, Swenson SL, Hill HT, Zimmerman JJ, Katz JB, Yaeger MJ, Chase CC, et al: Detection of porcine reproductive and respiratory syndrome virus in boar semen by PCR. J Clin Microbiol 1995;33:1730-1734.

Dr. W.M Nelson, Tetracore Inc., 11 Firstfield Rd, Suite C, Gaithersburg, MD 20878, USA.
E-mail: wnelson@tetracore.com

Vannier P, Espeseth D (eds): New Diagnostic Technology: Applications in Animal Health and Biologics Controls. Dev Biol (Basel). Basel, Karger, 2006, vol 126, pp 149-157.

Detection and Differentiation of Pathogenicity of Avian Paramyxovirus Serotype 1 (APMV-1) from Field Cases Using One-Step Real-Time RT-PCR

J.L. Creelan, S.J. McCullough

Veterinary Sciences Division, Department of Agriculture and Rural Development, Belfast, Northern Ireland, United Kingdom

Abstract: A real-time fluorescent RT-PCR assay was developed to amplify avian paramyxovirus serotype 1 (APMV-1)-specific nucleic acid fragments from field samples. Subsequent restriction endonuclease analysis (REA) using *BglI* was carried out to type strains according to their virulence. Primer sequences were used to amplify a 202 base-pair fragment, encompassing the fusion protein cleavage site, in a one-step RT-PCR test for detection of a range of field cases and reference strains of APMV-1. Subsequent restriction endonuclease analysis of the amplified fragments enabled differentiation of low virulent lentogenic field and vaccine strains from more virulent mesogenic and velogenic field strains of APMV-1, including pigeon PMV-1. In 2004, seven cases of pigeon PMV-1 in Northern Ireland were diagnosed and differentiated more rapidly using the fluorescent RT-PCR assay when compared with the use of virus isolation. We report the development and application of a one-step real-time RT-PCR test coupled with REA as a fast, specific method for both the detection and typing of APMV-1 from field samples.

Introduction

Newcastle Disease (ND) is caused by certain strains of avian paramyxovirus serotype 1 (APMV-1) and, since its first recognition in 1926, it has become a worldwide pathogen of poultry of considerable economic importance [1].

APMV-1 is a member of the genus *Rubulavirus* within the family Paramyxoviridae, with serological tests having distinguished nine avian paramyxovirus serotypes (designated APMV-1 to APMV-9) [2]. The APMV-1 viral genome consists of a single unsegmented, single-stranded negative sense RNA that encodes six gene products. One of these proteins, the fusion (F0) glycoprotein, is responsible for the adherence and subsequent penetration of the virus through the host cell membrane [2]. APMV-1 may be classified according to virulence based upon the ability to induce chicken embryo mortality [3] with strains described as lentogenic, mesogenic or velogenic in order of increasing virulence. The virulence of APMV-1 is also known to be associated with differences in the amino acid sequence surrounding the post-translational cleavage site of the F0 protein [4] with differences in the cleavage sites being directly related to the strain's virulence.

Within the EU, where a compulsory slaughter policy is implemented upon diagnosis, ND is defined as an infection of poultry caused by APMV-1 strains having an intracerebral pathogenicity index (ICPI) in day-old chicks of greater than or equal to 0.7 (Council Directive 92/66/EEC, Control of Newcastle Disease). However, the process of virus isolation and subsequent determination of ICPI are labour-intensive and time-consuming, with virus isolation sometimes requiring two six-day passages in eggs. Recently, the Office International des Epizooties (OIE) has adopted an expanded definition of ND with the inclusion of an alternative criterion of virulence. This criterion is the demonstration (directly or by deduction) of a characteristic pattern of amino acid residues in the region of the fusion protein cleavage site [5], thereby enabling molecular-based techniques to become integrated into the diagnosis of ND.

The potentially lengthy restrictions imposed on flocks while laboratory diagnosis is taking place can result in significant economic losses in both laying flocks and in broilers approaching commercial slaughter weight. In the latter group, restrictions can additionally result in significant welfare problems for the rapidly growing birds. Delays in confirmation of diagnosis may be compounded by the use of lentogenic vaccine strains, which may be adventitiously re-isolated for several weeks post-administration [6]. A more rapid, specific method for both detection and subsequent pathotyping of APMV-1 would have considerable advantages. Although sequence data are available which enable APMV-1 virulence prediction [7] through the sequence analysis of RT-PCR products, such methods are not suitable for routine, rapid diagnosis. Other methods more suited to routine laboratory diagnosis have been reported and include the use of RT-PCR, RT-PCR coupled with REA, dot-blot hybridisation and the use of strain-specific probes, although efficacy in the detection of all strains directly from field material has been limited.

This paper describes the development and application of a one-step real-time RT-PCR assay which, when combined with REA is capable of detecting APMV-1 from clinical samples and subsequently differentiating lentogenic strains from more virulent mesogenic and velogenic strains. The detection of pigeon APMV-1 from clinical material is particularly of note as research has demonstrated difficulties in both the amplification and characterisation of such strains.

Materials and Methods

Preparation of samples and RNA extraction

One hundred and sixty seven samples from 40 field cases, initially submitted with suspected ND, were characterised using RT-PCR and REA. The 167 samples examined comprised either 10% original tissue or swab homogenates (n=96) or from egg passaged material (n=71). In some cases both original clinical homogenates and egg passaged material were available.

In addition to these clinical samples, total RNA was extracted from three archived APMV-1 strains following egg passage, for use as positive controls. They included pigeon PMV-1 and the lentogenic vaccine strains Hitchner B1 and Ulster 2C. To evaluate specificity of amplification primers, RNA was also extracted from reference strains of avian PMV-2 to 9 and from additional avian RNA viruses including avian influenza type A (H7N7 and H5N3), turkey rhinotracheitis, infectious bronchitis virus and infectious bursal disease. For negative control purposes RNA was extracted from allantoic fluid from uninfected eggs.

Total RNA from field cases was isolated from 200 µl of allantoic fluid or 10% tissue or swab homogenate as appropriate using the RNeasy mini kit (Qiagen) according to the manufacturer's instructions.

RT-PCR

Two amplification primer sequences were generated using Genefisher primer design software following alignment of published APMV-1 sequences. The primers flanked the region of the fusion protein cleavage site and following RT-PCR produced a 202 base pair (bp) DNA fragment. The primer sequence and location were as follows: NDV F (sense) 4829 5'-GGTGAGTCTATCCGGARGATACAAG-3' 4893; NDV R (antisense) 5031 5'-TCATTGGTTGCRGCAATGCTCT-3' 5008. Sequence numbering is according to previously published sequences [8]. Both cDNA synthesis and subsequent DNA amplification were carried out in a single tube using these gene-specific primers and Quantitect SYBR Green RT-PCR kit (Qiagen). Amplification was carried out in an Opticon 2 DNA engine (MJ Research Inc.) real time thermal cycler using the following conditions. cDNA synthesis was carried out for one cycle at 45°c for 30 minutes, followed by one cycle at 95°c for 15 minutes. Amplification immediately followed with 40 cycles of 94°c for 15 seconds, 48°c for 30 seconds and 72°c for 30 seconds, followed by one final extension cycle of 72°c for seven minutes. Melt curve analysis was carried out by increasing PCR product temperature from 55°c to 95°c in 1°c increments.

Restriction endonuclease analysis and nucleic acid sequence analysis

PCR products were purified using QIAquick PCR purification kit (Qiagen) according to the manufacturer's instructions before restriction endonuclease digestion using *Bgl*I. This restriction endonuclease was selected because sequence analysis indicated that all lentogenic strains of APMV-1 selected for alignment contained a single *Bgl*I restriction site at the F0 cleavage region. This *Bgl*I site was missing from all virulent strains selected for alignment. Following digestion, RT-PCR products were electrophoresed on a 3% agarose gel. Nucleic acid sequencing of the 202 bp amplicons was carried out using the primers used for initial amplification and the BigDye Terminator cycle sequencing ready reaction kit (Applied Biosystems). Reactions were purified using the Montage SEQ_{96} Sequencing reaction cleanup Kit (Millipore), sequence reactions were run on an ABI 30100 Genetic Analyser (Applied Biosystems) and analysed using Vector NTI Advance 9 software (Informax, Invitrogen).

Results

Validation of reaction efficiencies and relative quantification

APMV-1 strains Ulster 2C, Hitchner B1 and pigeon PMV-1 produced a 202 bp product upon amplification. All other reference strains of APMV (2 to 9) and previously itemised unrelated avian RNA viruses were negative following amplification using these primers.

Identification of APMV-1 from clinical material

All cases (n=34) from which APMV-1 was identified by virus isolation produced an amplicon of the predicted size by RT-PCR, either from first pass allantoic fluid or original clinical material. All other cases found negative by virus isolation (n=6) failed to produce an amplicon by RT-PCR. The results of the RT-PCR on clinical samples and their comparison with virus isolation are summarised on both a case and sample basis in Table 1. One or more original field samples in 31/33 cases from which virus was subsequently isolated were positive by RT-PCR. Taking virus isolation as the gold standard gives RT-PCR a sensitivity of 93.94 % on a case basis and on a sample basis the sensitivity is 81.11%. Sixty-five of the 66 isolates available as egg passage material were positive by RT-PCR and all egg passage material isolates were RT-PCR positive on a case basis, thereby giving sensitivities of 98.48 % and 100 % on a sample and case basis respectively. Table 2 compares the ICPI and MAb typing results with REA profile results determined after RT-PCR was carried out.

Table 1: Comparison of RT-PCR results from original and egg passage material to virus isolation on a case and sample basis.

		RT-PCR results							
		Sample basis				Case basis			
		Original material		Egg passage material		Original material		Egg passage material	
		+	-	+	-	+	-	+	-
Virus isolation results	+	73	17	65	1	31	2	24	0
	-	0	6	0	5	0	1	0	5

Differentiation of APMV-1 strains using restriction endonuclease analysis and sequence analysis

APMV-1 specific RT-PCR products amplified from lentogenic vaccine strains and from field isolates of varying virulence, all produced a 202 bp fragment. Melt curve analysis of the amplified product indicated that the melting temperature of the fragment was 82.5°c +/- 0.5°c. Following amplification, the PCR products were digested with the restriction endonuclease BglI following purification as described previously. Figure 1 shows the restriction pattern following electrophoresis on a 3% agarose gel where the pigeon PMV-1 strain and four field cases that were shown to have an ICPI value greater than or equal to 0.7 show products that remain uncleaved by the endonuclease. The lentogenic strains Hitchner B1 and Ulster 2C, however, possess a BglI site in the region of the fusion protein cleavage site and are, therefore, cleaved into 2 fragments of 135 and 67 bp. The two field cases of APMV-1 that were typed as avirulent (ICPI < 0.7) also exhibit this restriction profile. Until 2004, differentiation of all field cases into virulent and avirulent strains using RT-PCR and REA matched the virulence as defined by ICPI, as all field cases identified as ND by virus isolation and ICPI remained uncleaved with BglI. In 2004, however, the seven cases of suspect ND in pigeons itemised in Table 2 were diagnosed as APMV-1 by virus isolation and RT-PCR. The restriction endonuclease profile indicated that the cases were virulent and this was further confirmed by sequence analysis of the fusion protein cleavage site where in all cases the amino acid residues were 113RQKRF117. The ICPI values, however, did not meet the EC criteria for ND as they were found to be less than 0.7 [9].

Potential identification of dual infection with pathogenic and non-pathogenic virus strains

Figure 2 shows the restriction profiles characterising an avirulent vaccine strain, Hitchner B1 and a virulent chicken field isolate (Group C1) of APMV-1 in lanes 1 and 2 respectively. No isolates from the field survey were identified with mixed populations of low virulent and virulent APMV-1. However, the possibility that a dual infection could occur in the field exists and has been previously reported. To test the discriminatory ability of the RT-PCR test under these circumstances, an aliquot of the virulent chicken field homogenate (as described for lane 2) was 'spiked' with an aliquot of the Hitchner B1 strain. Following extraction, amplification and restriction endonuclease digestion, an easily identifiable profile of an isolate containing a mixed population of avirulent and virulent virus was obtained.

Table 2: Comparison of ICPI, MAb and REA typing of APMV-1 virus isolates from field cases and controls.

Case number	ICPI	MAb typing	REA typing [a]
1	0.0	E	A
2	1.03	B	B
3	1.26	B	B
4	0.96	P	B
5	0.0	E	A
6	0.74	B	B
7	1.84	C1	B
8	1.82	C1	B
9	1.74	C1	B
10	0.0	E	A
11	0.0	E	A
12	1.89	C1	B
13	1.75	C1	B
14	1.89	C1	B
15	1.78	C1	B
16	1.66	C1	B
17	1.11	P	B
18	0.77	P	B
19	1.20	P	B
20	1.25	P	B
21	0.91	P	B
22	0.01	E	A
23	0.08	E	A
*24	0.33	P	B
*25	0.43	P	B
*26	ND	P	B
*27	0.6	P	B
*28	ND	P	B
*29	ND	P	B
*30	ND	P	B
Ulster 2C	0.0	E	A
Hitchner B1	0.0	E	A
Pigeon PMV-1	Not known	P	B

[a] REA typing: Profile A = RT-PCR product gave two fragments of 135 & 67 bp on digestion with *Bgl*I. Profile B = RT-PCR product of 202 bp remained uncleaved upon digestion with *Bgl*I.
ND = not done
*indicates pigeon PMV-1 cases in 2004

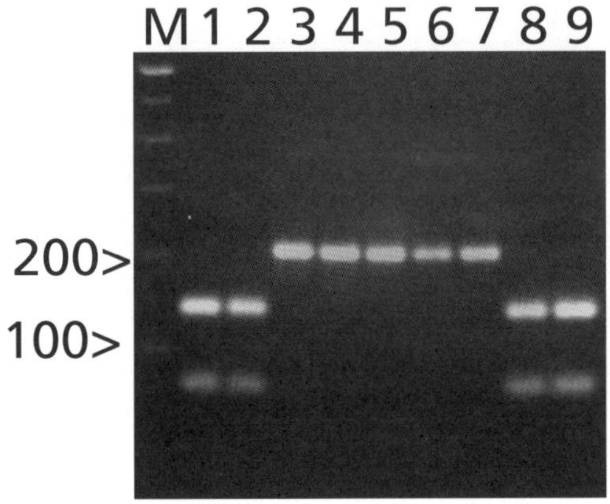

Fig. 1: BglI digested RT-PCR products amplified from RNA extracts derived from tissues and isolates with varying pathogenicity. Lane 1: lentogenic vaccine strain Hitchner B1; lane 2: lentogenic strain Ulster 2C; lane 3: pigeon PMV-1; lane 4: original chicken cloacal swab 10% homogenate; lane 5: original pheasant small intestine 10% homogenate; ICPI=1.74; lane 6: original pigeon kidney 10% homogenate, ICPI=0.74; lane 7: original pigeon faeces 10% homogenate, ICPI=1.03; lane 8: chicken trachea, 1 passage allantoic fluid, ICPI=0.00; lane 9: chicken trachea, 1 passage allantoic fluid, ICPI=0.00.

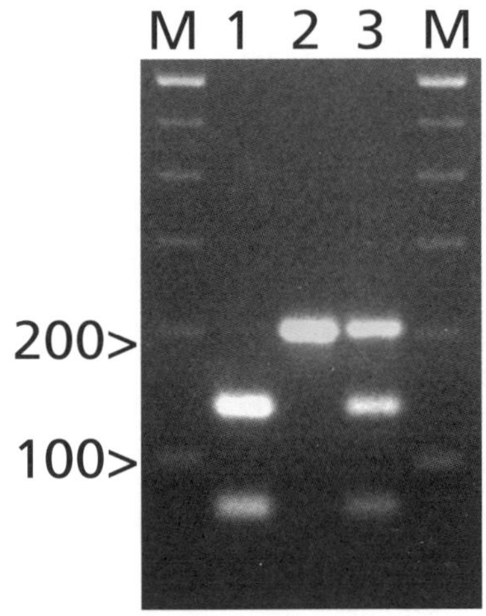

Fig. 2: Digestion of RT-PCR products with BglI. Lane 1: lentogenic vaccine strain Hitchner B1; lane 2: chicken original cloacal swab, Group C1; lane 3: original material as used in lane 2, spiked with lentogenic PMV-1 Hitchner B1 as described in Materials and Methods. M: 100 bp marker.

Discussion

The specificity of the primers to APMV-1 was demonstrated by amplification of a 202 bp product from all the lentogenic, mesogenic and velogenic cases tested, including pigeon PMV-1 cases and the absence of any amplification product for all other eight serotypes of APMV and unrelated avian RNA viruses as previously described. The specificity of the RT-PCR was also demonstrated by the negative RT-PCR results obtained from the six cases of suspect ND that were subsequently found to be negative by virus isolation. Correlation of the REA typing to the virulence (as defined by ICPI and sequence analysis) is demonstrated in Table 2. Avirulent strains have been observed to have an ICPI less than 0.7 [2], and all avirulent strains examined in this study were subsequently characterised as possessing a BglI restriction site within the amplification product.

The sensitivity of the test as applied to virus isolates that had received passage in eggs is 98.48% and 100% on both a sample and case basis respectively. However, lower sensitivities of 93.94% and 81.11% on a case and sample basis respectively were observed for the amplification of RT-PCR products from original clinical material. Given that multiple samples are usually tested from any suspect case, the former is a more realistic expectation of performance. Thirteen of these seventeen original field samples that failed to amplify a product were subsequently MAb typed as group E, with the sample types consisting of seven brain, three trachea, one kidney and two faeces samples. This reduced sensitivity, on a sample basis, may be explained by decreased viral load due to the tropism for the gut within avirulent strains [2], such as Group E, and is consistent with problems experienced by others in the amplification of avirulent strains from field isolates [10]. Four of the 17 original samples (23.53%) that were negative by RT-PCR but positive by virus isolation were faecal material. This is possibly explained by the presence of PCR inhibitors that have previously been reported in such clinical samples. This failure to detect viral nucleic acid in all samples from which viral isolates were made indicates that while this RT-PCR protocol is a useful adjunct to isolation in eggs, it cannot yet replace this procedure completely. Of note, however, is the fact that in one recent case of PPMV-1 all original clinical samples were positive by PCR, while virus isolation took two passages in eggs, thereby demonstrating the potential sensitivity of the RT-PCR assay.

Detection of APMV-1 from a range of tissues indicates that it is still difficult to recommend one particular tissue type for diagnostic purposes, but given the limited tropism of avirulent strains, the use of faeces, tracheal and cloacal swabs are possibly the best options. Direct detection of APMV-1 by RT-PCR from original faecal and cloacal swab material was achieved from all 4 MAb types studied, thereby allowing diagnosis from both systemic and non-systemic infections. This could enable diagnosis ante-mortem or post-mortem using samples from which virus can normally only be detected following passage in embryonated eggs.

Of particular note is the occurrence of the seven cases of PPMV-1 in 2004. According to the RT-PCR results followed by REA, the virus demonstrated a pattern indicative of virulence. This was confirmed by sequence analysis of the fusion protein cleavage site where, in all pigeon isolates, the presence of multiple basic amino acids at residues 113 to 116 plus the presence of a phenylalanine at residue 117 satisfies the OIE criteria for ND. The levels of mortalities and clinical signs shown in the birds, further establishes the typing of these cases as virulent. For example, the level of mortalities in one case was greater than 40% with most deaths occurring in juvenile unvaccinated birds. However, ICPI determination in all PPMV-1 cases tested (as indicated in Table 2) were found to be less than 0.7. Therefore, if the criteria according to the EC directive are applied, such cases would not be defined as ND. Although ICPI values in the case of PPMV-1 are normally greater than 1.0 [11], discrepancies between ICPI and molecular characterisation have previously been reported [12]. This reduction of virulence in chickens may be explained by adaptation to the host, as one of the recent cases of PPMV-1 where the ICPI was initially 0.6, upon re-isolation of the virus and repetition of ICPI, the value was increased to 1.05 [9].

Although the use of virus isolation and ICPI is still required under the EC directive, the use of a specific, DNA-based, faster diagnostic test shows potential for the detection and characterisation of APMV-1. Although other DNA-based tests have been reported that identify and type APMV-1, none has been able to amplify all virulent and avirulent strains of APMV-1 (including pigeon PMV-1) with one primer set.

The one-step real time RT-PCR method described here is of particular use for diagnosis as it reduces the possibility of cross-contamination between samples. On receipt of tissues or swabs in the laboratory, detection and characterisation could be carried out within 24-36 hours. This would not only allow control measures to be quickly implemented but may also prove beneficial to the welfare of birds under restriction and thereby reduce economic losses. Its use as a rapid real time test is also valuable with respect to differential diagnosis, as many clinical symptoms of the disease bear close resemblance to other important avian pathogens.

As the restriction endonuclease site used in this test is located in the region known to be associated with the molecular basis of pathogenicity, it makes progress towards fulfilling the OIE requirements of the alternative criteria for detection and characterisation of ND. The location and size of the fragment produced also lends itself to sequence analysis of the fusion protein cleavage site and the primer sequences described have been used for that specific purpose in the 2002-2003 outbreak of ND in California [13,14]. Additionally, further development and evaluation of such tests could potentially reduce and replace the number of embryos and chicks used for current diagnosis of ND by virus isolation and ICPI.

In this paper, we report a diagnostic method for a one-step RT-PCR test which has the ability to distinguish between virulent and low virulent isolates. Of note is the ability of the test to amplify pigeon PMV-1 isolates, as these strains can have variable virulence and can display ambiguities upon characterisation [12]. The test also has the potential to identify dual infections with virulent and avirulent strains, should they occur in the field. To our knowledge it is the first paper reporting the use of a real time RT-PCR and REA-based test that has the sensitivity to amplify nucleic acid directly from a range of virulent and avirulent strains (including pigeon PMV-1) derived from field cases.

Acknowledgements

APMV reference strains 2 to 9 were kindly supplied by Dr D.J. Alexander, VLA, Weybridge, Surrey, U.K. Monoclonal antibody typing and ICPI of some APMV-1 isolates were carried out by the VLA, Weybridge, Surrey, U.K. The authors acknowledge virus isolation and ICPI work carried out by DSIB staff at VSD, Stormont.

REFERENCES

1 Alexander DJ: Newcastle disease and other avian paramyxoviruses. Revue Scientifique et Technique, OIE 2000;19:443-462.

2 Alexander DJ: Newcastle Disease and other paramyxovirus infections, in Calnek BW (ed.): Diseases of Poultry London. Wolfe Ltd, 1991, 9th edition, pp 496-520.

3 Hanson RP, Brandly CA: Identification of vaccine strains of Newcastle Disease Virus. Science 1955;122:156-157.

4 Millar NS, Chambers P, Emmerson PT: Nucleotide sequence of the fusion and haemagglutinin-neuraminidase glycoprotein genes of Newcastle Disease virus, strain Ulster: Molecular basis for variations in pathogenicity between strains. J Gen Virol 1988;69:613-620.

5 Alexander DJ (ed): Definition of Newcastle Disease In Proceedings of the Joint Sixth Annual meetings of the National Newcastle Disease and Avian Influenza laboratories of countries of the European Union. Brussels, Belgium, 1999, p 92.

6 Van Eck JHH, van Wiltenburg N, Jaspers D: An Ulster 2C strain-derived Newcastle disease vaccine: efficacy and excretion in maternally immune chickens. Avian Pathol, 1991;20:481-495.

7 Seal BS, King DJ, Bennett JD: Characterisation of Newcastle Disease Virus isolates by reverse transcription PCR coupled to direct nucleotide sequencing and development of sequence database for pathotype prediction and molecular epidemiological analysis. J Clin Microbiol 1995;33:2624-2630.

8 De leeuw OS, Peeters B: Complete nucleotide sequence of Newcastle disease virus: evidence for the existence of a new genus within the subfamily Paramyxovirinae. J Gen Virol 1999;80:131-136.

9 Graham D, McCullough S: PPMV-1 viruses with low pathogenicity indices, in Alexander DJ (ed): Joint 10th Annual Meeting of National Laboratories for Newcastle Disease and Avian Influenza of EU Member States. 2004, pp 174-180.

10 Kant A, Koch G, van Roozelaar DJ, Balk F, Ter Huurne A: Differentiation of virulent and non-virulent strains of Newcastle Disease virus within 24 hours by polymerase chain reaction. Avian Pathol 1997;26:837-849.

11 Alexander DJ, Parsons G: Pathogenicity for chickens of avian paramyxovirus type 1 isolates obtained from pigeons in Great Britain during 1983-1985. Avian Pathol 1986;15:487-493.

12 Collins MS, Strong I, Alexander DJ: Evaluation of the molecular basis of pathogenicity of the variant Newcastle disease viruses termed 'pigeon PMV-1 viruses. Arch Virol 1994;134:403-411.

13 Crossley BM, Hietala SK, Shih L-M, Lee L, Skowronski EW, Ardans AA: High-throughput real-time RT-PCR assay to detect the exotic Newcastle Disease virus during the California 2002-2003 outbreak. J Vet Diag Invest 2005;17:124-132.

14 Hietala SK, Hullinger PJ, Crossley BM, Kinde H, Ardans AA: Environmental air sampling to detect exotic Newcastle Disease virus in two California commercial poultry flocks. J Vet Diag Invest 2005;17:198-200.

15 Nanthakumar T, Kataria RS, Tiwari AK, Butchaiah G, Kataria JM: Pathotyping of Newcastle Disease viruses by RT-PCR and restriction enzyme analysis. Vet Res Comm 2000;24:275-286.

Dr. J.L. Creelan, Veterinary Sciences Division, DARD, Stormont, Belfast, BT4 3SD, Northern Ireland, U.K.
E-mail: Julie.Creelan@dardni.gov.uk

Vannier P, Espeseth D (eds): New Diagnostic Technology: Applications in Animal Health and Biologics Controls. Dev Biol (Basel). Basel, Karger, 2006, vol 126, p 159.

SESSION IV

Certainties and Uncertainties of Agent or Toxin Identification by Molecular Methods

Chairman: *Steve Edwards,* United Kingdom

Vannier P, Espeseth D (eds): New Diagnostic Technology: Applications in Animal Health and Biologics Controls. Dev Biol (Basel). Basel, Karger, 2006, vol 126, pp 161-169.

SARS / Avian Coronaviruses

C. Monceyron Jonassen

Section for Virology and Serology, National Veterinary Institute, Oslo, Norway

Abstract: In the hunt for the aetiology of the SARS outbreak in 2003, a newly developed virus DNA micro-array was successfully used to hybridise PCR products obtained by random amplification of nucleic acids extracted from a cell culture infected with material from a SARS patient. The SARS agent was found to hybridise with micro-array probes from both coronaviruses and astroviruses, but one of the coronavirus probes and the four astrovirus probes contained redundant sequences, spanning a highly conserved motif, named s2m, found at the 3' end of the genomes of almost all astroviruses, one picornavirus, and the poultry coronaviruses. The three other coronavirus probes, that hybridised with the SARS agent, were located in the replicase gene, and it could be concluded that the SARS agent was a novel coronavirus, harbouring s2m. The presence of this motif in different virus families is probably the result of recombinations between unrelated viruses, but its presence in both poultry and SARS coronaviruses could suggest a bird involvement in the history of the SARS coronavirus. A recent screening of wild birds for the presence of coronaviruses, using a pan-coronavirus RT-PCR, led to the identification of novel coronaviruses in the three species studied. Phylogenetic analyses performed on both replicase gene and nucleocapsid protein could not add support to a close relationship between avian and SARS coronaviruses, but all the novel avian coronaviruses were found to harbour s2m. The motif is inserted at a homologous place in avian and SARS coronavirus genomes, but in a somewhat different context for the SARS coronavirus. If the presence of s2m in these viruses is a result of two separate recombination events, this suggests that its particular position in these genomes is the only one that would not be deleterious for coronaviral replication, or that it is the result of a copy-choice recombination between coronaviruses, following an ancestral introduction in the coronavirus family by an unrelated virus. In conclusion, the relative high frequencies of recombination observed both experimentally and in the natural evolution of RNA viruses, indicate that horizontal gene transfer does occur, even between unrelated viruses. This might represent a challenge in the rapid identification of novel pathogens with DNA micro-array techniques.

Introduction

The Severe Acute Respiratory Syndrome (SARS) emerged as a human disease in the Guandong province of China in late autumn 2002. It was introduced in Hong Kong in mid-February 2003, and spread from there to several parts of the world, infecting nearly 8,000 people, and killing about 800 [1-3]. The rapid identification of the agent responsible for SARS was a prerequisite to diagnostics and effective control of the epidemic.

Several agents were suspected at first as the aetiology of this syndrome. In particular, highly pathogenic avian influenza (HPAI) was considered, since a boy and his father died in Hong Kong from HPAI in February 2003, shortly after travelling in a province next to Guandong. HPAI was, however, rapidly ruled out as the cause of SARS. Chlamydia spp. were identified in samples from Chinese patients, while paramyxovirus was observed by electron microscopy in samples from SARS patients, both in Hong Kong and Germany [1,4]. In addition, several Canadian patients were found to be infected with metapneumovirus, by RT-PCR [5]. Metapneumovirus was a newly recognized human respiratory pathogen at that time [6], and the overall prevalence of this paramyxovirus was not known. It was therefore considered as one of the potential aetiological candidates for SARS.

In the hunt for the SARS agent, the use of a new diagnostic technology aimed at the discovery of novel viruses, was of special importance in identifying and characterizing the emerging pathogen [2,7,8], and by March 25th, the United States Center for Disease Control and Prevention (CDC) announced that a novel coronavirus was the probable aetiological cause of SARS.

Coronaviruses are enveloped viruses with a positive-strand RNA genome (27-32 kb) that have been associated with disease in several warm-blooded animals, including humans. The coronaviruses comprise three serologically and genetically distinct groups [9,10]. Coronaviruses from group I and II have been found to infect several mammalian species, including humans, pigs, cows, dogs, horses, cats and rodents, while group III coronaviruses have been found to infect poultry.

While the SARS coronavirus has been postulated to be an early split-off from the group II coronavirus, it was found to harbour an RNA motif at the 3' end of its genome that is also present in the poultry coronaviruses, but not in other coronaviruses [2,8,11]. A potential involvement of bird coronaviruses in the evolution of the SARS coronavirus was newly explored by molecular screening and characterizing coronavirus in wild bird populations, indicating that the use of a pan-coronavirus RT-PCR for the identification of novel coronaviruses was successful.

Use of Virus DNA Micro-Array in the Hunt for the SARS Agent - Certainties and Uncertainties

Sample material from a SARS patient was inoculated on VeroE6 cells. A virus was propagated that gave a cytopathic effect on the cells. The virus was confirmed as being the SARS agent in an immunocytochemical assay, using serum from a convalescent SARS patient [2].

Total nucleic acid was extracted from the positive cell culture, as well as from a negative control culture, and submitted to the University of California San Francisco (UCSF), U.S.A., for molecular analysis. A random amplification was performed, consisting of an initial step of reverse transcription and second strand synthesis, using a primer (A) that was a 26-mer with 9 Ns at its 3'-end, followed by PCR primed with the specific 17 bases of primer A. The amplicons were then hybridised to a virus DNA micro-array [2,8].

The DNA micro-array was developed by Wang and DeRisi, at the UCSF, for virus detection and potential identification of novel members of known virus families [7]. The array contains 70-mers from the 10 most conserved parts of every fully sequenced reference virus genome in the GenBank, comprising a total of about 10,000 oligonucleotides from approximately 1000 viruses. The initial hybridisation results showed evidence of sequence homology of the SARS agent with eight oligonucleotides on the array, from viruses belonging to two different families; the coronaviruses and the astroviruses. Three of the oligonucleotides were from the ORF1AB, encoding viral replicase, of one coronavirus from each of the three existing groups (the group I human 229E coronavirus, the group II bovine coronavirus, and the group III chicken infectious bronchitis virus (IBV)), one was from the 3' untranslated region (3' UTR) of IBV, while four were from the 3'-end of the genomes of different astroviruses, including avian nephritis virus, as well as human, turkey, and sheep astroviruses.

The surprisingly diverse hybridisation pattern could be explained by the fact that a conserved RNA stem-loop motif, named s2m, located near the 3' end of the genomes of all but one member of the astrovirus family, was also found in the 3' UTR of IBV, as observed earlier (12). As all five micro-array probes from viral genomic 3' ends that hybridised with the SARS viral sample spanned s2m, it could be concluded that the aetiological agent of SARS was probably a coronavirus harbouring s2m, rather than that the SARS patient from whom the virus was grown in cell culture was infected with both a coronavirus and an astrovirus. In addition to astroviruses and some coronaviruses, s2m is also found in the UTR of one member of the picornavirus family (horse rhinitis B virus), suggesting that the presence of s2m in some members of these virus families is probably the result of horizontal transfer between non-related viruses. The remarkable conservation of this RNA motif indicates an important function in the viruses posessing it, but the function of s2m is not yet known.

The use of a virus DNA micro-array proved to be a successful strategy for detecting novel pathogens, and later sequence information on the SARS coronavirus showed that the stretches of homology with the probes on the array leading to a clearly detectable hybridisation signal, were as short as 25 nucleotides. However, the recombination frequencies observed in several viruses, potentially involving non-related viruses, can make the interpretation of the results of hybridisation confusing. On the other hand, the use of a virus DNA micro-array for pathogen identification with several probes for each virus family could contribute to rapid identification of novel recombinant genomes.

Recombination in RNA Viruses

The ability of closely related RNA viruses to exchange genetic material through recombination has been noted in many animal and plant viruses, as well as bacteriophages, and among those, picornaviruses and coronaviruses display especially high frequencies of recombination [13]. Furthermore, viruses in several families, including alphaviruses, coronaviruses and luteoviruses, have been shown to contain genes that are closely related to genes from other virus families, suggesting relatively recent recombination events with unrelated viruses in their evolution [14]. In addition, bovine viral diarrhoea virus (BVDV) has been shown to evolve to cytopathogenic types in persistently infected animals by recombination events, resulting, on some occasion, to the insertion of cellular sequences in the viral genome [15].

The main recognized mechanism of RNA recombination involves a template-switch of the viral replicase during RNA synthesis, leading to dissociation of the replicase-nascent viral RNA strand complex from the donor template, and reassociation on an acceptor template for further polymerisation, in a copy-choice manner. Such discontinuous replication might be due, for example, to obstacles such as a strong RNA secondary structure or strand breakage in the donor template, or nucleotide misincorporation in the nascent strand. Replicase-driven RNA recombination has mostly been reported at homologous sites in related viral genomes, resulting in progeny molecule with no alteration in sequence frame at the cross-over point or genomic organization (homologous recombinants), or with deletions/insertions at the cross-over site (aberrant homologous recombinants), but such recombination might as well happen between non-related RNA molecules, resulting in non-homologous recombinants [13]. The replicative template-switch mechanism may require specific RNA-protein or protein-protein interactions too, and is mostly observed within members of the same viral genus, as for members of picornavirus families [16]. In coronaviruses, the discontinuous synthesis of sub-genomic RNAs involves template-switch of the replicase, which, in turn, has been associated with the high frequencies of coronavirus genomic recombinations observed [17].

In recent years, another mechanism has been proposed for viral RNA recombination, involving breakage/ligation of RNA molecules at non-homologous sites, through splicing-like mechanisms. Replicase-mediated transesterification reactions, as well as other types of spontaneous RNA rearrangements, have been proposed as mechanisms for recombination in bacteriophage Qβ following in vitro studies [18,19]. In addition, in vivo studies on both poliovirus and BVDV strongly suggest that recombination between viral RNAs can occur in infected cells independent of viral replicase activity for these viruses [20,21]. The discovery of non-replicative rearrangements of RNA molecules has an implication for our understanding of viral evolution, as two incompatible parental strains could give rise to novel genomes.

Even if replicase-independent RNA recombination can occur between RNA molecules, most of the viral recombinations observed are homologous or involve closely related viruses. This could be due to the fact that the vast majority of non-homologous recombinants involving incompatible parental strains would give rise to non-replicative genomes, and would therefore not be detected. In addition, non-homologous recombination events resulting in progeny viral genomes with no increase in fitness, that take place in one infected cell, on one occasion in a host, would hardly be fixed in a viral population. Besides, recombination involving unrelated viruses requires that a host is infected at the same time by two different viruses with the same cell tropism. However, successful non-homologous recombinations give viruses the opportunity to take huge evolutionary steps, and result in new viruses with altered properties. Such recombination events are not possible to foresee or control in nature, and novel viruses will probably emerge in the future, being a challenge for rapid molecular-based detection methods.

As recombination events involving unrelated viruses are rare, the preliminary results from the hybridisation identification of the SARS coronavirus suggested a stronger relationship of the novel virus to the poultry coronaviruses, since only these had previously been shown to harbour s2m. However, sequencing of the PCR products that had hybridised to the probes in the array, showed evidence that the SARS coronavirus was unique, but most closely related to group II coronaviruses. On the other hand, it has been proposed that recombination events with both group I and group III coronaviruses have taken place in the history of this virus. In particular, it has been argued that the 3'-part of the SARS coronavirus is more closely related to avian coronaviruses than to the other coronavirus groups [22]. However, antigenic reactivity of the SARS coronavirus nucleocapsid has been reported with polyclonal antisera against group I coronaviruses, but not with antisera against group II or group III coronaviruses [23].

Screening of Wild Bird Species for Coronavirus - s2m in Coronaviruses

To investigate the potential involvement of avian coronaviruses in the evolution of the SARS coronavirus further, a pan coronavirus RT-PCR [24] was recently performed to screen three wild bird species (greylag goose, feral pigeon, and mallard) for the presence of coronavirus in different sample types, including cloacal and tracheal swabs. Coronavirus could be found in cloacal swabs from all three bird species [25]. As the primers used had several degenerated positions, some degree of unspecific amplification was observed, especially in some organ samples. All PCR products had therefore to be sequenced for confirmation. Though demanding, the pan-coronavirus RT-PCR proved successful at identifying novel coronaviruses in different species. A rapid amplification of cDNA ends (5' RACE) strategy using s2m as a handle was used to obtain sequence information on the 3' part of the novel viruses, and confirmed the presence of s2m in the novel group III viruses. Phylogenetic analyses performed on the polymerase gene and nucleocapsid protein sequences, indicated that the novel wild bird coronaviruses all branch off from group III coronaviruses, and could not strengthen a bird coronavirus involvement in the history of the SARS coronavirus.

s2m is inserted at about the same place in the genomes of avian and SARS coronaviruses, a few nucleotides upstream of the conserved octamer sequence of the coronavirus 3' UTR (GGAAGAGC), but in a somewhat different context, (Fig. 1). Experimental replacement of the 3' UTR in the MHV genome with 3' UTR from the SARS coronavirus has been shown to yield replication-competent viruses, while this could not be achieved with 3' UTR from group I or group III coronaviruses [26]. This further suggests that the SARS coronavirus does not have a typical group III 3' UTR, but rather a group II-like 3' UTR, with s2m inserted.

An explanation for the presence of s2m in both SARS and group III coronaviruses could be that a common ancestor of both group II and III coronaviruses once harboured the motif, and that it was lost in the early branching of non-SARS group II coronaviruses. However, as s2m seems to be a readily transferable genomic element of RNA viruses, found in three virus families, its presence in both avian and SARS coronaviruses may rather be the result of two separate recombination events. The MHV sequence homologous to the sequence flanking s2m in the SARS coronavirus (underlined in Fig.1) has been shown to form a bulge, as was postulated for mammalian coronavirus sequences in group I, homologous of group III 3' UTR [12,27]. This indicates that independent insertions in two ancestor coronaviruses of s2m, that has a rigorously conserved structure itself [28], would not lead to a major conformational change of the acceptor RNA molecule, disturbing important secondary structures in the viral 3' UTR involved in replication. The presence of s2m in avian and SARS coronaviruses in a bulge structure right upstream of the conserved octamer could thus be the only position that would not be deleterious to coronavirus replication. On the other hand, if the two independent recombination events in coronavirus genomes involve first a recombination event with an unrelated virus, and second a recombination event between coronaviruses, then the latter could be of a replicative copy-choice type involving related sequences, and would most probably result in insertion of s2m in the novel recombinant at a homologous place as in the parental strain. Screening of several wildlife species for coronavirus using a pan-coronavirus RT-PCR, could lead to identification of yet undiscovered coronaviruses in group II, including the SARS coronavirus lineage, and this could shed light on whether s2m is present in several group II coronaviruses.

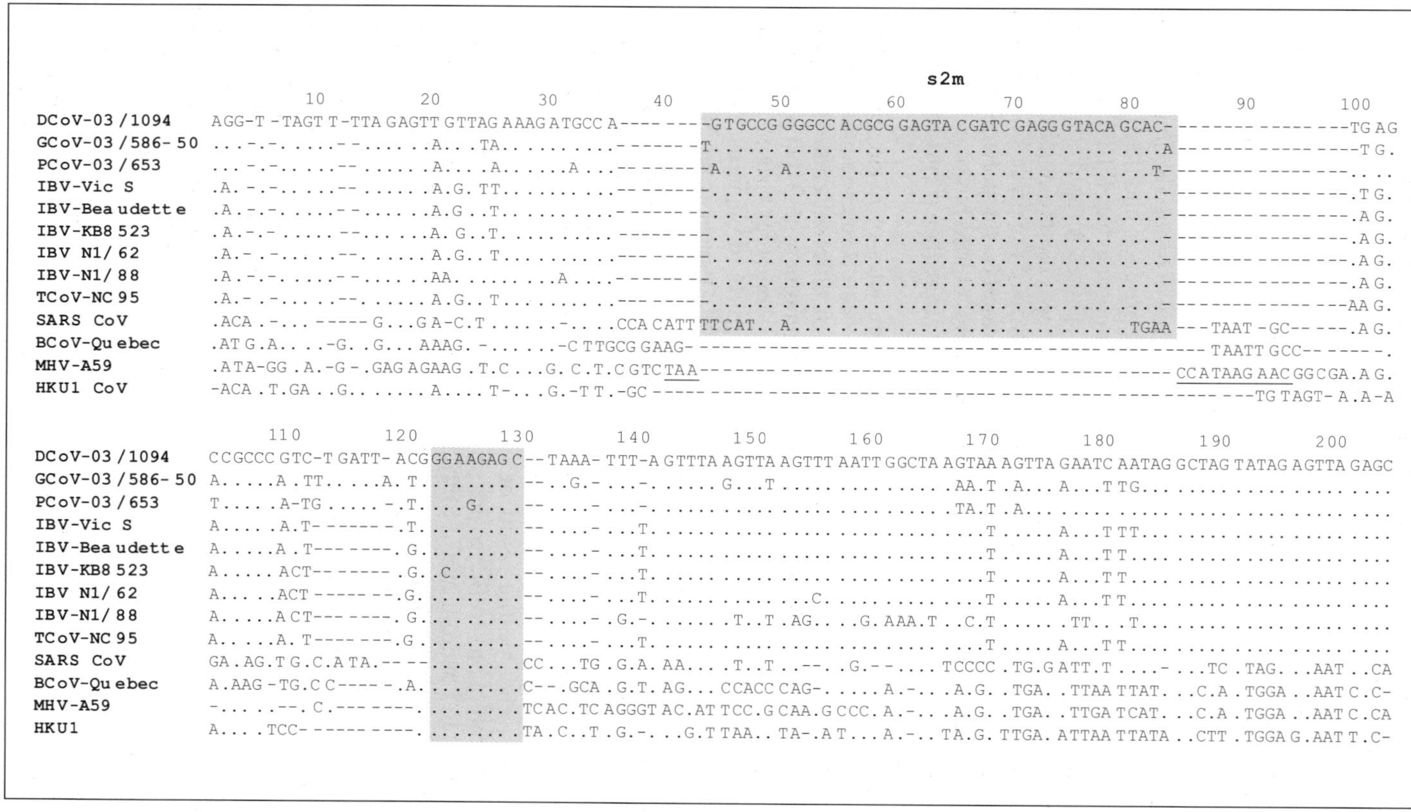

Fig. 1: Nucleotide alignment of the downstream part of the 3' UTR, ending immediately upstream of the poly(A) tail, of several coronaviruses from group II and III, as well as the SARS coronavirus. The coronaviruses are duck coronavirus (DCoV), goose coronavirus (GCoV), pigeon coronavirus (PCoV), several infectious bronchitis virus (IBV) strains, turkey coronavirus (TCoV), SARS coronavirus (SARS CoV), bovine coronavirus (BCoV), murine hepatitis virus (MHV), and human HKU1 coronavirus (HKU1 CoV). Dots indicate nucleotides identical to DCoV sequence, and gaps are shown as dashes. s2m and the conserved octamer in the 3' UTR of all coronaviruses are shown as shaded boxes. The bulge forming nucleotides in the MHV genome that are homologous to the s2m flanking sequence in the SARS coronavirus are underlined.

Conclusion

The relative high frequencies of recombination observed both experimentally and in the natural evolution of RNA viruses, indicate that horizontal gene transfer does occur, even between unrelated viruses. Such transferable genes might be short motifs that may not be easily detected when analysing viral genomes, or designing conserved probes for rapid pathogen detection and identification. Especially, the existence of mobile elements in the RNA viruses, such as the highly conserved s2m found in three separate virus families, further emphasizes the importance of being aware of the dynamics of viral genomic evolution when dealing with novel pathogens.

References

1 Drosten C, Gunther S, Preiser W, van der Werf S, Brodt HR, Becker S, Rabenau H, Panning M, Kolesnikova L, Fouchier RA, Berger A, Burguiere AM, Cinatl J, Eickmann M, Escriou N, Grywna K, Kramme S, Manuguerra JC, Muller S, Rickerts V, Sturmer M, Vieth S, Klenk HD, Osterhaus AD, Schmitz H, Doerr HW: Identification of a Novel Coronavirus in Patients With Severe Acute Respiratory Syndrome. N Engl J Med 2003;348(20):1967-1976.

2 Ksiazek TG, Erdman D, Goldsmith CS, Zaki SR, Peret T, Emery S, Tong S, Urbani C, Comer JA, Lim W, Rollin PE, Dowell SF, Ling AE, Humphrey CD, Shieh WJ, Guarner J, Paddock CD, Rota P, Fields B, DeRisi J, Yang JY, Cox N, Hughes JM, LeDuc JW, Bellini WJ, Anderson LJ: A Novel Coronavirus Associated With Severe Acute Respiratory Syndrome. N Engl J Med 2003;348(20):1953-1966.

3 Kuiken T, Fouchier RAM, Schutten M, Rimmelzwaan GF, van Amerongen G, van Riel D, Laman JD, de Jong, T, van Doornum G, Lim W, Ling AE, Chan PK S, Tam JS, Zambon MC, Gopal R, Drosten C, van der Werf S, Escriou N, Manuguerra JC, Stohr K, Peiris JSM, Osterhaus ADME: Newly Discovered Coronavirus As the Primary Cause of Severe Acute Respiratory Syndrome. Lancet 2003;362(9380):263-270.

4 Cyranoski D: China Joins Investigation of Mystery Pneumonia. Nature 2003;422(6931):459.

5 Poutanen SM, Low DE, Henry B, Finkelstein S, Rose D, Green K, Tellier R, Draker R, Adachi D, Ayers M, Chan AK, Skowronski DM, Salit I, Simor AE, Slutsky AS, Doyle PW, Krajden M, Petric M, Brunham RC, McGeer AJ: Identification of Severe Acute Respiratory Syndrome in Canada. N Engl J Med 12003;348(20):1995-2005.

6 Kahn JS: Human Metapneumovirus: a Newly Emerging Respiratory Pathogen. Cur Opin Infect Dis 2003;16(3):255-258.

7 Wang D, Coscoy L, Zylberberg M, Avila PC, Boushey HA, Ganem D, Derisi JL: Micro-array-Based Detection and Genotyping of Viral Pathogens. Proc Natl Acad Sci U S A 2002;99(24):15687-15692.

8 Wang D, Urisman A, Liu YT, Springer M, Ksiazek TG, Erdman DD, Mardis ER, Hickenbotham M, Magrini V, Eldred J, Latreille JP, Wilson RK, Ganem D, Derisi JL: Viral Discovery and Sequence Recovery Using DNA Micro-arrays. PLoS.Biol. 2003;1(2):E2.

9 Enjuanes L Brian D, Cavanagh D, Holmes K, Lai MMC, Laude H, Masters P, Rottier P, Siddell SG, Spaan WJM, Tagushi F, Talbot P: *Coronaviridae*. van Regenmortel MHV, Fauquet CM, Bishop DHL, Carsten EB, Estes MK, Lemon SM, McGeoch DJ, Maniloff J, Mayo MA, Pringle CR, Wickner RB (eds): Virus Taxonomy. Classification and Nomenclature of Viruses. San Diego, Academic Press, 2000, pp 835-849.

10 Gonzalez JM, Gomez-Puertas P, Cavanagh D, Gorbalenya AE, Enjuanes LA: Comparative Sequence Analysis to Revise the Current Taxonomy of the Family Coronaviridae. Arch Virol 2003;148(11):2207-2235.

11 Marra MA, Jones SJ, Astell CR, Holt RA, Brooks-Wilson A, Butterfield YS, Khattra J, Asano JK, Barber SA, Chan SY, Cloutier A, Coughlin SM, Freeman D, Girn N, Griffith OL, Leach SR, Mayo M, McDonald H, Montgomery SB, Pandoh PK, Petrescu AS, Robertson AG, Schein JE, Siddiqui A, Smailus DE, Stott JM, Yang GS, Plummer F Andonov A, Artsob H, Bastien N, Bernard K, Booth TF, Bowness D, Czub M, Drebot M, Fernando L, Flick R, Garbutt M, Gray M, Grolla A, Jones S, Feldmann H, Meyers A, Kabani A, Li Y, Normand S, Stroher U, Tipples GA, Tyler S, Vogrig R, Ward D, Watson B, Brunham RC, Krajden M, Petric M, Skowronski DM, Upton C, Roper RL: The Genome Sequence of the SARS-Associated Coronavirus. Science 2003;300(5624):1399-1404.

12 Jonassen CM, Jonassen TO, Grinde B: A Common RNA Motif in the 3' End of the Genomes of Astroviruses, Avian Infectious Bronchitis Virus and an Equine Rhinovirus. J Gen Virol 1998;79(Pt 4):715-718.

13 Lai MM: RNA Recombination in Animal and Plant Viruses. Microbiol.Rev. 1992;56(1):61-79.

14 Miller WA, Koev G: Getting a Handle on RNA Virus Recombination. Trends Microbiol 1998;6(11):421-423.

15 Nagai M, Sakoda Y, Mori M, Hayashi M, Kida H, Akashi H: Insertion of Cellular Sequence and RNA Recombination in the Structural Protein Coding Region of Cytopathogenic Bovine Viral Diarrhoea Virus. J GenVirol 2003;84(Pt 2):447-452.

16 Kirkegaard K, Baltimore D: The Mechanism of RNA Recombination in Poliovirus. Cell 1986;47(3):433-443.

17 Zhang X, Lai MM: Unusual Heterogeneity of Leader-MRNA Fusion in a Murine Coronavirus: Implications for the Mechanism of RNA Transcription and Recombination. J Virol 1994;68(10):6626-6633.

18 Chetverin AB, Chetverina HV, Demidenko AA, Ugarov VI: Nonhomologous RNA Recombination in a Cell-Free System: Evidence for a Transesterification Mechanism Guided by Secondary Structure. Cell 1997;88(4):503-513.

19 Chetverina HV, Demidenko AA, Ugarov VI, Chetverin AB: Spontaneous Rearrangements in RNA Sequences. FEBS Lett. 1999;450(1-2):89-94.

20 Gallei A, Pankraz A, Thiel HJ, Becher P: RNA Recombination in Vivo in the Absence of Viral Replication. J Virol 2004;78(12):6271-6281.

21 Gmyl AP, Belousov EV, Maslova SV, Khitrina EV, Chetverin AB, Agol VI: Nonreplicative RNA Recombination in Poliovirus. J Virol 1999;73(11):8958-8965.

22 Stavrinides J, Guttman DS: Mosaic Evolution of the Severe Acute Respiratory Syndrome Coronavirus. J Virol 2004;78(1):76-82.

23 Sun ZF, Meng XJ: Antigenic Cross-Reactivity Between the Nucleocapsid Protein of Severe Acute Respiratory Syndrome (SARS) Coronavirus and Polyclonal Antisera of Antigenic Group I Animal Coronaviruses: Implication for SARS Diagnosis. J Clin Microbiol 2004;42(5):2351-2352.

24 Stephensen CB, Casebolt DB, Gangopadhyay NN: Phylogenetic Analysis of a Highly Conserved Region of the Polymerase Gene From 11 Coronaviruses and Development of a Consensus Polymerase Chain Reaction Assay. Virus Res 1999;60(2):181-189.

25 Jonassen CM, Kofstad T, Larsen IL, Løvland A, Handeland K, Follestad A, Lillehaug A: Molecular Identification and Characterization of Novel Coronaviruses Infecting Graylag Geese (Anser Anser), Feral Pigeons (Columbia Livia) and Mallards (Anas Platyrhynchos). Journal of General Virology 2005;86:1597-1607.

26 Goebel SJ, Taylor J, Masters PS: The 3' Cis-Acting Genomic Replication Element of the Severe Acute Respiratory Syndrome Coronavirus Can Function in the Murine Coronavirus Genome. J Virol 2004;78(14):7846-7851.

27 Liu Q, Johnson RF, Leibowitz JL: Secondary Structural Elements Within the 3' Untranslated Region of Mouse Hepatitis Virus Strain JHM Genomic RNA. J Virol 2001;75(24):12105-12113.

28 Robertson MP, Igel H, Baertsch R, Haussler D, Ares M, Jr, Scott WG: The Structure of a Rigorously Conserved RNA Element Within the SARS Virus Genome. PLoS Biol 2005;3(1):e5.

Dr. C. Monceyron Jonassen, Section for Virology and Serology, National Veterinary Institute, P.O. Box 8156 Dep., NO-0033 Oslo, Norway.
E-mail: christine.monceyron-jonassen@vetinst.no.

Vannier P, Espeseth D (eds): New Diagnostic Technology: Applications in Animal Health and Biologics Controls. Dev Biol (Basel). Basel, Karger, 2006, vol 126, pp 171-177.

Rapid Diagnosis of Avian Influenza (AI) and Assessment of Pathogenicity of Avian H5 and H7 Subtypes by Molecular Methods

D.A. Senne[1], J.C. Pedersen[1], D.L. Suarez[2], B. Panigrahy[1]

[1] U.S. Department of Agriculture, Animal and Plant Health Inspection Service, Veterinary Services, National Veterinary Services Laboratories, Ames, Iowa, USA

[2] U.S. Department of Agriculture, Agriculture Research Service, Southeast Poultry Research Laboratory, Athens, Georgia, USA

Key words: Avian influenza, low pathogenicity avian influenza, highly pathogenic avian influenza, sequencing, virulence marker, cleavage site, haemagglutinin, H5, H7, poultry, real time RT-PCR, RT-PCR

Abstract: Avian influenza (AI) is a highly contagious viral disease of poultry that is found worldwide. There are two forms of AI: a mild form called low pathogenicity avian influenza (LPAI), and a severe form called highly pathogenic avian influenza (HPAI). HPAI is associated with the H5 and H7 subtypes of AI virus (AIV) and is subject to Federal control and International reporting. A real-time reverse transcription-polymerase chain reaction (rRT-PCR) assay has been developed and validated that can help in the early detection of AI outbreaks. The rRT-PCR assay can also be used to identify infections caused by H5 and H7 subtypes of AIV. New isolates of AIV must be characterized as LPAI or HPAI for reporting and control purposes. The criteria for classification of an AI virus as HPAI are defined by the World Organization for Animal Health (OIE); the definition includes a virulence and a molecular criterion. The virulence requirement for HPAI is defined as an AIV killing 75% or more of eight inoculated chickens within 10 days. The molecular criterion is the presence of multiple dibasic amino acids at the proteolytic cleavage site of the haemagglutinin (H) protein. All HPAI viruses isolated before 2002 fulfilled both the virulence and molecular criteria. Consequently, nucleotide sequencing of the H gene and deduction of the amino acid motif at the H cleavage site has been successfully used to assess the virulence of H5 and H7 AIVs rapidly. Since 2002, however, there have been three outbreaks of HPAI where the viruses responsible for the outbreaks have either fulfilled the virulence criterion or the molecular criterion, but not both.

INTRODUCTION

Avian influenza (AI) is caused by influenza A viruses belonging to the family Orthomyxoviridae. All influenza type A viruses share common and highly conserved internal proteins, namely the matrix and nucleoprotein. Influenza A viruses are further divided into subtypes based on the antigenically variable surface proteins, the haemagglutinin (H) and neuraminidase (N). There are currently 16 H and 9 N subtypes recognized that can occur in any combination, giving the potential for 144 combinations of H and N subtypes such as H1N1, H1N2, N5N1, H9N2 etc. Infections of avian influenza (AI) in commercial poultry in the United States are rare, but occasionally the virus is introduced into domestic poultry through activities associated with the marketing of birds through the live bird marketing system (a man-made reservoir for AI viruses) or by direct exposure to the wild aquatic bird reservoir [1]. Wild aquatic birds are the primary reservoir of avian influenza virus (AIV) and have been shown to harbour all 16 haemagglutinin (H) and nine neuraminidase (N) subtypes of the virus [2-4]. Although poultry are susceptible to infections from most of the H and N AIV subtypes, infections in poultry caused by the H5 and H7 subtypes are of great concern because these subtypes have been associated with the highly pathogenic form of the disease [5]. In poultry, avian influenza infections result in one of two clinicopathological forms of disease: low pathogenicity AI (LPAI) and highly pathogenic AI (HPAI). With LPAI, infections are localized in epithelial cells lining the respiratory tract and/or the intestinal tract with little or no clinical disease unless complicated by secondary viral or bacterial infections. In contrast, HPAI infections result in an acute systemic disease with high mortality [5]. The criteria for classifying AI viruses as LPAI or HPAI are defined in the Manual of Standards for Diagnostic Tests and Vaccines published by the World Organization for Animal Health (OIE) [6]. To date, highly pathogenic avian influenza (HPAI) has only been associated with some strains of H5 and H7 subtypes; however, the vast majority of H5 and H7 subtype viruses are of low pathogenicity for domestic poultry. Studies have shown that HPAI arise through mutation of LPAI strains after circulation in domestic poultry [7]. Therefore, it is important to detect and eliminate infections of LPAI H5 and H7 AIV in poultry to reduce the chances of the virus mutating to HPAI. Advances in the molecular basis for pathogenicity of AIV viruses have made it possible to predict whether some strains of H5 and H7 subtype viruses have the genetic make-up to be highly pathogenic to poultry. It has been shown that HPAI viruses have multiple dibasic amino acids at the cleavage site of the haemagglutinin protein and that this characteristic can serve as a virulence marker for HPAI H5 and H7 viruses. Therefore, RT-PCR assays have been developed to aid in the rapid assessment of virulence of these viruses [8,9].

Diagnostic tests that are currently used to diagnose infections of AI in poultry include detection of specific antibodies in serum or yolk and isolation of the virus in embryonating chicken eggs. Isolates of AIV are characterized by subtype identification of the H and N surface proteins followed by the inoculation of susceptible chickens to determine its virulence [5]. These procedures are time-consuming, expensive, require a diverse set of reagents, and laboratory facilities with a high level of biosecurity (BSL-3AG) to handle these viruses safely. Early detection of AI infections in poultry and rapid assessment of the virulence of AIV is critical for the development of appropriate strategies to control the spread of avian influenza. This is especially important for infections with LPAI where clinical signs cannot be used as definitive evidence of infection. Recently, a molecular assay was developed to detect AIV RNA in clinical specimens which reduces the time for diagnosis to fewer than three hrs [10]. This paper will review the reliability of the method to identify AIV in clinical specimens rapidly and molecular methods to identify virulent H5 and H7 viruses isolated from poultry.

MATERIALS AND METHODS

Rapid diagnosis of AI by molecular methods

A real time reverse transcription-polymerase chain reaction (rRT-PCR) developed by Spackman et al has been shown to be a reliable assay in the detection of AIV in clinical specimens [10]. The assay is a one step procedure that uses specific primers designed to amplify the target sequence. Non-extendible fluorogenic hydrolysis (Taqman) probes are used to monitor amplification of the target sequence after each PCR cycle, thus providing results in real-time. The rRT-PCR procedure comprises three separate assays; one that targets the matrix (M) gene and two that are subtype specific for H5 and H7 AIVs. The primers and probe for M assay are designed to detect the highly conserved M gene of AIV and, as such, will detect most if not all strains of AIV, regardless of subtype [10]. The H5 and H7 assays are designed to detect most North American H5 and H7 viruses, respectively; however, the H5 primers and probes also are shown to detect the Asian H5N1 HPAI viruses successfully (D. Suarez, personal communication).

During an outbreak of LPAI H7N2 in Virginia in 2002, 3,628 tracheal swabs specimens were tested by rRT-PCR and virus isolation in embryonating chicken eggs to evaluate the usefulness of the rRT-PCR assay in an outbreak situation. In this study, tracheal swabs were obtained from 20 birds/house and pooled 5/tube and submitted to the laboratory under refrigeration. The swabs specimens were extracted with Qiagen RNeasy mini-columns (Qiagen, Valencia, CA) before testing by the rRT-PCR assay.

Rapid assessment of pathogenicity of H5 and H7 viruses

The presence of multiple basic amino acids (arginine [R] and lysine [K]) at the cleavage site of the H protein has been a characteristic of highly pathogenic H5 and H7 subtypes of AIV and has been incorporated into the definition of HPAI by the OIE [6]. The method most commonly used to determine the presence or absence of the virulence marker has been RT-PCR amplification of the coding sequence at the cleavage site of the H protein, followed by sequencing of the amplicon [8,9]. The procedure involves the isolation of the RNA using Qiagen RNeasy mini columns (Qiagen, Valencia, CA), conversion of the viral RNA to cDNA by the use of reverse transcriptase, amplification of the cDNA by PCR, evaluation of the PCR product (amplicon) in gels stained with ethidium bromide, and sequence determination of the amplicon by automated sequencing techniques. The isolates from diagnostic specimens were H5 or H7 AIV in infectious egg allantoic amniotic fluid. The isolates represented a wide range of temporal, geographical and host origins [8, D. Senne, unpublished data].

RESULTS

Rapid diagnosis of AI by molecular methods

The AI rRT-PCR assay developed by Spackman et al was initially validated by testing more than 1,500 specimens collected during surveillance studies in the live bird marketing system in the northeast United States [10]. The assay was further validated with 3,628 tracheal swab specimens collected during the 2002 outbreak of LPAI H7N2 in Virginia, USA [11]. Results obtained by rRT-PCR were compared with virus isolation in embryonating chicken eggs for sensitivity and specificity. Specimens from the VA outbreak were evaluated at the specimen level and at the flock level (consensus of all specimens collected from a single flock). The diagnostic sensitivity (DxSN) and specificity (DxSP) of the assay at the specimens level were 88.2 and 99.5 respectively, whereas, on a flock basis, the DxSN and DxSP were 95.1 and 99.1, respectively (D. Senne, unpublished data).

Table 1: Amino acid sequences at the HA0 cleavage site of H5 and H7 AIV strains (modified from Senne et al, [8]).

Virus strain	Subtype	Pathotype	HA0 cleavage site sequence
A/Chicken/Hidalgo/26654-1368/94	H5N2	LPAI	P Q-----------------------------**R** E T **R***G L F
A/Emu/NY/12716-67/94	H5N9	LPAI	P Q-----------------------------**R** E T **R***G L F
A/Turkey/CA/6878/79	H5N3	LPAI	P Q-----------------------------**R** E T **R***G L F
A/Turkey/MN/10734/94	H5N2	LPAI	P Q-----------------------------**R** E T **R***G L F
A/Chicken/Queretaro/14588-19/95	H5N2	HPAI	P Q---------------------- **R K R K** T **R***G L F
A/Chicken/Yamaguchi/7/02	H5N1	HPAI	P Q----------------------**R ER R KK R***G L F
A/Turkey/Ireland/83	H5N8	HPAI	P Q----------------------**R K R K K R***G L F
A/Chicken/Hong Kong/220/97	H5N1	HPAI	P Q-----------------**R** E **R R R K K R***G L F
A/Chicken/TX/04	H5N2	HPAI**	P Q ----------------------------**R K K R***G L F
A/Chicken/NY/13142-5/94	H7N2	LPAI	P E-------------------------- N P **K** T **R***G L F
A/Turkey/Italy/977/99	H7N1	LPAI	P E---------------------------I P **K** G **R***G L F
A/Quail/AR/16309-7/94	H7N3	LPAI	P E--------------------------N P **K** T **R***G L F
A/Chicken/Chile/176822/02	H7N3	LPAI	P E--------------------------**K** P **K** T **R***G L F
A/Chicken/Canada/AVFV1/04	H7N3	LPAI	P E--------------------------N P **K** T **R***G L F
A/Chicken/Victoria/85	H7N7	HPAI	P E------------------I P **K K R** E **K R***G L F
A/Chicken/The Netherlands/04	H7N7	HPAI	P E---------------------I P.**K R R R R***G L F
A/Turkey/Italy/4580/99	H7N1	HPAI	P E----------------I P **K** G S **R** V **R R***G L F
A/Chicken/Chile/4325/02	H7N3	HPAI	P E **K** P **K** T C S P L S **R** C **R KT** **R***G L F
A/Chicken/Canada/AVFV2/04	H7N3	HPAI	P E------N P **K** Q A Y Q **K R** M T **R***G L F

* indicates cleavage site, dashes are used for alignment. Basic amino acids arginine (R) and lysine (K) are shown in bold.

** classified as HPAI but the virus did not produce disease in experimentally inoculated chickens

Rapid assessment of pathogenicity of H5 and H7 viruses

The RT-PCR has been used to predict the virulence characteristics of a large number of H5 and H7 AIVs representing a wide range of temporal, geographical and host origins and is routinely used at the National Veterinary Services Laboratories (NVSL), Ames, Iowa. Selected examples of the H cleavage site sequences for LPAI and HPAI H5 and H7 viruses are shown in Table 1. The molecular assessment of virulence for all H5 and H7 isolates recovered and tested before 2002 were in agreement with the chicken pathogenicity, i.e. all chicken LPAI viruses had few dibasic amino acids at the H cleavage site while all HPAI viruses had multiple dibasic amino acids at the cleavage site (results not shown). The proposed minimum amino acid sequence for HPAI is B-X-B-R/G, where B = basic amino acids arginine or lysine, X = non-basic amino acid, and R

= arginine [8]. Since 2002 there have been three AIVs isolated from the Americas (one H5 and two H7 subtypes), where the molecular marker for virulence was not consistent with in vivo test results (Table 1). One of the atypical viruses was an H5N2 virus isolated from Texas in 2004; the virus met the molecular criterion for HPAI but did not produce disease when inoculated into experimental chickens [12]. The remaining two were H7N3 viruses, one from Chile in 2002 and one (genetically unrelated to the Chilean H7N3 virus) from British Columbia, Canada, in 2004. Both viruses were highly pathogenic to chickens but did not meet the molecular criterion for HPAI. The H7N3 virus from Chile had a 30 nucleotide insert (derived from the nucleoprotein gene) near the cleavage site on the H gene, whereas the H7N3 virus from Canada had a 21 nucleotide insert (derived from the matrix gene) near the cleavage site on the H gene. The insertions were believed to be examples of recombination events, the first such events seen in natural outbreaks of AI [13,14].

Discussion

Molecular tests are being widely used in veterinary diagnostic laboratories to diagnose a variety of animal diseases including AI. Clinical specimens generally submitted for AI diagnosis include tissues and tracheal and cloacal swabs from avian species. For rRT-PCR, particular attention should be given to the selection of an appropriate method for RNA extraction to maximize recovery and avoid PCR inhibiting substances that can lead to false negative findings. In these studies, RNeasy membrane kits (Qiagen, Valencia, CA) were used for extraction of tracheal swabs. However, this method was shown to have lower sensitivity for cloacal swabs and tissue suspensions. The reduction in sensitivity is attributed to the heavy organic load in these samples as extraneous RNA compete for binding sites in the extraction column. Organic extraction methods such as Trizol (Invitrogen, Carlsbad, CA) have been shown to be better than silica gel membrane kits for specimens with a high organic content, but the extra time for processing and use of toxic chemicals make this method less desirable. The cloacal swab and tissue samples may also have a problem with PCR inhibitors, so the use of a positive internal control is also recommended for these samples. Preliminary studies with the MagMAX kit (Ambion, Austin, TX), that uses special magnetic beads to adsorb RNA may be more resistant to RNA saturation problems associated with silica gel membrane kits. The MagMAX is also suitable for high throughput testing because the extraction is performed in 96-well microtitre plates.

In this paper we present "fitness for purpose" data for RT-PCR assays that can be used to detect AIV in clinical samples and to assess pathogenicity of H5 and H7 subtypes of AIV rapidly. Both traditional and real-time RT-PCR assays have been used for this purpose. In general, these assays have shown levels of sensitivity that are comparable to in vivo assays such as virus isolation (VI), and the chicken pathogenicity test, both of which are the gold standard tests for detection of AIV and pathogenicity testing, respectively. Because of the ease and speed of using molecular tests in the laboratory, the rRT-PCR assay has been used as the front line test for surveillance and during outbreaks of AIV in Virginia (H7N2, 2002), Delaware (H7N2, 2004), and Texas (H5N2, 2004) and in outbreaks in Hong Kong (H5N1, 2003) and Canada (H7N3, 2004) [11,12,14, D. Suarez, personal communication]. The rRT-PCR assay is now used as a screening test in 39 laboratories in the United States that are part of the National Animal Health Laboratory Network.

Although molecular tests have greatly increased the efficiency of diagnostic laboratories in diagnosing AI infections and assessing the virulence of H5 and H7 viruses, periodic testing of outbreak isolates by virus isolation and chicken pathogenicity should continue to monitor for the emergence of new strains with unusual characteristics and to detect genetic changes that molecular tests may fail to detect.

References

1 Senne DA, Suarez DL, Stallnecht DE, Pedersen JC, Panigrahy B: Ecology and epidemiology of avian influenza in North and South America, in Schudel A, Lombard M (eds): OIE/FAO International Scientific Conference on Avian Influenza. Dev Biol (Basel). Basel, Karger, 2006, vol 124, pp 37-44.

2 Slemons RD, Johnson DC, Osborn JS, Hayes F: Type-A influenza viruses isolated from wild free-flying ducks in California. Avian Diseases 1974;18:119-124.

3 Kawaoka Y, Chambers TM, Sladen WL, Webster RG: Is the gene pool of influenza viruses in shorebirds and gulls different from that in wild ducks? Virology 1988;3:247-250.

4 Fouchier, RA, Munster V, Wallensten A, Bestebroer TM, Herfst S, Smith D, Rimmelzwaan GF, Olsen B, Osterhaus AD: Characterization of a novel influenza A virus haemagglutinin subtype (H16) obtained from black-headed gulls. J Virol 2005;79(5):2814-2822.

5 Swayne DE, Halvorson DA: Influenza, in Saif YM, Barnes J, Glisson JR, Fadley AM, McDougald LR, Swayne DE (eds): Diseases of Poultry, 11th ed. Ames, Iowa, Iowa State Press, Blackwell Publishing, 2003, pp 135-160.

6 OIE Standards Commission. Manual of Diagnostic Tests and Vaccines for Terrestrial Animals. World Organization for Animal Health (OIE), Paris, France. Fifth edition, 2005.

7 Webster RG: Influenza: an emerging disease. Emerging Infectious Diseases 1998;4(3):436-441.

8 Senne DA, Panigrahy B, Kawaoka Y, Pearson JE, Suss J, Lipking M, Kida H, Webster RG: Survey of the hemagglutinin (HA) cleavage site sequence of H5 and H7 avian influenza viruses: Amino acid sequence at the HA cleavage site is a marker of pathgenicity potential. Avian Diseases 1996;40:425-437.

9 Wood GW, McCauley JW, Bashiruddin JB, Alexander DJ: Deduced amino acid sequences at the haemagglutinin cleavage site of avian influenza A viruses of H5 and H7 subtypes. Arch Virol 1993;130:209-217.

10 Spackman E, Senne DA, Myers TJ, Bulaga LL, Garber LP, Perdue M, Lohman K, Daum LT, and Suarez DL: Development of a real-time reverse transcriptase PCR assay for type A influenza virus and the avian H5 and H7 hemagglutinin subtypes. J Clin Microbiol 2002;40(9):3256-3260.

11 Akey BL: Low-pathogenicity H7N2 avian influenza outbreak in Virginia during 2002. Avian Diseases 2003;47(3 Suppl):1099-1103.

12 Lee CW, Swayne DE, Linares JA, Senne DA, Suarez DL: 5N2 avian influenza outbreak in Texas in 2004: the first highly pathogenic strain in the United States in 20 years? J Virol 2005;79(17):11412-11421.

13 Suarez DL, Senne DA, Banks J, Brown IH, Essen SC, Lee CW, Manvell RJ, Mathieu-Benson C, Moreno V, Pedersen JC, Panigrahy B, Rojas H, Spackman E, Alexander DJ: Recombination resulting in virulence shift in avian influenza outbreak, Chile. Emerging Infectious Diseases 2004;10(4):693-699.

14 Hirst M, Astell CR, Griffith M, Coughlin SM, Moksa M, Zeng T, Smailus DE, Holt RA, Jones S, Marra MA, Petric M, Krajden M, Lawrence D, Mak A, Chow R, Skowronski DM, Tweed SA, Goh S, Brunham RC, Robinson J, Bowes V, Sojonky K, Byrne SK, Li Y, Kobasa D, Booth T, Paetzel M: Novel avian influenza H7N3 strain outbreak, British Columbia. Emerging Infectious Diseases 2004;10(12):2192-2195.

D.A. Senne, Diagnostic Virology Laboratory, National Veterinary Services Laboratories, 1800 Dayton Avenue, Ames, IA 50010, USA.
E-mail: dennis.a.senne@aphis.usda.gov

Vannier P, Espeseth D (eds): New Diagnostic Technology: Applications in Animal Health and Biologics Controls. Dev Biol (Basel). Basel, Karger, 2006, vol 126, pp 179-186.

Validation of a Real-Time RT-PCR Assay for Rapid and Specific Diagnosis of Classical Swine Fever Virus

M.F. Le Potier, M. Le Dimna, G. Kuntz-Simon, S. Bougeard, A. Mesplède

AFSSA, Ploufragan, France

Key words: Classical swine fever, real-time RT-PCR, diagnosis validation

Abstract: Two new real-time RT-PCR kits developed by LSI (TaqVet CSF) and ADIAGENE (Adiavet CSF) obtained a manufacturing agreement in France during the past year. For that purpose, the Classical Swine Fever (CSF) National Reference Laboratory (NRL) planned a schedule of conditions to be fulfilled by commercial real-time RT-PCR assays. The producers were asked to introduce an internal control to check the RNA extraction efficacy. The different criteria assessed were: sensitivity, specificity, especially "pestivirus specificity", reproducibility and easy handling, using 187 different samples distributed in four different panels. These samples were either CSFV inactivated strains or organs collected from CSF experimental SPF infected pigs, or naturally infected wild boars. All these samples were previously tested for genome detection using an RT-nested PCR assay and for virus isolation on cell culture.

The LSI TaqVet kit was used for the CSF surveillance of wild boars in an area known to be infected, during the winter of 2004-2005. This field evaluation was carried out on 4710 spleen samples. In summary, the new CSF real-time RT-PCR assays have a higher predictive value than the current diagnostic standard, Virus Isolation.

Introduction

Classical swine fever (CSF), previously named Hog Cholera, is a highly contagious viral disease of worldwide importance. CSF disease is registered on OIE list A, and is endemic in many countries of Eastern Europe, South-Eastern Asia, and Southern and Central America. Nowadays, even if CSF has been eradicated in the domestic pigs of Western Europe, it is still endemic in some wild boar populations, increasing the risk of outbreaks on farms. There are, however, other pestiviruses than Classical Swine Fever virus to which pigs might be exposed, including Bovine Viral Diarrhoea (BVD) virus and Border Disease (BD) virus. The main problem in CSF diagnosis had for a long time been the lack of specificity regarding pestiviruses, as demonstrated by the presence of BVD-BD antibodies in pig sera interfering with CSF eradication programmes.

Several CSF epidemics in Europe during the past decade proved that the key control during an outbreak is the early identification of CSF infected animals and their elimination [1]. There was therefore a real need for the availability of a diagnostic method permitting the early detection of infected pigs. During the past ten years, different RT-PCR have been described, generally focusing on the 5' non-coding region, whether followed by genetic typing or not. Viral RNA detection can be performed on blood samples collected on live animals as well as on organ samples. Even if these RT-PCR methods have been evaluated in two European ring-tests [2], no official test is currently recommended, as some improvement was required to obtain a rapid, reliable and automated test, before being used in routine diagnosis laboratory. The aim of this study was to define validation procedures and assess the performances of new real-time RT-PCR assays.

Materials and Methods

Real-time RT-PCR

The two commercially available real-time single tube RT-PCR assays were developed by LSI (TAQVET PPC) and ADIAGENE (ADIAVET CSF). These tests were based on a first reverse-transcription (RT) of the CSF Virus RNA into complementary DNA. Then cDNA was amplified (PCR) by a DNA polymerase using primers and Taqman probes labelled with FAM specific from CSF virus. Both enzymatic reactions occur in the same tube (one step RT-PCR). This mix also contained an internal control, with primers and probes labelled with a fluorophore (VIC) specific from a porcine cell RNA permitting RNA extraction, reverse transcription and further DNA amplification to be checked. The passive reference was ROX. Reagents provided by QUIAGEN (Courtaboeuf, France) are used for RNA extraction.

Table 1: Sensitivity and specificity of the real-time RT-PCR on β-propiolactone inactivated strains

Virus	Strain	Genogroup	Virus titre* ($TCID_{50}$/ml)	Positive until dilution**:
CSF	Alfort 187	1.1	$10^{6.9}$	10^{-7}
	Brescia	1.2	$10^{6.1}$	10^{-4}
	Diepholz	2.3	$10^{5.7}$	10^{-4}
	Viesbeck	2.3	$10^{7.1}$	10^{-4}
	Moselle 1	2.3	$10^{5.5}$	10^{-4}
	Paderborn	2.1	$10^{6.5}$	10^{-4}
	Eystrup	1.	$10^{6.7}$	10^{-4}
BDV	Aveyron	/	/	negative
BVDV	C24 Oregon	/	/	negative
BD/PRRS	Aveyron/Espagnol	/	/	negative
BD/AUJ/PPV	Aveyron/75V19/NADL8	/		negative

*: before inactivation with β-propiolactone

**: positive in real-time RT-PCR after inactivation

Choice of samples for validation studies

Virus strains

The selected CSF virus strains were either reference or field strains, representing different CSF genogroups. Other pestivirus or porcine virus were used to evaluate the specificity of the real-time RT-PCR. The strains were inactivated with β-propiolactone and the limit of detection was previously assessed by RT-nested PCR (Table 1).

Clinical samples

Organs sampled from Specific Pathogen-Free (SPF) pigs experimentally infected with different CSF strains or other porcine virus were chosen. On these samples, virus isolation for CSF was performed by standard techniques, based on the OIE Manual of Standards for Diagnostic Tests and Vaccines (2000) and the presence of the genome was detected using a single tube RT-nested PCR adapted from the method previously described [3]. Others organ samples collected from CSF and pestivirus-free herds were included in the panels as well as organs sampled from an area containing CSF-infected wild boars (Table 2).

Agreement procedure

First, the industrial company had to present a dossier with a description of the method and the results regarding specific detection of different CSF genogroups and sensitivity tests. Then the first panel was tested with the kit and in the event of success, the second and third panels are submitted to the Real-time RT-PCR assay kit (Table 2). Some samples, especially the strains, are repeated in different panels to assess the reproducibility of the results. The fourth panel of 22 samples was specially designed for the kit batch-to-batch control and for inter-laboratory ring tests.

Field validation

Since the 1st October 2004, the first agreed kit (LSI Taqvet) was used for the epidemiological survey carried out in two wild boar-infected areas in the North-Eastern part of France. In these two outbreaks, blood and spleen were sampled from each hunted wild boar. Between the 1st October 2004 and the 1st April 2005, 4690 spleen were collected from wild boars and tested by real-time RT-PCR, but blood for specific CSF antibody detection was obtained from only 3646 wild boars. Out of the 4690 spleen samples, all those positive by real-time RT-PCR were further assessed by virus isolation. Within the 4690 spleen samples, the first 120 obtained were tested both by virus isolation and real-time RT-PCR.

Data analysis

Performances of the real-time RT-PCR were compared to virus isolation in terms of specificity, sensitivity, and negative and positive predictive values [4]. The 95% confidence interval (CI) was calculated as:

$$p \pm 1.98 \sqrt{\frac{p(1-p)}{n}}$$

where p is the proportion of animals detected and n the number of animals tested. Negative and positive predictive values are corrected, taking into account the infection prevalence in the target population (4690 wild boars).

Table 2: 189 samples/4 panels.

Species	Organs	Strain	Results	Number	Panel1	Panel2	Panel3	Panel4
Experimental pigs	tonsil	circovirus	neg	2			*	
	lymph nodes	circovirus	neg	2			*	
	ileo-caecal valves	circovirus	neg	2			*	
	kidney	CSFV (Eystrup)	Pos f	1		*	*	
		CSFV (Eystrup)	+/-	2		*	*	*
	spleen	CSFV (Eystrup)	Pos	1		*	*	*
		CSFV (paderborn)	Pos	4		*	*	
		/	neg	2	*	*	*	*
	lymph nodes	CSFV (Eystrup)	Pos	3		*	*	*
	ileo-caecal valves	CSFV (Eystrup)	Pos	3		*	*	
	blood	CSFV (Eystrup)	Pos	1		*	*	
	hyper immune serum	CSFV	neg	1			*	
	lymph nodes	/	neg	2	*	*	*	
	serum	/	neg	1			*	
Herd swine	tonsil	CSFV (Moselle1)	Pos	2		*	*	
		/	neg	2		*	*	
Wild boar	lymph nodes	/	neg	4			*	*
	spleen	/	neg	3			*	*
	blood	/	neg	4			*	
	spleen	CSFV	Pos f	9		*	*	
		CSFV	+/-	5	*	*	*	
		/	neg	17	*	*	*	
	kidney	/	neg	1	*			
Viral strain	supernatant	CSFV Alfort	Pos /+/-	3	*	*	*	*
		CSFV Moselle1	Pos/neg	3	*	*	*	*
		CSFV Bas rhin	Pos/neg	3	*	*	*	*
		CSFV Eystrup	Pos/neg	3	*	*	*	*
	strain culture	CSFV Brescia	Pos	1	*	*	*	
		CSFV Visbeck	Pos	1	*	*	*	
		CSFV Diepholz	Pos	1	*	*	*	
		CSFV Paderborn	Pos	1	*	*	*	*
		BDV	neg	1	*	*	*	*
		BVDV	neg	1		*	*	
	mixed strain	BD/SDRP	neg	1			*	*
		BD/AUJ/PPV	neg	1			*	*
					20	58	89	22

Results

Laboratory validation

All the selected CSF strains were scored positive by real-time RT-PCR assays, when others pestivirus or porcine virus were negative. In most of the cases, the sensitivity of the real-time RT-PCR assay on the dilution of inactivated strains was reduced approximately 10- to 100- fold when compared with virus isolation, except for the Alfort 187 strain (Table 1).

CSFV experimental infection:

Six SPF pigs were oro-nasally infected with the CSFV Eystrup strain. Three pigs were euthanised 24 hours post-infection (hpi) and the three others at 48hpi. Virus was only isolated from organs sampled at 48hpi. Spleen and mesenteric lymph nodes were CSFV positive by RT-nested PCR and by real-time RT-PCR on 24 hpi as well as 48 hpi. Two of the three ileo-caecalis valves were positive at 24 hpi and three at 48 hpi. Only one of the kidneys was positive at 48 hpi. This experimental infection confirmed that CSFV genome can be detected as early as 24hpi in organs.

In addition, the results obtained using the kits had to match the expected results of the three panels. Finally 189 samples were tested by virus isolation, RT-nested PCR and real-time PCR. The sensitivity of the real-time RT-PCR was equal or even higher as obtained with those which were RT-nested. No specificity problem was noted.

The two kits developed by LSI and ADIAGENE were used in a ring test with height laboratories. The results obtained on panel n°4 (Table 2) confirmed the high reproducibility of these tests.

Field validation

The results of the comparison of performances between virus isolation and real-time RT-PCR made on 120 spleen samples are summarised in Table 3.

Of 4690 spleen samples, 62 were positive by the CSFV real-time RT-PCR assay, that permitted to calculate an infection prevalence of 1.3% in the wild boars population. Of these positive spleen, 14 were virus isolation (VI) positive.

When considering VI as the diagnostic standard for CSF and applying data correction taking into account the infection prevalence in the targeted population, the positive predictive value calculated was 2,8% (IC 95%: 0-6) and the negative predictive value was 100%. To try to understand the origin of this discrepancy, analysis of available serological data was carried out. Out of the 48 wild boars scored "PCR+/VI-", only 38 had been blood-sampled and 35 were positive in ELISA. RT-nested PCR performed as described by McGoldrick [3] confirmed the presence of CSF genome in the spleen of the three wild boars "PCR+/VI-/ELISA-".

When the cycle threshold (Ct) values of the real-time RT-PCR positive samples was below 26, virus was always isolated. When Ct values were between 26 and 42, virus isolation was not systematic depending most of the time of antibodies presence, as shown in Figure1.

Table 3: Performance of the real-time RT-PCR compared to virus isolation made on 120 spleen samples.

		Real-time RT-PCR	
		negative	positive
Virus isolation	negative	58	48
	positive	0	14

	Percentage	Confidence interval 95%
Sensitivity	100	-
Specificity	54,7	46-64
Positive predictive value	22,6	15-30
Negative predictive value	100	-
False positives	77,4	70-85
False negatives	0	-

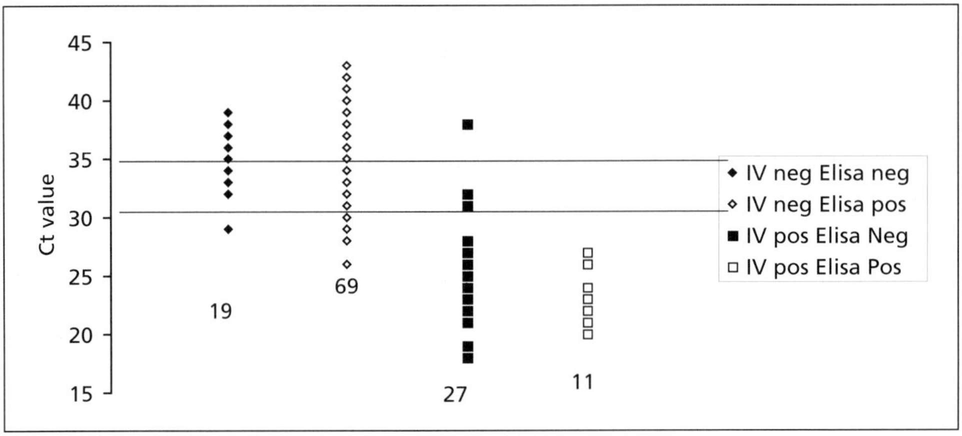

Fig. 1: Relationship between (Ct) values of real-time PCR positive samples and VI and ELISA status of 126 samples

Discussion

The validation procedure applied to these real-time RT-PCR assays was designed for our diagnostic purposes. We needed a very high sensitivity to detect the viral genome in every pig or wild boar infected and also a high specificity to be confident in negative results. For this purpose we defined a four-step procedure that had been applied in France to deliver the official agreement for two commercial kits already.

The use of the real-time RT-PCR assay for virological surveillance of wild boars permitted the real incidence of the infection in infected or vaccinated areas to be followed. As all positive real-time RT-PCR spleen were further assessed for virus isolation and the higher sensitivity of the real-time RT-PCR was demonstrated when compared to virological isolation. When neutralising antibodies were already present, virus could not be isolated, even if viral genome was detectable. This property was very interesting to follow the diffusion of the wild virus within a vaccinated population when serological differentiation was not possible.

Another RT-nested PCR assay [3] targeting the same highly conserved region of the viral genome confirmed the positive results for the wild boars that were positive by real-time RT-PCR but negative for virus isolation and antibody detection. In these cases, it was confirmed that the wild boars had been infected but maybe the infection was too recent to isolate viral particles. The results obtained with these real-time RT-PCR assays were very reliable as no false negative was found. On this basis, we can now imagine using this tool when clinical specimens are submitted to the laboratory for suspected swine fever. The short time needed to obtain the first results can permit containment measures applied to the suspected herd in the event of negative results to be lifted quickly. Moreover, because of the high sensitivity of the test, infected herds could be detected where virus isolation failed, as described in the Caribbean [5].

To summarize, this diagnostic tool, using EDTA treated blood on live animals, allowed the rapid identification of CSF infected pigs and could save considerable time when control measures had to be applied to prevent virus dissemination between herds, after the first detection of the virus. The following validation step could be a sensitivity assessment of the detection of a viral genome in a pool of blood or sera with these real-time RT-PCR assays.

Acknowledgements

The authors thank Eric Selal and Cécile Dives from LSI and Béatrice Blanchard and Blandine Civelli from ADIAGENE for their very kind collaboration.

References

1 Elbers ARW, Stegeman A, Moser H, Ekker HM, Smak JA, Pluimers FH: The classical swine fever epidemic 1997-1998 in the Netherlands: descriptive epidemiology. Preventive Veterinary Medicine 1999;42(3-4):157-184.

2 Paton DJ, McGoldrick A, Bensaude E, Belak S, Mittelholzer C, Koenen F, et al: Classical swine fever virus: a second ring test to evaluate RT-PCR detection methods. Veterinary Microbiology 2000;77(1-2):71-81.

3 Mcgoldrick A, Bensaude E, Ibata G, Sharp G, Paton DJ: Closed one-tube reverse transcription nested polymerase chain reaction for the detection of pestiviral RNA with fluorescent probes. Journal of Virological Methods 1999;79(1):85-95.

4 Rumeau-Rouquette C, Blondel B, Kaminski M: Evaluation de la valeur diagnostique d'une méthode de dépistage. Epidémiologie, méthodes et pratiques. 1995;20:294-306.

5 Risatti G, Holinka L, Lu Z, Kutish G, Callahan JD, Nelson WM, et al: Diagnostic evaluation of a real-time reverse transcriptase PCR assay for detection of classical swine fever virus. Journal of Clinical Microbiology 2005;43(1):468-471.

Dr. M.-F. Le Potier, Agence Française de Sécurité Alimentaire, Laboratoire d'études et de recherches avicoles et porcines, Unité Virologie Immunologie Porcines, BP53, FR-22440 Ploufragan, France.
E-mail: mf.lepotier@afssa.fr

Vannier P, Espeseth D (eds): New Diagnostic Technology: Applications in Animal Health and Biologics Controls. Dev Biol (Basel). Basel, Karger, 2006, vol 126, pp 187-196.

Bluetongue: Characterization of Virus Types by Reverse Transcription-Polymerase Chain Reaction

S. Zientara, E. Bréard, C. Sailleau

Agence Française de Sécurité Sanitaire des Aliments, Afssa/INRA/ENVA, Maisons-Alfort, France

Key words: Bluetongue, diagnosis, type-specific-polymerase chain reaction.

Abstract: Bluetongue virus (BTV) is the prototype of the member of the *Orbivirus* genus within the family *Reoviridae*. The BTV serogroup contains 24 serotypes. Traditionally, viruses have been isolated in cultured cells, suckling mouse brain or embryonated chicken eggs before their identification and biochemical, antigenic and biological characterization. These procedures are time-consuming and may fail to detect low levels of infectious virus or strains of BTV which fail to replicate in eggs, mice or tissue culture. In the past decade, traditional procedures for virus characterization, such as ELISA and serum neutralisation with serotype-specific antisera, have been supplemented by reverse transcriptase-polymerase chain reaction (RT-PCR) and sequencing. A number of procedures have been developed to detect the presence of BTV antigens or nucleic acids. RT-PCR technique has appeared to be a powerful tool in the field of BTV diagnosis. Polymerase chain reaction techniques may be used not only to detect the presence of viral nucleic acid but also to 'serogroup' orbiviruses and provide information on the serotype and possible geographical source (topotype or genotype) of BTV isolates within a few days of receipt of a clinical sample such as infected sheep blood. Real-time PCRs have recently been developed.

INTRODUCTION

Bluetongue is a non-contagious, infectious viral disease of domestic and wild ruminants, although clinical signs are usually only seen in sheep and certain species of deer. The bluetongue viruses (BTV) are arthropod-borne and constitute the type species of the genus *Orbivirus* within the *Reoviridae* family. The genome is composed of ten double-stranded RNA segments which encode at least ten viral proteins. Seven of the viral proteins are structural and form the double-shelled virus particle. At least three non-structural proteins (NS1, NS2 and NS3) have been identified [1,2]. The inner capsid of BTV is composed of five polypeptides: three minor proteins (VP1, VP4 and VP6) and two major proteins (VP3 and VP7) [2]. The outer capsid is composed of two major viral proteins, VP2 and VP5, which determine the antigenic variability of the BTV. To date, 24 antigenically distinct serotypes of the virus have been identified.

Virus is transmitted almost entirely by the bites of certain species of *Culicoides* midges and as a result is restricted to areas where these vectors are present. *C. Imicola* is considered to be the most efficient vector in Europe, although other species of *Culicoides*, of lesser importance, have been identified in these areas. The global distribution of BTV lies approximately between latitudes 35°S and 40°N, although in parts of Western North America and in China, it may extend up to almost 50°N.

Since 1998, five different serotypes (1, 2, 4, 9 and 16) have circulated in the Mediterranean Basin) and in particular in Greece, Italy, Tunisia, Spain (Balearic Islands) and France (Corsica). In 2004, serotype 4 spread to Morocco, Spain and Portugal.

Molecular tools (in particular polymerase chain reaction) can be used for the detection of early infection to allow early rapid implementation of sanitary measures to prevent the occurrence of an outbreak.

Interest of PCR for Bluetongue Diagnosis

BTV routine diagnosis is based primarily on serological methods that detect virus-specific antibodies in serum [3]. A number of other procedures are also currently used to detect BTV from the blood or tissues of infected animals. These include direct inoculation of cultured mammalian or insect cells, or intravenous inoculation into 10-12 embryonated chicken eggs, followed by one passage in insect cell cultures and up to three passages in mammalian cell cultures [4]. In particular, the inoculation of embryonated chicken eggs and passaging through cell culture is the generally accepted method for testing of animals for export and other regulatory purposes [4]. This is, however, a laborious and time-consuming protocol that may take up to five weeks for completion. Consequently, alternative methods of virus detection have been sought. These include antigen capture enzyme-linked immunosorbent assay (ELISA), dot immunobinding assay (DIA), immuno-electron microscopy and polymerase chain reaction (PCR) [5-11]. The use of antigen capture ELISA for the detection of bluetongue virus in the blood of infected ruminants has either been unsuccessful [10], has detected antigen only in animals with high viraemias [12], or was not consistent enough to allow for the reliable diagnosis of BTV [6].

To avoid these problems, RT-PCR-based assays were developed and evaluated for the detection of BTV serotypes based on nucleotide sequences of different genome segments [8,11,13-21]. The polymerase chain reaction (PCR), first described in 1985, is a highly sensitive and specific technique used for the detection of nucleic acids [22]. As far as BTV diagnosis is concerned, the application of polymerase chain reaction technology has permitted very rapid amplification of BTV RNA in clinical samples and PCR-based procedures are now available to provide information on virus serogroup and serotype. The remarkable ability of PCR to amplify specific DNA sequences has, along with its obvious benefits, some practical pitfalls that require careful attention. First among these is the ability of PCR to amplify DNA inadvertently introduced into the reaction. Precautions against contamination are especially important in clinical applications, but must be considered in every laboratory using the technique. Some principles used in sterile culture of micro-organisms are applicable, but additional precautions such as the strict segregation of sample preparation, reaction assembly, thermocycler, and analysis work areas, and the use of positive displacement pipettes or aerosol preventive tips, may also be necessary. The use of ultraviolet light and chemical decontamination procedures as well as of enzymatic methods to prevent the amplification of "carryover" templates should be used in some situations.

The high level of conservation of the nucleic acid sequences of some BTV genome segments among the different BTV serotypes has permitted primers which allow specific amplification of BTV genes to be selected. Oligonucleotides have been derived from VP7 gene, NS1 gene, VP3 gene and NS3 gene [8,15,16,18,23,24].

The different RT-PCR assays are shown to be sensitive as they could detect quantities of BTV as low as about 10-100 TCID50. Mainly, the diagnostic sensitivity of group-specific RT-PCR is greater than the virus isolation in embryonated chicken eggs or assays of cytopathic effects on cultured cells. The conditions of stringency, the choice of the primers, the different PCR parameters (the denaturation, annealing and extension conditions, the number of cycles performed, the RT and Taq polymerase concentrations, ...) have been studied by different laboratories to optimize the parameters which influence DNA amplification efficiency and specificity. In particular, pairs of primers are now available which did not cross-react with the genome of other orbiviruses (like EHDV) [11,25].

Determination of the nucleic acid sequence of portions of genome segment 3 (VP3) gene may provide information on whether the virus came from Australia, North America or South Africa [26,27].

Nested-PCRs were also developed. For example, the nested-PCR based on NS1 gene [20] is very sensitive, detecting the equivalent to one plaque-forming unit. Moreover, reverse-transcriptase multiplex PCR have been developed but needed to be validated [28,29].

The RT-PCR diagnosis outlined is rapid and specific for the detection of the BT serogroup, which holds promise as an alternative to the time-consuming, costly and cumbersome conventional protocols. However, virus isolation is still needed for positive identification or confirmation of BTV infection.

By the determination of the nucleotide sequences of the PCR-amplified products, it is now easy to confirm the specificity of the PCR reaction rapidly. As shown, the nucleic acid sequences of known BTV genes appear to differ with the geographical area of isolation; thus the comparison of nucleotide sequences can give useful data to complement BTV epidemiology [8,27].

Bluetongue Virus Typing by RT-PCR

The low level of conservation of the nucleic acid sequences of some BTV genome segments among the different BTV serotypes has permitted primers which allow specific amplification of BTV genes to be selected. Oligonucleotide primers have been designed for the amplification and sequence analysis of genome segment 2 (VP2 gene) of BTV 2, 4, 9 and 16. Segment 2 has a single, open reading frame encoding the VP2 protein, which comprises 962 amino acid residues. These pairs of primers did not amplify segments 2 of the other 23 serotypes. These can be used in a multiplex RT-PCR reaction to generate different size products from each of the five serotypes that are currently present in Europe (1,2,4,9,16) and thereby identify their serotype.

Specific primers, based on the published nucleotide sequence of the genome S2 of

BTV 2 (GenBank accession number M21946) [30,31] were designed. A BTV 9 specific RT-PCR is also available using oligonucleotide primers designed and published by McColl and Gould [8]. These BTV 9 serotype specific primers amplified a 805 bp fragment of the BTV 9 vaccine strain.

Differentiation of Wild and Vaccine BTV Strains

BT serotype 2 was first suspected in Tunisia in December 1999. Virus was isolated and confirmed as BTV serotype 2 (BTV 2) early in 2000. The virus spread westward to Algeria and northward into Southern Italy. In October 2000, BTV 2 was confirmed on the Spanish islands of Majorca and Minorca and on the French island of Corsica. Following the isolation and confirmation of BTV 2 in Corsica 2000, the French Veterinary Authorities decided to vaccinate all the sheep on the island systematically during the winter of 2000-2001 using the South African attenuated BTV 2 vaccine virus. The South African attenuated BTV 2 vaccine virus (batch 7, Onderstepoort Biological Product, South Africa) was used for the vaccination campaign against BTV 2 on the Corsican island during the winter of 2000/2001 (January to April 2001). About 80% of sheep (non-pregnant or over the age of three months) were vaccinated on the island during the winter of 2000-2001 using the South African attenuated BTV 2 vaccine virus [31].

Coincidentally, outbreaks of a second BTV serotype, BTV 9, were being reported and confirmed on mainland Italy around this time. Vaccination with a bivalent attenuated vaccine (2 and 9) was decided in some regions of Italy.

Alignment of the S2, S7 and S10 of the BTV 2 vaccine and Corsican virus isolates showed that the nucleotide sequence of S10 presented the highest degree of divergence between these strains. Primers that incorporated nucleotide sequence differences in genome S10 of the vaccine and wild-type strains of BTV 2 were designed and used to help differentiate between vaccination and infection [32]. The S10 gene has two in-phase and overlapping open reading frames, which encode the two non-structural proteins of NS3 and NS3A, containing 229 and 216 amino acids, respectively. The nucleotide sequence of S10 from the BTV 2 vaccine strain showed 82.1% homology with that of the S10 from the wild-type strains isolated in 2000 and 2001. The NS3 predicted amino acid sequence of the BTV 2 vaccine strain showed 94.3% homology with the wild-type BTV 2 NS3 protein.

Primers showing a high level of nucleotide substitution within S10 of the wild-type and vaccine viruses were selected. The specific vaccine or wild strain primers have been tested and validated on RNAs extracted from red blood cells, spleen and lymph nodes samples from lambs experimentally vaccinated, from Corsican vaccinated sheep blood, from Corsican infected sheep in 2001 or from wild BTV 2 isolates [33].

Similar work has been performed by Aguero et al [34]. To perform epidemiological studies in the Balearic islands, these authors developed a PCR method (based on the use of an RT-PCR, followed by restriction endonuclease analysis) to differentiate between the NS1 genes of BTV-2 and that of the Onderstepoort commercial live virus vaccine (monovalent, serotype 2). Concerning serotype 9, primers have been designed from the nucleotide sequence of the BTV-9 segment 2; these primers amplify the genome segment 2 of the wild BTV-9 strain (present in Italy, Greece, ...) but do not detect the vaccine virus. The primers designed by McColl and Gould [8] only amplify the BTV-9 vaccine strain. However, it is necessary to validate the primers already designed (BTV-2 and BTV-9) and to select primers for the other 22 serotypes.

21 strains	Type-specific primers					
	1	**2**	**4a**	**4b**	**9 wild**	**16**
BT1 NIG82/01	+	-	-	-	-	-
BT1 GRE01/01	+	-	-	-	-	-
BT1 MAL87/01	+	-	-	-	-	-
BT1 SUD87/01	+	-	-	-	-	-
BT1 Australia	+	-	-	-	-	-
BT1 IND92/02	+	-	-	-	-	-
BT1 IND88/01	+	-	-	-	-	-
BT1 IND92/01	+	-	-	-	-	-
BT1 ref strain	+	-	-	-	-	-
BT2SA Vaccine	-	+	-	-	-	-
BT2 SA ref strain	-	+	-	-	-	-
BT2 ITL01/02	-	+	-	-	-	-
BT2 ITL01/01	-	+	-	-	-	-
BT2 FRA 01/03	-	+	-	-	-	-
BT2 TUN 00/01	-	+	-	-	-	-
BT4 SA Vaccine	-	-	+	-	-	-
BT4 TURKEY	-	-	+	+	-	-
BT9 BUL99/01	-	-	-	-	+	-
BT9 Bos02/02	-	-	-	-	[illegible]	
BT9 GRE00/02	-	-	-	-	+	-
BT16 N°434	-	-	-	-	-	+

Fig. 1: Validation of the serotype-specific RT-PCR (serotypes 1, 2, 4, 9 and 16) using a coded panel of BTV reference strains and field isolates (P Mertens, IAH, Pirbright, U.K.).

Validation

The specificity of the serotypes (2, 4, 9 and 16)-specific pairs of primers has been partially validated. Comparative testing by serotype-specific RT-PCR and by virus neutralisation test for 21 BTV samples (Dr. Peter Mertens, Institute for Animal Health, Pirbright, U.K.) showed total agreement (Fig. 1). These included recent strains isolated in 2001 and 2002 and reference strains, in a number of different countries, often where epidemiological links would be unlikely. The success of the RT-PCR in correctly identifying the serotype of isolates of 4 BTV serotypes that were from different 12 different geographical areas emphasises the choice of the sequenced regions from which the primers were selected. It was also noted that differences in passage history did not appear to interfere with the ability of the RT-PCR to serotype the viruses. The same validation procedure has been followed for the validation of the primers which allow the differentiation between the vaccine and wild strains of BTV-9 (Fig. 2).

	9 vacc	**9 wild**
BT1 NIG82/01	-	-
BT1 GRE01/01	-	-
BT1 MAL87/01	-	-
BT1 SUD87/01	-	-
BT1 Australia	-	-
BT1 IND92/02	-	-
BT1 IND88/01	-	-
BT1 IND92/01	-	-
BT1 ref strain	-	-
BT2SA Vaccine	-	-
BT2 SA ref strain	-	-
BT2 ITL01/02	-	-
BT2 ITL01/01	-	-
BT2 FRA 01/03	-	-
BT2 IND82/01	-	-
BT2 SUD85/01	-	-
BT2 NIG82/02	-	-
BT2 TUN 00/01	-	-
BT9 BUL99/01	-	+
BT9 SER01/01	-	+
BT9 Bos02/02	-	+
BT9 GRE00/02	-	+
BT9 TUR00/03	-	+

Fig. 2: Validation of the vaccine or wild-specific RT-PCR (for serotype 9) using a coded panel of BTV vaccine or field isolates (P Mertens, IAH, Pirbright, U.K.).

Real-Time PCR

Recently, quantitative RT-PCR have been developed. The primers and probes used in the real-time PCR described by Orru et al [35] were designed using the genome segment 2 sequences of the Italian field strain of BTV serotype 2. This FRET (fluorescence resonance energy transfer) real-time PCR has a limit of detection of 10^2 TCID50/ml and is able to differentiate between the vaccine and wild-type strains.

In the near future, real-time PCR will probably become the method of choice for molecular diagnosis of BTV. This method is sensitive, limits laboratory contaminations, is faster than conventional PCR, allows the simultaneous identification of different BTV types in a same sample and can be automated.

However, it is absolutely necessary to validate the different protocols and to make international positive controls available.

Conclusion

Traditional approaches, which rely on virus isolation followed by virus identification, may require at least three to four weeks to generate information on serogroup and serotype and yield no data on the possible geographical origin of the isolated virus. On the contrary, the BTV RT-PCR can provide a rapid, sensitive and specific viral identification for BTV infections. PCR and sequencing provides information on 'type', genotype and topotype very rapidly [36]. Segments coding for VP2, VP5, VP3, NS1 and NS3 are currently relevant. However, successful identification of BTV around the world depends on availability of relevant sequence data for primer development.

However, in particular in areas which are BTV-free, the sole interpretation of positive BTV PCR results must be analysed carefully. Before officially reporting BTV cases in the absence of clear epidemiological data, virus isolation is absolutely recommended to complement the molecular diagnosis.

The nucleotide sequence divergence between field vaccine strains has permitted the design of primers which allow them to be differentiated. These tools are very useful for monitoring the disappearance of a wild virus in a vaccinated population. Implementing such assays is important and will help in our understanding of the epidemiology and spread of BTV in vaccinated areas. The information obtained will also prove invaluable in control and eradication programmes.

Efforts should be made to send viruses or PCR products to all OIE reference laboratories or other competent laboratories to be sequenced and primer information made available (via the web) to facilitate characterization at the source laboratory. An excellent start has been made in the development of a website (http://www.iah.bbsrc.ac.uk/dsRNA_virus) in which a collection of relevant sequence data can be found (phylogenetic tree analysis of BTV isolates based on RNA2, segment 2 sequencing primers and primers which differentiate between vaccine and wild strains,...).

At last, real-time PCR technology which is faster, less susceptible to contamination problems but more expensive than traditional PCR methods will certainly be more frequently used in the near future. But, as there may be problems attempting to identify new isolates with already-existing 'real-time' probes, it is necessary to improve efforts to facilitate exchanges of strains and probes to validate the newly designed primers and probes.

Acknowledgements

The authors wish to thank Dr. P. Mertens, Institute for Animal Health, Pirbright (U.K.) for supplying BTV strains. The financial support of the European Project (Contract number:QLK2-CT-2000-00143) is gratefully acknowledged.

References

1 Mertens PPC, Brown F, Sangar DV: Assignment of the genome segments of bluetongue virus type 1 to the proteins they encode. Virology 1984;135:207-217.

2 Roy P : Bluetongue virus proteins. J General Virology 1992;73:3051-3064.

3 Pearson JE, Gustafson GA, Shafar AL, Alstad AD: Diagnosis of bluetongue and epizootic hemorrhagic diseases, in Walton TE, Osburn BI (eds): Bluetongue, African Horsesickness and related Orbiviruses. Boca Raton, FL, CRC press, 1992, pp 533-546.

4 Charalambos B, Koumbati M, Spyrou V, Nomikou K, Mangana O, Panagiotidis CA, Papadopoulos O: Bluetongue virus diagnosis of clinical cases by a duplex reverse transcription–PCR : a comparison with conventional methods. J Virol Methods 2001;98:77-89.

5 Afshar A, Thomas FC, Wright PF, Shapiro JL, Anderson J: Comparison of competitive ELISA, indirect ELISA and standard AGID tests for detecting bluetongue antibodies in cattle and sheep. Vet Record 1989;124:136-141.

6 Hawkes RA, Kirkland PD, Sanders DA, Zhang F, Li Z, Davis RJ, Zhang N: Laboratory and field studies of an antigen capture ELISA for bluetongue virus. J Virol Methods 2000;85:137-149.

7 Katz JB, Gustafson GA, Alstad AD, Adler KA, Moser KM: Colorimetric diagnosis of prolonged bluetongue viremia in sheep using an enzyme-linked oligonucleotide sorbent assay of amplified viral nucleic acid. Am J Vet Res 1993b;54:2021-2026.

8 McColl KA, Gould AR: Detection and characterisation of bluetongue virus using the polymerase chain reaction. Virus Res 1991;21(1):180-188.

9 Lunt RA, White JR, Blacksell SD: Evaluation of a monoclonal antibody blocking ELISA for the detection of group-specific antibodies to bluetongue virus in experimental and field sera. J Gen Virol 1988;69:2729-2740.

10 Mecham JO: Detection of bluetongue virus from blood of infected sheep by use of an antigen-capture enzyme-linked immunosorbent assay after amplification of the virus in cell culture. Am J Vet Res 1993;54:370-372.

11 Shad G, Wilson WC, Mecham JO, Evermann JF: Bluetongue virus detection : a safer reverse-transcriptase polymerase chain reaction for prediction of viremia in sheep. J Vet Diagn Invest 1997;9:118-124.

12 Stanislawek WL, Lunt RA, Blacksell SD, Newberry KM, Hooper P, White JR: Detection by ELISA of bluetongue antigen directly in the blood of experimentally infected sheep. Vet Microbiol 1996;52:1-12.

13 Akita GY, Chinsangaram J, Osburn BI, Ianconescu M, Kaufman R: Detection of bluetongue virus serogroup by polymerase chain reaction. J Vet Diagn Invest 1992;4:400-405.

14 Akita GY, Glenn J, Castro AE, Osburn BI: Detection of bluetongue virus in clinical samples by polymerase chain reaction. J Vet Diagn Invest 1993;5:154-158.

15 Aradaib IE, Schore CE, Cullor JS, Osburn BI: A nested PCR for detection of North American isolates of bluetongue virus based on NS1 genome sequence analysis of BTV-17. Vet Microbiol 1998;59:99-108.

16 Dangler CA, de Mattos CA, de Mattos CC, Osburn BI: Identifying bluetongue virus ribonucleic acid sequences by the polymerase chain reaction. J Virol Methods 1990;28:281-292.

17 Parsonson IM, Mc Coll KA: Retrospective diagnosis of bluetongue virus in stored frozen and fixed tissues samples using PCR. Vet Microbiol 1995;46:143-149.

18 Wade-Evans AM, Mertens PPC, Bostock CJ: Development of the polymerase chain reaction for detection of bluetongue virus in tissue samples. J Virol Methods 1990;30:15-24.

19 Wilson WC: Preliminary description of a polymerase chain reaction test for bluetongue and epizootic hemorrhagic disease viral RNA in bovine semen. J Vet Diagn Invest 1999;11:377-379.

20 Wilson WC, Chase CC: Nested and multiplex polymerase chain reaction for the identification of bluetongue virus infection in the biting midge, Culicoides variipennis. J Virol Methods 1993;45:39-47.

21 Billinis C, Koumbati M, Spyrou V, Nomikou K, Mangana O, Panagiotidis CA, Papadopoulos O. Bluetongue virus diagnosis of clinical cases by a duplex reverse transcription-PCR : a comparison with conventional methods. J Virol Methods 2001;98(1):77-89.

22 Saiki RK, Scharf S, Faloona F, Mullis KB, Horn GT, Erlich HA, Arnheim N: Enzymatic amplification of beta-globin genomic sequences and restriction site analysis for diagnosis of sickle cell anemia. Science 1985;230:1350-1354.

23 Katz JB, Alstad AD, Gustafson GA, Moser KM: Sensitive Identification of bluetongue virus serogroup by a colorimetric dual oligonucleotide sorbent assay of amplified viral nucleic acid. J Clin Microbiol 1993a;31:3028-3030.

24 Harding MJ, Prud'homme I, Rola J: Identification of the major North America bluetongue viruses using nucleic acid amplification techniques. Molecular and Cellular probes 1995;9:223-231.

25 Wilson WC: Development of nested-PCR tests based on sequence analysis of epizootic hemorrhagic disease viruses non-structural protein 1 (NS1). Virus Res 1994;31:357-365.

26 Gould AR, Pritchard LI: Relationships amongst bluetongue viruses revealed by comparisons of capsid and outer coat protein nucleotide sequences. Virus Res 1990;17:31-52.

27 Gould AR, Mc Coll KA, Pritchard LI: Phylogenetic relationships between bluetongue viruses and other Orbiviruses, in Walton TE, Osburn BI (eds): Bluetongue, African Horsesickness and related Orbiviruses. Boca Raton, FL, CRC press, 1992, pp 452-460.

28 Johnson DJ, Wilson WC, Paul PS: Validation of a reverse-transcriptase multiplex PCR test for the serotype determination of U.S. isolates of bluetongue virus. Vet Microbiol 2000;76(2):105-115.

29 Aradaib IE, Smith WL, Osburn BI, Cullor JS: A multiplex PCR for simultaneous detection and differentiation of North American serotypes of bluetongue and epizootic hemorrhagic disease viruses. Comp Immunol Microbiol Infect Dis 2003;26(2):77-87.

30 Yamaguchi S, Fukusho A, Roy P: Complete sequence of neutralization protein VP2 of the recent US isolate bluetongue virus serotype 2: Its relationship with VP2 species of other US serotypes. Virus Res 1988 ;11 :49-58.

31 Zientara S, Sailleau C, Dauphin G, Roquier C, Remond M, Lebreton F, Hammoumi S, Dubois E, Agier C, Merle G, Bréard E: Identification of the Bluetongue virus serotype 2 (Corsican strain) by RT-PCR of the genome segment 2. The Veterinary Record 2002;150(19):598-601.

32 Bréard E, Sailleau C, Coupier H, Mure Ravaud K, Hammoumi S, Gicquel B, Hamblin C, Boutrand S, Dubourget P and Zientara S: Molecular epidemiological analysis of genome segments 2, 7 and 10 of bluetongue viruses in Corsica, and differentiation between field isolates and the vaccine strain by RT-PCR. Vet Research 2003;34(6):777-789.

33 Hammoumi S, Bréard E, Sailleau C, Russo P, Grillet C, Cetre-Sossah C, Albina E, Sanchis R, Pépin M, Guilbert JM and Zientara S: Innocuity and activity of the monovalent bluetongue serotype 2 vaccine. Specific detection of the genome of the vaccine strain by RT-PCR. J Vet Medicine 2003;B50:316-321.

34 Aguero M, Arias M, Romero LJ, Zamora MJ, Sanchez-Vizcaino JM: Molecular differentiation between NS1 gene of a field strain Bluetongue virus serotype 2 (BTV-2) and NS1 gene of an attenuated BTV-2 vaccine. Vet Microbiol 2002;86(4):337-341.

35 Orru G, Santis PD, Solinas F, Savini G, Piras V, Caporale V: Differentiation of Italian field and South African vaccine strains of bluetongue virus serotype 2 using real-time PCR. J Virol Methods. 2004;122(1):37-43.

36 Bréard E, Sailleau C, Hamblin C, Zientara S: Bluetongue virus in the French Island of Reunion. Veterinary Microbiology 2005;106:157-165.

Dr. S. Zientara, Agence Française de Sécurité Sanitaire des Aliments, UMR 1161 Afssa/INRA/ENVA, 23 Avenue du Général de Gaulle, FR-94703, Maisons-Alfort, France.
E-mail : s.zientara@afssa.fr.

Vannier P, Espeseth D (eds): New Diagnostic Technology: Applications in Animal Health and Biologics Controls. Dev Biol (Basel). Basel, Karger, 2006, vol 126, pp 197-201.

Molecular Epidemiology of West Nile Virus in Humans

C. Pfleiderer[1], C. König[1], M. Chudy[1], M. Schmidt[2], W.K. Roth[2], C.M. Nübling[1]

[1] Paul-Ehrlich-Institut, Division of Virology, Langen, Germany
[2] Institute of Transfusion Medicine and Immunohematology, Johann Wolfgang Goethe University, Frankfurt am Main, Germany

Abstract: After the introduction of West Nile Virus into the United States of America in 1999 followed by annual WNV epidemics during the mosquito seasons and spreading of the virus from the East (New York; 1999) to the West of the U.S. (California; 2003/2004) there appeared the question of whether a similar scenario could happen in Europe, too. To be able to answer this question the German Ministry of Health decided to investigate the prevalence and incidence of WNV infections in German blood donors. First a test algorithm was established taking into account the high level of cross-reactivity between different flavivirus infections in serological test systems. AntiWNV-suspicious specimens were further investigated for their neutralisation capacity and by an antiWNV confirmation assay developed in-house. As a preliminary result of our studies a very low prevalence of WNV infections in healthy German blood donors was measured. Development of highly specific test systems is necessary for accurate and reliable differential diagnosis of flavivirus infections.

INTRODUCTION

West Nile Virus (WNV) is a mosquito-transmitted flavivirus, widely distributed throughout Africa, Asia, Middle East, and Europe causing epidemics of human meningo-encephalitis [1]. The viral replication cycle includes mosquitos, mainly of the species Culex, and birds as hosts, with humans, horses and other mammalian species being "dead-end" hosts. The vast majority of WNV infections in humans takes an asymptomatic course. However, some WNV infections are associated with mild symptoms such as fever, headache or rash. Only few WNV infections (<0,2 %) result in more serious symptoms such as meningitis or encephalitis, sometimes with fatal outcome, with a significantly higher risk for elderly or immunocompromised individuals.

The unexpected emergence of WNV 1999 in New York [2] and its rapid spread throughout North America during the next few years caused about 20,000 human cases of West-Nile fever and more than 500 fatalities so far in the USA. Screening of the US blood supply since 2003 by nucleic acid amplification techniques (NATs) resulted in a yield of more than 1.000 WNV-RNA positive blood donations up to the middle of 2005 which otherwise would have been transfused to patients [3,4]. These data underline the ability of WNV to become a major health threat.

Despite multiple reports about sporadic WNV infections in humans and horses across Europe, e.g. in France, Italy, and Romania [5], there are no data available on the existence or spread of WNV in Germany. Therefore, a study was initiated to investigate the prevalence and incidence of WNV in human and animal populations in Germany. We started with the screening of healthy blood donors for WNV specific antibodies to obtain an impression of the extent of past infections. As a first step a test algorithm had to be developed taking into account the high level of cross-reactivity [6] of different flavivirus infections in serological test systems.

AntiWNV IgG Test Algorithm

As a first step we determined the sensitivity of three antiWNV-IgG ELISAs potentially useful for blood donor screening. Two of the assays (Focus, PanBio) recognized all 26 well-characterized antiWNV-positive sera from U.S. patients (sensitivity 100%) while a third assay (prototype ELISA, not yet marketed) missed three specimens resulting in a sensitivity of 88% in this study. However, ranking of the three assays according to their specificity gave a different picture: cross-reactivity rates of the anti-WNV assays with panels comprising 14 antiDengue virus positive and 39 antitick-borne-encephalitis virus (TBEV)-positive specimens were highest in the PanBio test (92% and 80%, respectively) followed by the Focus antiWNV test (92% and 56%, respectively). Highest specificity was achieved by the prototype ELISA with cross reactivity of 36% and 18%, respectively. Therefore we decided to screen the blood donors with Focus antiWNV IgG test as one of the most sensitive assays, followed by analysis of all Focus reactive specimens in the prototype antiWNV test as the most specific assay. Furthermore all specimens reactive in the Focus test were also analysed in an antiTBEV-IgG ELISA. All blood donations reactive in the two antiWNV ELISAs and negative in the antiTBEV ELISA were preferentially tested in the WNV neutralisation assay.

Blood Donor Screening

More than 14,000 healthy blood donors from Hesse/Germany and around 1,000 from Austria (Vienna and Carinthia) were screened for WNV specific IgG antibodies. Around 6% and 70% of the donors, respectively, were tested antiWNV-reactive in the screening assay (Table 1). Most of the reactive test results can be explained by cross-reacting antiTBEV-antibodies, originating either from natural TBEV infection or TBEV vaccination. The high percentage of reactive blood donors from Austria is in accordance with the immunisation level against TBEV of approximately 80% for the adult population. The rate of reactive blood donors in Hesse may also reflect the TBEV immunisation and /or infection level in a TBE low-risk region.

Table 1: AntiWNV (IgG) reactivity of blood donor samples (Focus ELISA).

Origin of blood donors Country (Region)	Tested n	AntiWNV-reactive n (%)
Germany (Hesse)	14,437	852 (5,9 %)
Austria (Vienna, Carinthia)	928	669 (72 %)

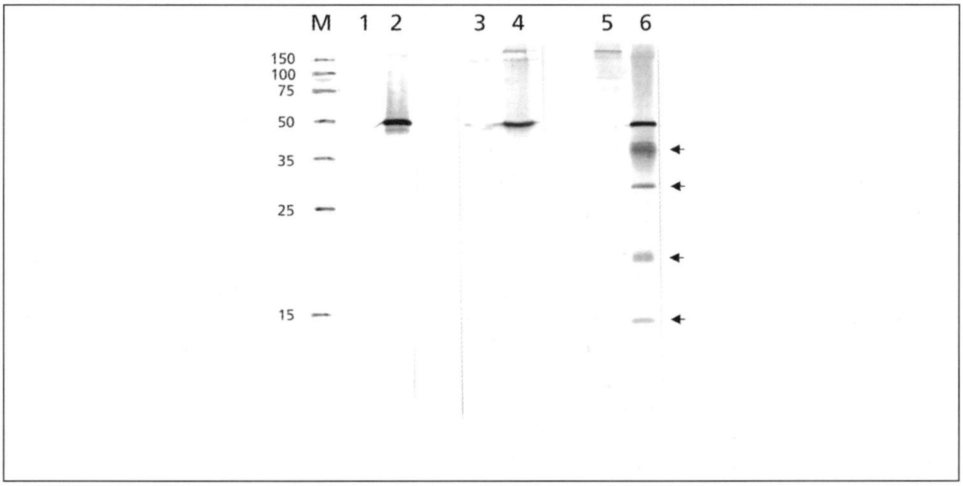

Fig. 1: Western Blot analysis of antiDengue virus-positive (lanes 1-2), anti-TBEV-positive (lanes 3-4), and anti-WNV-positive (lanes 5-6) sera. Each serum was tested with Western blot strips loaded with proteins from mock-infected (lanes 1, 3, 5) and WNV-infected Vero cells (lanes 2, 4, 6). All sera react with the 50 kD WNV-specific E protein. The arrows indicate WNV-specific proteins of different molecular weights recognized only by anti WNV-positive sera. The sizes of the molecular weight marker (lane M) are shown in kilodaltons

The antiWNV-reactive German blood donations were further tested in the second antiWNV and in the antiTBEV-assay. About 1.4% of the screened population were also reactive in the second antiWNV ELISA. We analysed at least 202 samples in our in-house WNV neutralization assay. 148 of the antiWNV-preselected specimens from German blood donors (73%) did not inhibit WNV infection in cell culture at all or only to a very low extent (log $ND_{50} < 1$), 50 specimens (25%) neutralized WNV infection at a titre between log $ND_{50} \geq 1$ and < 2. Only four of the 202 antiWNV IgG pre-selected specimens (2%) exhibited higher neutralisation titres (log $ND_{50} = 2$) comparable to those titres we observed with the antiWNV IgG positive well-characterized control sera obtained from U.S. patients.

However, several antiDengue virus IgG positive sera tested in the WNV neutralization assay also showed inhibition of WNV infection comparable to antiWNV control sera. Cross-reactivity was also observed to a lower extent in the WNV neutralization assay with antiTBEV IgG positive sera. The test algorithm was therefore complemented by a more specific confirmation assay for antiWNV-reactive specimens. This Western

Blot based assay using viral structural and non-structural (NS) antigens produced in WNV-infected Vero cells is obviously capable of differentiating between antibodies directed against WNV, Dengue virus or TBEV (Fig. 1). Using this test format we were able to confirm WNV infection in a total of four blood donors (three with high and one with medium NT titres).

Discussion

There is currently no indication that WNV could cause an epidemic in Europe comparable to the course in the U.S.A. during the recent years. However, we have to recognize that the available diagnostic tools for detection of past WNV infections are not optimal for investigations in Europe, where highly cross-reacting flaviviruses such as TBEV or Dengue virus are much more prevalent compared to WNV [6]. Therefore there is a need for developing more specific in vitro diagnostic assays. One approach could be the Western Blot-based confirmation test which in our hands is able to differentiate between the humoral response directed against different flaviviruses. Our studies on WNV prevalence in Germany by testing for WNV-specific antibodies are being completed by NAT testing of healthy individuals donating blood during the 2005 mosquito season in Germany (WNV incidence). Furthermore, we are comparing the WNV-RNA contents in plasma pools used for the fractionation of plasma derivatives (e.g. immune globulin preparations, albumin, factor concentrates) where the source material has been collected in different countries at different times of the year. However, preliminary results of our virus validation studies prove that WNV should not pose a specific risk for these virus-inactivated biologicals.

Acknowledgement

For this study a grant was received from the German Ministry of Health (BMGS; 115-1720-1/31) in the project "Prevalence and incidence of West Nile Virus infections in Germany".

References

1 Solomon T, Ooi MH, Beasley DW, Mallewa M : West Nile encephalitis.BMJ 2003 ;326 :865-869.

2 Lanciotti RS, Roehrig JT, Deubel V, Smith J, Parker M, Steele K, Crise B, Volpe KE, Crabtree MB, Scherret JH, Hall RA, MacKenzie JS, Cropp CB, Panigrahy B, Ostlund E, Schmitt B, Malkinson M, Banet C, Weissman J, Komar N, Savage HM, Stone W, McNamara T, Gubler DJ : Origin of the West Nile virus responsible for an outbreak of encephalitis in the northeastern United States. Science 1999 ;286 :2333-2337.

3 Stramer SL, Fang CT, Foster GA, Wagner AG, Brodsky JP, Dodd RY : West Nile virus among blood donors in the United States, 2003 and 2004. N Engl J Med 2005 ;353:451-459.

4 Busch MP, Caglioti S, Robertson EF, McAuley JD, Tobler LH, Kamel H, Linnen JM, Shyamala V, Tomasulo P, Kleinman SH : Screening the blood supply for West Nile virus RNA by nucleic acid amplification testing. N Engl J Med 2005 ;353 :460-467.

5 Zeller HG, Schuffenecker I: West Nile virus: an overview of its spread in Europe and the Mediterranean basin in contrast to its spread in the Americas. Eur J Clin Microbiol Infect Dis 2004;23 :147-156.

6 Allwinn R, Doerr HW, Emmerich P, Schmitz H, Preiser W: Cross-reactivity in flavivirus serology: new implications of an old finding? Med Microbiol Immunol 2002;190:199-202.

Dr. C.M. Nübling, Paul-Ehrlich-Institut, Division of Virology, Paul-Ehrlich-Str. 51-59, DE-63225 Langen, Germany.
E-mail: nuemi@pei.de

Vannier P, Espeseth D (eds): New Diagnostic Technology: Applications in Animal Health and Biologics Controls. Dev Biol (Basel). Basel, Karger, 2006, vol 126, pp 203-210.

The Use of DNA Microarray Technology for Detection and Genetic Characterisation of Chlamydiae

K. Sachse[1], *H. Hotzel*[1], *P. Slickers*[2], *R. Ehricht*[2]

[1] Friedrich-Loeffler-Institut (Federal Research Institute for Animal Health), Institute of Bacterial Infections and Zoonoses, Jena, Germany

[2] Clondiag Chip Technologies, Jena, Germany

Key words: DNA microarray technology, species identification, *Chlamydia, Chlamydophila*, clinical samples, sensitivity, specificity.

Abstract: Due to its highly parallel approach, DNA microarray technology opens up new possibilities that may be particularly beneficial for laboratory diagnosis of infectious diseases. We developed a microarray assay for detection and differentiation of all currently defined chlamydial species belonging to the genera *Chlamydia* and *Chlamydophila* using the ArrayTube™ system, which we found to be particularly user-friendly and economical. The test includes PCR amplification of a 23S rDNA target region with concurrent biotinylation and subsequent hybridisation in the ArrayTube, a micro-reaction tube carrying the microarray chip on the bottom. In addition to high specificity, the assay was shown to allow detection and genetic characterisation of single PCR-amplifiable target DNA copies.

INTRODUCTION

The diagnostic test of the future will be expected to provide complex information on the causative agent of an infection beyond the mere identification of the species. As soon as the diagnostician is confronted with this demand for "multi-dimensional" evidence on the pathogenic micro-organism the limitations of present PCR-based tests become obvious. Detection of subtle differences among strains, such as single-nucleotide polymorphisms and other intra-species variations, is problematic with existing amplification assays, and even in multiplex PCR only a few target regions can be examined in parallel.

DNA microarray technology opens up new possibilities that may be particularly beneficial for laboratory diagnosis of infectious diseases. This highly parallel approach allows any DNA sample to be simultaneously examined using a large number of probes, which may be derived from polymorphic gene segments of one or more genomic regions. In essence, this amounts to sequencing the respective target site.

Thus, DNA microarray-based tests can attain far higher specificity than PCR.

Although microarrays have become a widely accepted tool for mRNA expression monitoring in gene transcription analysis, their use in rapid diagnosis of bacterial and viral pathogens is only emerging. Recent applications include identification and genotyping of gram-positive bacteria [1], staphylococci [2], *Escherichia coli* [3], mycobacteria, *Listeria* spp., *Clostridium perfringens, Vibrio* spp., as well as neuro-invasive viruses.

In a recent study, we developed a microarray assay for detection and differentiation of *Chlamydia* spp. and *Chlamydophila* spp. [4]. We used the commercially available ArrayTube™ (AT) system which represents a less expensive platform allowing the use of both *in situ* synthesised and spotted DNA microarrays. It involves chips of 3.4 x 3.4 mm size, which are assembled to form the bottom of 1.5-ml plastic micro-reaction tubes. Due to this unique geometry, all steps of the hybridisation reaction including staining of duplexes can be conducted in the ArrayTube without the need for a separate hybridisation chamber. As visualisation of specific duplexes is based on enzyme-catalysed precipitation rather than fluorescence, a simple transmission reader can be used to measure signal intensities dynamically. This renders the system particularly user-friendly and economical.

Materials and Methods

Probe design

Species-specific and genus-specific probes for chlamydiae were designed on the basis of a multiple sequence alignment that included 16S rRNA genes, 16S-23S intergenic spacer regions and 23S rRNA genes from 44 chlamydial strains entered in the GenBank database (http://www.ncbi.nlm.nih.gov/entrez/query.fcgi?db=Nucleotide). To identify genomic segments of high discriminatory power within an existing alignment, we developed a sequence analysis algorithm designated the "most variable window approach". Using this procedure, species-specific nucleotide polymorphisms in a region of generally high sequence similarity in the 23S rRNA gene were identified.

Microarray fabrication

The selected 26-nt probe sequences were used on different versions of customized microarrays: (i) combinatorial high-density in situ synthesised arrays (one version, 1024 spots) and (ii) low-density spotted arrays in three versions.

Sample preparation

Target DNA was extracted from chlamydia-infected cell cultures or clinical specimens using the High Pure PCR Template Preparation Kit (Roche Diagnostics, Mannheim, Germany). Chlamydial DNA was amplified and biotin-labelled in 40 cycles of 94°C/30s, 60°C/60s, and 72°C/30s, using primers U23F-19 (5'-ATTGAMAGGCGAWGAAGGA-3') and 23R-22 (5'-biotin-GCYTACTAAGATGTTT-CAGTTC-3').

Hybridisation

One µl of heat-denatured PCR product in 99 µl hybridisation buffer (Hybridisation buffer 1, Clondiag Chip Technologies, Jena, Germany) was transferred to the pre-conditioned ArrayTube and incubated at 60 °C for 60 min. After washing and blocking of vacant binding sites (Blocking Reagent, Roche Diagnostics), hybridising spots were visualised using the Array Staining Kit (Clondiag), which includes a 3,3,'5,5'-tetramethyl benzidine (TMB) derivative as precipitation substrate for streptavidine-conjugated horseradish peroxidase to be coupled to the biotin-labelled DNA strand that hybridised to the array.

Microarray processing

Hybridisation signals were measured using the ATR-01 array tube reader (Clondiag) at 25°C for 10 min, recording one image per 20 s. A CCD camera integrated in the ATR-01 reader is used to monitor positive hybridization signals by kinetic measurement of the precipitation reaction at each spot via specific changes in red light transmission. Signal intensity data with local background correction were obtained for each spot of each picture using the Iconoclust software, version 2.3 (Clondiag).

Results

In a preliminary study, a panel of chlamydial strains from our own collection, which included all nine species of the genera *Chlamydia (C.)* and *Chlamydophila (Cp.)*, was tested on the in situ synthesised combinatorial array. As this array comprised all possible nucleotide variations (n=512) within the most variable window, analysis of hybridisation patterns allowed selection of the most efficient probes. The latter were included in further trials on a series of low-density spotted microarrays. The third version of the spotted microarray carried 28 probes for species identification, i.e. four probes each for *C. trachomatis, Cp. pneumoniae* and *Cp. psittaci*, three probes for *C. muridarum, Cp. pecorum, Cp. caviae* and *C. suis*, two probes for *Cp. abortus* and *Cp. felis*, three genus-specific probes, five probes for the closest relatives, i.e. *Simkania negevensis* and *Waddlia chondrophila*, as well as four positive controls (consensus probes), and one internal staining control (biotin marker). The number of species-specific probes chosen depended on the degree of intra-species genetic heterogeneity. Each probe was spotted five-fold, yielding a total of 289 spots. The results are presented in Figure 1.

To evaluate the sensitivity of the microarray assay [5], we examined decimal dilution series of a recombinant plasmid carrying the intergenic spacer and adjacent regions of the 16S and 23S rRNA genes of *C. trachomatis*, i.e. the insert included the target region of the microarray assay and primer binding sites of the consensus biotin PCR. Results showed that a species-specific hybridization pattern was obtained from a single copy of target DNA. However, the detection limit for chromosomal DNA of *C. trachomatis* was near 56 genomic copies or 1.87 inclusion-forming units (Fig. 2).

During a validation trial that included testing a number of clinical tissue samples for the presence of chlamydiae we repeatedly observed amplicon bands of apparently correct size on agarose gels from samples found negative by other tests such as nested PCR and real-time PCR, as well as in some non-template controls (Fig. 3). Notably, these PCR products failed to generate meaningful hybridisation signals in the microarray assay. To clarify this discrepancy, we sequenced these PCR products, and BLAST analysis of the 172-bp fragment revealed a complete match with the *E. coli* 23S rRNA gene and significantly less homology with the analogous sequences of *Chlamydia* and *Chlamydophila spp*. Subsequently, when testing commercially available thermostable DNA polymerases for residual *E. coli* DNA using *blaZ* and 23S rRNA PCR assays, 18 out of 21 were found to contain detectable amounts of DNA (data not shown).

Nevertheless, even the formation of such non-specific amplicons from truly negative samples did not hamper the performance of the microarray hybridisation assay for chlamydiae. As an example, the identification of *Cp. psittaci* from broncho-alveolar lavage of a patient suffering from psittacosis is shown in Figure 4.

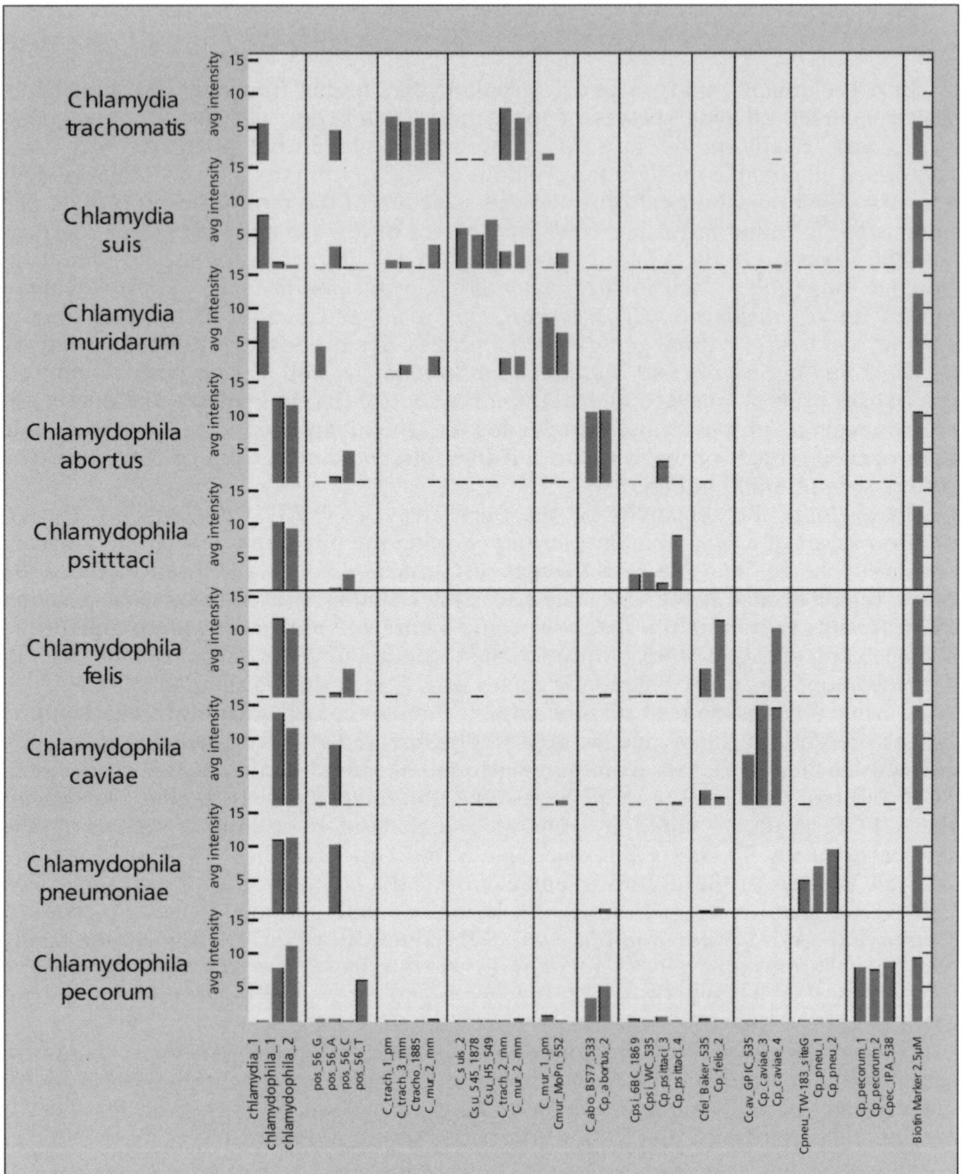

Fig. 1: Barplot showing specific identification of all nine species of *Chlamydia* and *Chlamydophila* using the present DNA microarray.

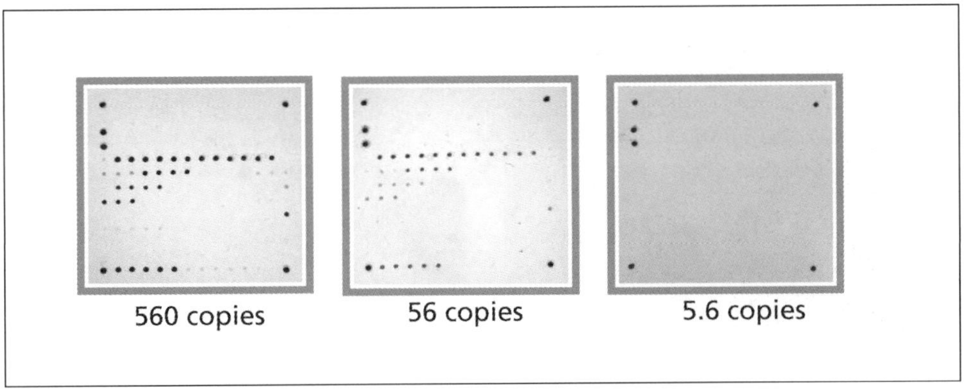

Fig. 2: Sensitivity of microarray detection. Hybridisation patterns of a dilution series of *C. trachomatis* genomic DNA showing the detection limit of 56 genome copies.

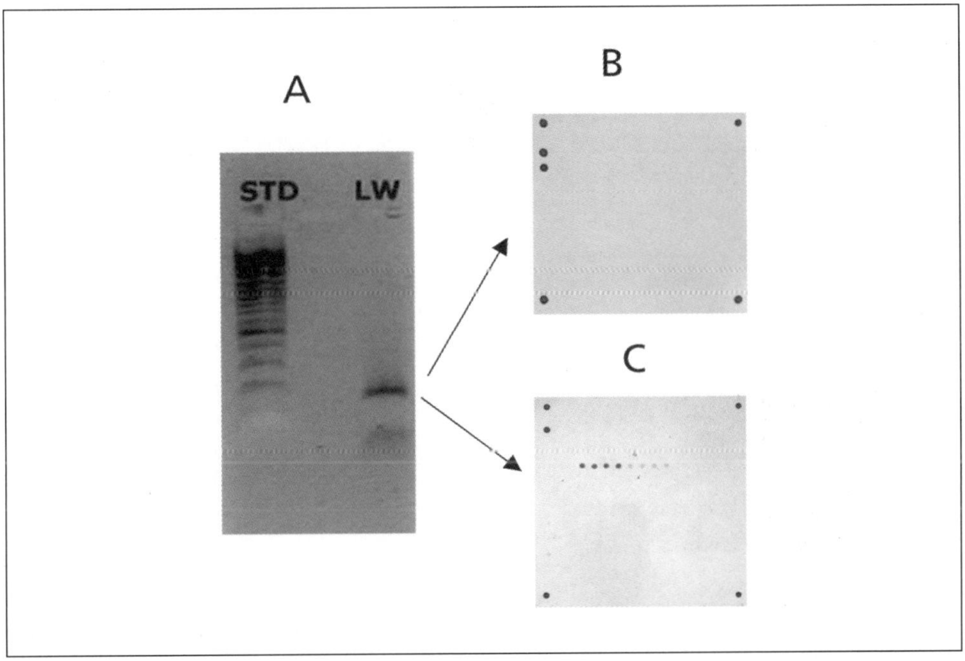

Fig. 3: A) Agarose gel electrophoresis of chlamydiae-negative samples showing a non-specific PCR product (same size as expected from a specific amplicon) due to the presence of residual *E. coli* DNA in thermostable DNA polymerase used for amplification. B) Hybridisation of this PCR product on the chlamydia microarray: The absence of specific hybridisation signals demonstrates that the sample was truly negative. C) Hybridisation of this PCR product on an *E. coli* microarray showing positive hybridisation signals.

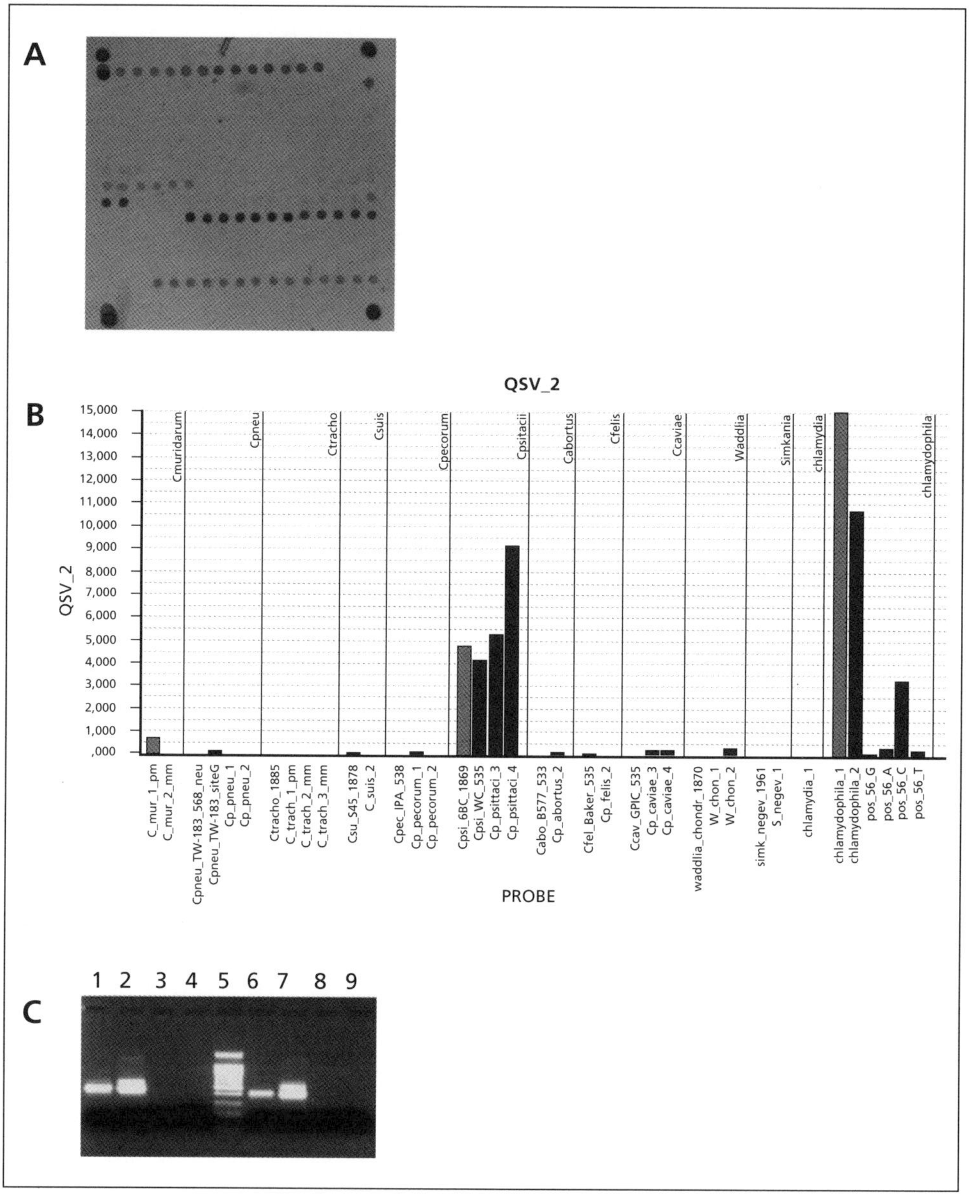

Fig. 4: Use of the DNA microarray assay for testing clinical samples. Identification of *Cp. psittaci* from bronchalveolar lavage of a patient suffering from psittacosis. A) Microarray hybridisation pattern, B) Barplot showing hybridisation signal intensities of the probes present on the microarray (highest intensity with *Cp. psittaci*-specific probes), C) Confirmation of the microarray result by nested *ompA*-PCR of the same sample: 1-4 *Chlamydiaceae*-specific amplification, 5 DNA size marker, 6-9 *C. psittaci*-specific amplification, 1 and 6 patient sample, 2 and 7 positive amplification controls, 3,4 and 8,9 negative (reagent) controls.

Discussion

We were able to define a panel of species-specific probes for all nine species of the family *Chlamydiaceae* and develop a DNA microarray that produced specific hybridisation patterns.

After proving the suitability of the present assay for unambiguous species identification of chlamydial cell cultures we used it for direct detection of chlamydiae from clinical tissue. As many clinical samples contain only low numbers of bacteria, sensitivity becomes the crucial parameter in this application.

In a series of experiments we were able to show that the sensitivity of the microarray assay was equivalent to that of real-time PCR, thus rendering it suitable for use in the diagnostic laboratory [5]. The fact that we found a lower detection limit for plasmid DNA (carrying a defined genomic insert of chlamydia) compared to extracted chlamydial chromosomal DNA in both microarray and real-time PCR indicates that, in the case of chromosomal DNA, the actual number of target copies available for amplification was lower than the number of genome copies present in the sample. We assume that this is due to cleavage and partial degradation effects during pre-PCR DNA extraction. Interestingly, these findings imply that in PCR detection tests 50-100 genome copies need to be present in a clinical sample for one copy to be amplified. In any case, the inferior sensitivity of detection in the case of chromosomal DNA was not associated with the hybridisation or visualisation reactions on the AT microarray.

The truly "two-dimensional" specificity of the present DNA microarray assay became obvious when we were confronted with the occurrence of apparently correct amplicons from chlamydiae-negative samples in the pre-hybridisation PCR. Using DNA sequencing we were able to demonstrate that the false-positive amplicons detected by agarose gel electrophoresis were from residual *E. coli* DNA in thermostable DNA polymerase. This problem was described more than once in the early years of the emergence of PCR as a diagnostic technique. However, the present findings reminded us that most of the commercially available polymerases for amplification tests are not entirely free of DNA stemming from the recombinant production strain. Although the DNA microarray test was not hampered by this phenomenon, it is important to note that in the development of PCR detection assays for eubacterial species, primer matches with *E. coli* sequences should be avoided to ensure specificity.

We conclude that the present DNA microarray assay and the ArrayTube™ platform in general have the potential to be used in routine diagnosis.

References

1 Perreten V, Vorlet-Fawer L, Slickers P, Ehricht R, Kuhnert P, Frey J: Microarray-based detection of 90 antibiotic resistance genes of gram-positive bacteria. J Clin Microbiol 2005;43:2291-2302.

2 Monecke S, Ehricht R: Rapid genotyping of methicillin-resistant Staphylococcus aureus (MRSA) isolates using miniaturised oligonucleotide arrays. Clin Microbiol Infect 2005;11:825-833.

3 Korczak B, Frey J, Schrenzel J, Pluschke G, Pfister R, Ehricht R, Kuhnert P: Use of diagnostic microarrays for determination of virulence gene patterns of Escherichia coli K1, a major cause of neonatal meningitis. J Clin Microbiol 2005;43:1024-1031.

4 Sachse K, Hotzel H, Slickers P, Ellinger T, Ehricht R: DNA microarray-based detection and identification of Chlamydia and Chlamydophila spp. Mol Cell Probes 2005;19:41-50.

5 Ehricht R, Slickers P, Goellner S, Hotzel H, Sachse K: Optimized DNA microarray assay allows detection and genotyping of single PCR-amplifiable target copies. Mol Cell Probes 2006;20:60-63.

Dr. K. Sachse, Friedrich-Loeffler-Institut, Institut für Bakterielle Infektionen und Zoonosen, Naumburger Str. 96a, DE-07743 Jena, Germany.
E-mail: konrad.sachse@fli.bund.de.

Vannier P, Espeseth D (eds): New Diagnostic Technology: Applications in Animal Health and Biologics Controls. Dev Biol (Basel). Basel, Karger, 2006, vol 126, pp 211-212.

Molecular and Biological Method for Detection and Confirmation of Different Fish Viruses

S.M. Bergmann, D. Fichtner

Friedrich-Loeffler-Institut, National Reference Laboratory for Fish Diseases, Greifswald – Insel Riems, Germany

Introduction

EU legislation allows, in contrast to OIE recommendations, the use of virological and serological methods as cell cultivation, antigen ELISA, neutralization assays or immunofluorescence for detection and confirmation of fish disease agents only. First the lack of cell cultures, detection and confirmation methods in the case of Infectious Salmon Anaemia virus (ISAV), occurring in Atlantic salmon (*Salmo salar*), led to a situation where e.g. RT-PCR was included in the legislation. Fast, high-sensitive and specific RT-PCRs or PCRs, respectively according to nucleic acids, for detection and confirmation of salmonid harming agents as Viral Haemorrhagic Septicaemia virus (VHSV), Infectious Haematopoietic Necrosis virus (IHNV) or emerging cyprinid diseases as Spring Viremia of Carp virus (SVCV) or even Koi Herpesvirus (KHV) or other disease agents as Epizootic Haematopoietic Necrosis virus (EHNV) in different percid species have recently not been included because of a lack of method validation.

Development and Testing

We developed, took over and established a panel of molecular biological assays, e.g. PCR, RT-PCR, nested PCR or in-situ hybridization, for detection, differentiation and confirmation of these and other viruses, bacteria and parasites in the German National Reference Laboratory for Fish Diseases and combined them with animal experiments.

The safest detection method for ISAV consists of a combination of RT-PCR recognizing different genes, e.g. NS gene, or for differentiation of European and American viruses in HA or NA genes and cell cultivation in SHK-1 or AS cells.

Beside the use of classical-virological and biological techniques, RT-PCRs for the detection and differentiation of all genotypes of VHSV, IHNV and SVCV, based on published and our own sequences of the G and N genes of these viruses, were carried out in parallel with the aim of validation. These results were confirmed by immunofluorescence using commercially available and our own monoclonal antibodies against these viruses.

For KHV detection it is very difficult to cultivate the virus in cell cultures. The only possible method consists of the application of different PCR procedures. The confirmation can be done by sequencing of the amplicons, immunofluorescence with polyclonal antisera or monoclonal antibodies, immunohistochemistry also with polyclonal antisera, monoclonal antibodies or in-situ hybridization with probes recognising different genes of the viruses. Currently there is no safe method for detection of latently infected animals which seem to be healthy.

Current Policy to Eradicate Viral Fish Diseases

Based on Council Directive 93/53/EEC (24 June 1993 introducing minimum Community measures for the control of certain fish diseases), viral fish diseases listed in CD 91/67/EEC are:

- ISAV (list 1)
- IHNV (list 2)
- VHSV (list 2)
- SVCV (list 3)

EHNV is a listed disease by the OIE but not by the EU; KHV is not listed at all but is a threat for the cyprinid industry all over the world.

The aim of all the measurements is to combat certain fish diseases effectively, to avoid their spread in the Community.

References

Anonymous: Concil Directive 91/67/EEC, 1991.

Anonymous: Concil Directive 93/53/EEC, 1993.

Anonymous OIE: International Aquatic Animal Health Code and Diagnostic Manual for Aquatic Animal Diseases, 2000-2003.

Dr. S.M. Bergmann, Friedrich-Loeffler-Institut, National Reference Laboratory for Fish Diseases, Boddenblick 5a, DE-17493 Greifswald – Insel Riems, Germany.
E-mail: bergmann@rie.bfav.de

Vannier P, Espeseth D (eds): New Diagnostic Technology: Applications in Animal Health and Biologics Controls. Dev Biol (Basel). Basel, Karger, 2006, vol 126, pp 213-218.

Development of a New Integrated Diagnostic Test for Identification and Characterization of Pathogens

G. Martinez[1], *G. Bruant*[1], *R. Brousseau*[2], *L. Masson*[2], *J. Harel*[1]

[1] Groupe de Recherche sur les Maladies Infectieuses du Porc, Faculté de Médecine vétérinaire, Université de Montréal, Saint-Hyacinthe, Qc, Canada

[2] Biotechnology Research Institute, National Research Council of Canada, Montreal, Qc, Canada

Abstract: Animal diseases directly cause multi-million dollar losses world-wide. Therefore a rapid, highly specific, cost-effective diagnostic test for detecting a large set of bacterial virulence and antimicrobial resistance genes simultaneously is necessary. Hence, our group, the BCBG (Bacterial Chips Bacterial Genes) group, proposes **developing a powerful molecular tool (DNA microarray) to detect a broad range of infectious agents, their endogenous main virulence factors and antibiotic resistance genes simultaneously**. Effectively, a 70-mer oligonucleotide microarray capable of detecting the presence or absence of 169 *Escherichia coli* virulence genes or virulence marker genes as well as their variants, in addition to 30 principal antimicrobial resistance genes previously characterized in *E. coli* strains was developed by our group. This microarray was validated with a large collection of well characterized pathogenic and reference *E. coli* strains. Moreover, we are developing a new powerful clinical diagnostic microarray tool, to identify pathogenic bacteria of veterinary interest. The commercialization of this assay would allow same day diagnosis of infectious agents and their antibiotic resistance resulting in early treatment. In addition, this technology is also applicable to microbial quality control of food and water.

INTRODUCTION

Animal diseases directly cause multi-million dollar losses world-wide. Therefore a rapid, highly specific, cost-effective diagnostic test to detect a plethora of bacterial virulence and antibiotic resistance genes simultaneously is necessary. The development of a micro-array-based methodology for analysis of clinical specimens would provide a rapid means of directly identifying, at the same time, the most important agents associated with particular pathologies without the need for conventional culture or diagnostic methods that would be cost-prohibitive. Further, the specific virulence or antimicrobial resistance gene profiles of certain pathogens could be concomitantly ascertained. These data could provide important epidemiological information for tracking the source and spread of particular infectious diseases and determine the novel or emerging pathotypes of particular pathogens. Additionally, the detection of the antimicrobial resistance gene make-up based on microarray analyses can provide a quick and guided choice of potential therapeutic treatments of certain bacterial infections that are becoming increasingly resistant to multiple antibiotics. The concept and principle of using microarrays for diagnostics have already been validated on some bacterial pathogens by our work and by that of others [1-4]. The BCBG group proposes to develop a DNA microarray to detect a broad range of infectious agents, their endogenous main virulence factors and antibiotic resistance genes simultaneously. The results obtained with the *E. coli* «atb-vir» microarray, in which a chip is able to detect the known virulence genes of *Escherichia coli* as well as the most importantly acquired antibiotic resistance genes, which was developed by our group, are presented here. We hope to use this model to develop other microarrays.

Materials and Methods

The microarray is an earlier prototype of that developed by our group and is based on a 70-mer oligonucleotides capable of detecting the presence or absence of: (i) 169 *Escherichia coli* virulence genes or virulence marker genes as well as their variants and (ii) 30 principal antimicrobial resistance genes belonging to six antimicrobial families (Beta-lactams, aminoglycosides, tetracylines, phenicols, trimethoprim, sulfonamides) previously characterized in *E. coli* strains [5]. In addition, the negatives (*gfp* and a gene from *Arabidopsis thaliana*) and positive controls such as *lacZ* and *lacY* were included. The oligonucleotides were printed in triplicate for a technical report.

The *E. coli* DNA was directly labelled with Cy5-dCTP, hybridized overnight at 50°C and analysed with ScanArray Express software. Furthermore, biological and technical replicates were carried out. The reference strains: MG1655, EDL933 and CFT073; 26 human and animal extra-intestinal strains resistant to at least one antimicrobial and 18 well-characterized strains belonging to various pathotypes were used to validate the microarray.

Two techniques, membrane hybridization and PCR were used to confirm or negate the microarray results.

Results

This microarray was validated with a large collection of well characterized pathogenic and reference *E. coli* strains. In general, a good correlation between membrane hybridization and microarray results was observed (Tables 1 and 2). Just a few discrepancies were detected, and in most of the cases PCR and DNA sequencing confirmed the microarray results. Only two false-positive results for class 1 integron and *bfpA* and one false-negative result for *eae* gamma variant were obtained (Tables 1 and 2). New oligonucleotides were designed for these cases and added in a new prototype which was validated. No positive hybridization with the negative control strain MG1655 for antibiotic oligonucleotides was obtained (Table 1).

Excellent reproducibility was obtained with our microarray. Effectively, identical results were obtained for two hybridizations at two different times for each strain (biological and technical replicates) (Fig. 1).

Table 1: Determination of the virulence profile of the sequenced *E. coli* used for the validation of the virulence oligonucleotides.

Strain identification	Pathotypes[a]	Microarray results	Reference or source
EDL933[b]	EHEC (O157H7)	*stx1A, stx1B, stx2A, stx2B-1, eae, espA2, espB1, tir2, ehxA, paa, wzy(O157H7), artJ, astA, ccdB, chuA, csgE, Ecs1282, efa1, espP, etpD, fepC, fimA, fimH, fliC, gad, hlyE, ibeB, iha, katP, l7095, lacY-Ec, lacZ, lpfA(O157), mviM, mviN, ompA, ompT, rfbE, rtx, set, tlrA, tnaA, uidA, ureD*	Perna et al [7]
CFT073[b]	UPEC	*papA(7-1), papA(7-2), papC, papGII, chuA, fepC, irp1, irp2, fyuA, iroN, usp, agn43, artJ, astA, b1121, b1432, cela, csgE, fimA, fimH, fliC, focA, gad, hlyA, hlyE, ibeB, iha, iss, iucD, kpsM-II, lacY-Ec, lacZ, malX, mchB, mviM, mviN, ompA, ompT, pic, sat, sfaA, tnaA, uidA, wzx(O6), yjaA*	Welch et al [8]
MG1655[b]	K12	*agn43, artJ, b1121, csgE, fimA, fimH, fliC, gad, hlyE, ibeB, lacY-Ec, lacZ, mviM, mviN, ompA, ompT, tnaA, uidA, yjaA*	Blattner et al [9]

[a] a pathotype is attributed to clinical strains according to their set of virulence genes or markers: EHEC (shiga-like toxin encoding genes, genes from the LEE, *ehxA*), EPEC (genes from the LEE, *bfpA*), ETEC (heat-stable and heat-labile toxin encoding genes, F4 and F18 fimbriae encoding genes), EIEC (*ipaB, invX, senB*), EAEC (*capU, shf, virK*, aggregative adherence fimbriae encoding genes), DAEC (*aidaI*), UPEC (P pili encoding genes, *chuA, fepC, irp1, irp2, fyuA, iroN, usp*), NMEC (*ibeA, neuA, neuC*), APEC (P pili encoding genes variants *papA11, papGII and papGIII, tsh, iutA, iucD, iroN*).

[b] reference strains for which the genome has been entirely sequenced [7-9].

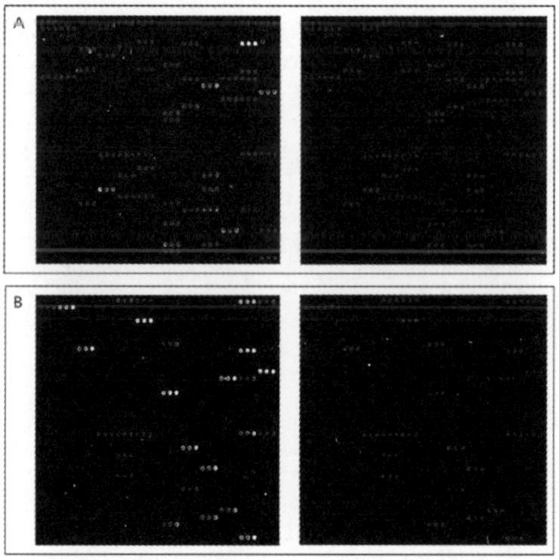

Fig. 1: Examples of the reproducibility obtained with our oligonucleotide microarray. Two biological replicates performed at different times with the *E. coli* reference strains EDL933 (A) and MG1655 (B).

Table 2: Determination of the antimicrobial resistance profile of 26 clinical *E. coli* extra-intestinal isolates used for the validation of the antimicrobial resistance oligonucleotides.

Strain identification	Origin	Microarray results[a, b]	Membrane hybridizations[a, b]
01-6288-1122	pig	*aadA1*[d], *tet*(A), *sulI*, class 1 integron	*tet*(A), ***tet*(C)**, *sulI*, class 1 integron
01-8344-0611	caribou	bla_{TEM}, *aadA1*[d], *tet*(A), *tet*(B), *dhfrI*, *sulI*, *sulII*, class 1 integron	bla_{TEM}, *tet*(A), *tet*(B), ***tet*(C)**, *dhfrI*, ***dhfrXV***, *sulI*, *sulII*, class 1 integron
EcL1329 (01-D913)[c]	pig	bla_{TEM}, *aadA1*[d], *aphA1*, *tet*(A), *dhfrV*, *sulII*, ***class 1 integron***	bla_{TEM}, *aphA1*, *tet*(A), ***tet*(C)**, *dhfrV*, *sulII*
EcL10038 (02-779-175)[c]	pig	bla_{TEM}, *aadA1*[d], *tet*(A), *sulI*, class 1 integron	bla_{TEM}, *tet*(A), ***tet*(C)**, *sulII*, class 1 integron
EcL10040 (02-839)[c]	pig	*aadA1*[d], *tet*(A), *sulI*, class 1 integron	*tet*(A), ***tet*(C)**, *sulI*, class 1 integron
EcL10044 (02-1273-175)[c]	cattle	bla_{TEM}, *tet*(A)	bla_{TEM}, *tet*(A), ***tet*(C)**
EcL1335 (02-1902-2)[c]	pig	bla_{TEM}, *aadA1*[d], *aphA1*, *tet*(A), *dhfrV*, *sulI*, *sulII*, class 1 integron	bla_{TEM}, *aphA1*, *tet*(A), ***tet*(C)**, *dhfrV*, *sulI*, *sulII*, class 1 integron
02-1926	avian	bla_{TEM}, *aadA1*[d], *tet*(A), *sulI*, *sulII* class 1 integron	bla_{TEM}, *tet*(A), ***tet*(C)**, *sulI*, *sulII*, class 1 integron
EcL10106 (02-1940-1)[c]	pig	bla_{TEM}, *aadA1*[d], *aphA1*, *tet*(A), *dhfrV*, *sulI*, *sulII*, class 1 integron	bla_{TEM}, *aphA1*, *tet*(A), ***tet*(C)**, *dhfrV*, *sulI*, *sulII*, class 1 integron
02-6386	cattle	bla_{TEM}, *tet*(A), *sulII*	bla_{TEM}, *tet*(A), ***tet*(C)**, *sulII*
1182768	human	bla_{TEM}, *aadA1*[d], *tet*(A), *sulI*, class 1 integron	bla_{TEM}, *tet*(A), ***tet*(C)**, *sulI* class 1 integron
67101013-1	human	bla_{TEM}, *aadA1*[d], *tet*(A), *catI*, *sulI*, class 1 integron	bla_{TEM}, *tet*(A), ***tet*(C)**, *catI*, *sulI*, class 1 integron
67170713-3	human	*aadA1*[d], *tet*(A), *sulI*, class 1 integron	*tet*(A), ***tet*(C)**, *sulI*, class 1 integron
67170713-4	human	*aadA1*[d], *tet*(A), *sulI*, class 1 integron	*tet*(A), ***tet*(C)**, *sulI*, class 1 integron
67170751-1	human	bla_{TEM}, *aadA1*[d], *aadB1*[d], *aphA1*, *tet*(A), *sulI*, *sulII*, class 1 integron	bla_{TEM}, *aphA1*, *tet*(A), ***tet*(C)**, *sulI*, *sulII*, class 1 integron
67300471-1	human	bla_{TEM}, *aadA1*[d], *tet*(A), *dhfrI*, *sulI*, *sulII*, class 1 integron	bla_{TEM}, *tet*(A), ***tet*(B)**, ***tet*(C)**, *dhfrI*, ***dhfrXV***, *sulI*, *sulII*, class 1 integron
67310973-1	human	*aadA1*[d], *tet*(A), *sulI*, class 1 integron	*tet*(A), ***tet*(C)**, *sulI*, class 1 integron
68030679-1	human	bla_{TEM}, *tet*(A), *sulI*, **class 1 integron**	bla_{TEM}, *tet*(A), ***tet*(C)**, *sulI*
68070861-1	human	bla_{TEM}, *tet*(A), *tet*(B), *catI*, *sulII*	bla_{TEM}, $\boldsymbol{bla_{SHV}}$, *tet*(A), *tet*(B), ***tet*(C)**, *catI*, *sulII*
68080537-1	human	*aadA1*[d], *tet*(A), *sulI*, class 1 integron	*tet*(A), ***tet*(C)**, *sulI*, class 1 integron
68090616-1	human	*aadA1*[d], *tet*(A), *sulI*, class 1 integron	*tet*(A), ***tet*(C)**, *sulI*, class 1 integron
EcL8134 (98-2453)[c]	dog	*tet*(A)	*tet*(A), ***tet*(B)**, ***tet*(C)**
EcL1344 (P01-8342-1)[c]	pig	bla_{TEM}, *tet*(A), *sulII*	bla_{TEM}, *tet*(A), ***tet*(C)**, *sulII*
EcL1332 (P01-8342-é)[c]	pig	bla_{TEM}, *tet*(A), *sulII*	bla_{TEM}, *tet*(A), ***tet*(C)**, *sulII*
EcL1333 (P02-014-1)[c]	pig	bla_{TEM}, *aadA1*[d], *tet*(A), *sulI*, class 1 integron	bla_{TEM}, *tet*(A), ***tet*(B)**, ***tet*(C)**, *sulI*, class 1 integron
EcL1340 (P02-014-2)[c]	pig	bla_{TEM}, *aadA1*[d], *tet*(A), *sulI*, class 1 integron	bla_{TEM}, *tet*(A), ***tet*(C)**, *sulI* class 1 integron

[a] all antimicrobial resistance genes targeted in this study are described in [5].

[b] discrepancies are indicated in bold characters. For these genes, PCR and DNA sequencing were performed and validated the microarray results.

[c] previous nomenclature.

[d] gene not tested by membrane hybridization.

Conclusion

The results confirm that our oligonucleotide microarray is suitable for the detection of multiple antimicrobial resistance and virulence genes in *E. coli* isolates. This oligonucleotide microarray thus represents a powerful and valuable tool for the assessment of the pathotype of an *E. coli* strain and its genetic characterization and should find widespread use in diagnostic studies, for monitoring antimicrobial resistance among *E. coli* strains, as well as for environmental, epidemiological and phylogenetic studies [5].

Perspective

We are now developing a powerful new clinical diagnostic microarray tool to identify pathogenic bacteria of veterinary interest, their principal virulence genes or genes which are markers of virulence together with most antimicrobial genes. The use of this array would allow same day diagnosis of infectious agents and their antibiotic resistance resulting in early treatment. In addition to veterinary and medical diagnosis, this technology is also applicable to the microbial quality control of food and water [6].

Acknowledgements

We thank Miria Elias and Tracey Rigby for their excellent technical support.

References

1 Bekal S, Brousseau R, Masson L, Prefontaine G, Fairbrother J, Harel J: Rapid identification of *Escherichia coli* pathotypes by virulence gene detection with DNA microarrays. J Clin Microbiol 2003;41:2113-2125.

2 Perreten V, Vorlet-Fawer L, Slickers P, Ehricht R, Kuhnert P, Frey J: Micro-array-based detection of 90 antibiotic resistance genes of gram-positive bacteria. J Clin Microbiol 2005;43:2291-2302.

3 Sergeev N, Volokhov D, Chizhikov V, Rasooly A: Simultaneous analysis of multiple staphylococcal enterotoxin genes by an oligonucleotide microarray assay. J Clin Microbiol 2004;42:2134-2143.

4 van Ijperen C, Kuhnert P, Frey J, Clewley JP: Virulence typing of *Escherichia coli* using microarrays. Mol Cell Probes 2002;16:371-378.

5 Bruant G, Maynard C, Bekal S, Gaucher I, Masson L, Brousseau R, Harel J: Development and validation of an oligonucleotide microarray for the detection of antimicrobial resistance genes in *Escherichia coli*. Appl Environ Microbiol 2006;72(5): In Press.

6 Maynard C, Berthiaume F, Lemarchand K, Harel J, Payment P, Bayardelle P, Masson L, Brousseau R: Waterborne Pathogens Detection Using Oligonucleotide-Based Microarrays. Appl Environ Microbiol 2005;71:8548-8557.

7 Perna NT, Plunkett G, 3rd, Burland V, Mau B, Glasner JD, Rose DJ, Mayhew GF, Evans PS, Gregor J, Kirkpatrick HA, Posfai G, Hackett J, Klink S, Boutin A, Shao Y, Miller L, Grotbeck EJ, Davis NW, Lim A, Dimalanta ET, Potamousis KD, Apodaca J, Anantharaman TS, Lin J, Yen G, Schwartz DC, Welch RA, Blattner FR: Genome sequence of enterohaemorrhagic *Escherichia coli* O157:H7. Nature 2001;409:529-533.

8 Welch RA, Burland V, Plunkett G, 3rd, Redford P, Roesch P, Rasko D, Buckles EL, Liou SR, Boutin A, Hackett J, Stroud D, Mayhew GF, Rose DJ, Zhou S, Schwartz DC, Perna NT, Mobley HL, Donnenberg MS, Blattner FR: Extensive mosaic structure revealed by the complete genome sequence of uropathogenic *Escherichia coli*. Proc Natl Acad Sci U S A 2002;99:17020-17024.

9 Blattner FR, Plunkett G, 3rd, Bloch CA, Perna NT, Burland V, Riley M, Collado-Vides J, Glasner JD, Rode CK, Mayhew GF, Gregor J, Davis NW, Kirkpatrick HA, Goeden MA, Rose DJ, Mau B, Shao Y: The complete genome sequence of *Escherichia coli* K-12. Science 1997;277:1453-1474.

Dr. J. Harel, Groupe de Recherche sur les Maladies Infectieuses du Porc, Faculté de Médecine Vétérinaire, Université de Montréal, Saint-Hyacinthe, Québec J2S 7C6, Canada.
E-mail : josee.harel@umontreal.ca

Vannier P, Espeseth D (eds): New Diagnostic Technology: Applications in Animal Health and Biologics Controls. Dev Biol (Basel). Basel, Karger, 2006, vol 126, pp 219-226.

Validation of an Internal Method for the Diagnosis of Infections with *Chlamydophila abortus* and *Coxiella burnetii* by Real-Time Multiplex PCR

C. Pelletier[1], S. Chartier[2], J. Berthillier[1], E. Dégletagne[1], C. Rigaud[1], H. Berthet[1], A. Valognes[2], A. Reynaud[2], P. Véry[1]

[1] Laboratoire Départemental d'Analyses de Saône-et-Loire, Mâcon, France
[2] Laboratoire Départemental d'Analyses de l'Allier, Moulins, France

Key words: *Chlamydophila abortus, Coxiella burnetii*, real-time multiplex PCR, abortions, cattle, validation.

Abstract: *Chlamydophila abortus* and *Coxiella burnetii* are one of the major pathogens implicated in abortion in cattle. Their characteristic of obligate intracellular bacteria, and of zoonotic agents, makes their culture difficult in diagnostic laboratories, and the traditional tools of diagnosis (detection of sera antibodies by ELISA, Stamp's coloration) encounter specificity, sensitivity and interpretability limits. Individual PCR have recently been developed. Nevertheless, their income/cost is a limiting factor for breeders. As the symptoms are not specific, the request for analysis often concerns the two valences. Consequently, the development and the validation of an internal multiplex PCR appears to be a suitable solution.

INTRODUCTION

Although often suspected in the case of serial abortions in domestic ruminants, the diagnosis of certainty of the infections with *Coxiella burnetii* and *Chlamydophila abortus* until recently were limited to specialized research laboratories. With the recent provision of individual commercial PCR, the veterinary laboratories can now propose a significant, specific and reliable diagnosis. However, the analytical cost is still a limiting factor owing to the fact that the request for analysis generally associates the two valences, because of a similar clinical expression (epizootic late abortions), and also because of a possible association of the two pathogenic agents.

We then decided to develop and validate a real-time multiplex PCR associating three valences: *Coxiella burnetii, Chlamydophila abortus* and GAPDH as internal positive control. After having characterized the multiplex PCR in term of specificity, fidelity and applicability, we compared it with two individual PCR [1,2] developed by the team of A.Rodolakis, INRA, to check its accuracy and to determine its relative threshold of detection. Lastly, we compared the results of the multiplex PCR with those obtained in Stamp's coloration or in detection of sera antibodies by ELISA.

Materials and Methods

The reference PCR described by the team of A.Rodolakis [1,2] are respectively targeting *htp*AB-associated repetitive element *IS111* for the detection of *C.burnetii* and *pmp* gene for the detection of *C.abortus, psittaci* and *caviae*. The triplex PCR developed by LDA71 on ABI PRISM 7000, co-amplifies specific fragments of the repetitive element *IS111* for *C.burnetii* (detector VIC), *omp2* gene for *C.abortus* [3] (detector FAM) and a common sequence of *GAPDH* gene (detector NED) from *Bos taurus, Capra hircus* and *Ovis aries*. Real-time PCR is performed with QuantiTect Multiplex PCR kit (Qiagen) in a final volume of 25µl during 45 cycles as described in the handbook. Cycle threshold (C_t) is used for qualitative interpretation: C_t under 40 is considered as Positive, and a C_t equal to or above 40 is considered as a Negative result.

The triplex PCR was first characterized in terms of specificity on 18 positive samples (LDA03 collection) and 36 bacterial strains. Its fidelity was also verified on 20 positive samples (10 for each valence) with a different level of intensity (low, medium and high). The same samples were analysed in duplicates in the same try (repeatability), or in another run by a different operator (run and laboratory effect), or by LDA03 (in-between laboratory effect).

The triplex PCR was then validated by comparison with the two individuals PCR used by LDA03 and considered as reference method. Its accuracy was determined on 40 samples tested in duplicates and formerly found positive (10 samples for each valence) or negative (20 samples) with reference PCR. The relative threshold of detection of each PCR was determined on successive dilutions of three positive samples for *C.burnetii*, three positive samples for *C.abortus* and one sample naturally co-infected by the two bacteria.

Lastly, the multiplex PCR was compared with more traditional tools of diagnosis such as Stamp's coloration and ELISA (Intervet-Bommeli).

Results

The specificity of the multiplex PCR was tested and checked with respect to 36 bacterial strains. These strains were isolated from placenta, vaginal swabs and milk (*Arcanobacterium pyogenes, Bacillus sp., Brucella abortus, Clostridium perfringens, E.coli, Listeria monocytogenes, Mycoplasma sp., Proteus sp., Pseudomonas sp., Salmonella Typhimurium, Staphylococcus aureus, Streptococcus uberis*) or are close to the genus *Coxiella* or *Chlamydophila* on the genetic level (*Legionella pneumophila*). For each of these strains a negative result was found for the three valences of the multiplex PCR.

To check the specificity of the *C.abortus* target more particularly, we tested various strains belonging to the *Chlamydophila* genus. The concentrations in total DNA being relatively equivalent (about 10^6 copies of genomes per PCR), a comparison of the C_t presented in Table 1 is thus possible. We noted that for the three identified strains *C.abortus* and a mixed strain *C.pecorum* + *C.abortus*, C_t obtained are low (Ct from 19.82 to 22.13 for the mixed strain). The PCR *C.abortus* seems on the other hand to be less effective with respect to strains very close to *C.abortus* but isolated from other animal species: *C.psittaci* in the birds (C_t from 25.67 to 34.20) and *C.caviae* in the guinea-pig (C_t 25.69). In an interesting way, the results obtained for the various identified strains *C.pecorum* are either negative, or close to the negative threshold. For one strain, however, a lower C_t (34.31) was obtained. This strain, tested with the PCR of reference *C.abortus, C.psittaci* and *C.caviae,* provides a slightly positive result. A contamination of the initial culture by *C.abortus* thus seems possible.

The applicability of the multiplex PCR was checked on 18 samples resulting from the collection of the LDA03 and found positive for one or both valences with the PCR of reference. The diversity of the samples provided by the LDA03 in term of animal species (bovine, caprine, ovine) and of origins (vaginal swabs, placenta, milk) initially made it possible to validate internal positive control selected to check the integrity of the samples, the quality of the DNA extract and the absence of inhibitors of the PCR. Indeed, a positive result was obtained for target GAPDH on the totality of the samples tested.

Table 1: C_t obtained with the valence *C.abortus* of the multiplex PCR for different strains belonging to the *Chlamydophila* genus (no signal of amplification was observed with the valences *C.burnetii* and GAPDH – data not shown)

Isolate	**Host**	**Identification**	C_t ***C.abortus***
AB10-14632		*C.pecorum*	34.31*
AC1-14154		*C.abortus*	20.86
AB7-10402		*C.abortus*	19.82
Loth	Bird	*C.psittaci*	30.31
VS1	Bird	*C.psittaci*	31.46
CP3	Bird	*C.psittaci*	30.44
L2A	Bird	*C.psittaci*	34.20
GR9	Bird	*C.psittaci*	25.67
NJ1	Bird	*C.psittaci*	29.29
TT3	Bird	*C.psittaci*	27.20
Cal 10	Bird	*C.psittaci*	29.11
VS225	Bird	*C.psittaci*	31.77
C8	Swine	*C.suis*	>40
C14	Bovine	*C.pecorum*	39.22
C18	Sheep	*C.abortus*	21.43
DC13	Calf	*C.pecorum*	38.62
DC25	Guinea-pig	*C.caviae*	25.59
DC34	Bovine	*C.abortus*	20.08
DC39	Mouse	*C.muridarum*	37.30
CM-1ATCC VR1360	Human	*C.pneumoniae*	Undet.
ZC53	Sheep	*C.pecorum+ C.abortus*	22.13
W73	Sheep	*C.pecorum*	39.35
C95/38	Sheep	*C.pecorum*	Undet.

* For this particular isolate, a positive result was found with the reference PCR *C.abortus, C.psittaci, C.caviae.*

With regard to the targets *Chlamydophila abortus* and *Coxiella burnetii*, a positive result was obtained for the whole of the initially positive samples with the respective PCR of reference. Two naturally co-infected samples were also confirmed positive for the two valences. On the other hand, a sample, found only positive for *Chlamydophila abortus* by the LDA03 with the reference PCR, proved also to provide a positive result for *Coxiella burnetii* with the multiplex PCR (C_t 32.51). Two hypotheses could explain such a result: a lack of specificity of the multiplex PCR or a lack of sensitivity of the PCR of reference *Coxiella burnetii* for values of C_t close to 32.51. This second assumption seems most probable if one refers on the one hand to the results obtained with the multiplex PCR on the various bacterial strains, and on the other hand with the results obtained with four particular samples. These four samples initially provided negative results in PCR of reference *Coxiella burnetii* but were found positive with a commercial real-time PCR (ADIAVET Cox, AES Laboratoires). A positive result was found with the multiplex PCR with C_t of the same order as those found with the commercial kit (C_t from 31.08 to 34.80).

The specificity of the three targets of the multiplex PCR is thus checked on the collection of positive samples from the LDA03.

The fidelity of the PCR was studied on 10 samples positive for each valence *C.abortus* and *C.burnetii*, the samples being distributed in three levels of intensity (low, medium or high). These samples were analysed in duplicates in the same series of analysis (repeatability), at least once in another series of analysis by a different manipulator (run and laboratory effect), and once at the LDA03 (in-between laboratory effect). Figures 1 and 2 represent the C_t obtained for each sample under the various conditions of variability. It is noted that for each target and each level of intensity (low, medium, high), the values of C_t obtained in conditions of repeatability or reproducibility are at an interval of low amplitude. The fidelity of the multiplex PCR can be also materialized by the coefficients of variation shown in Tables 2 and 3.

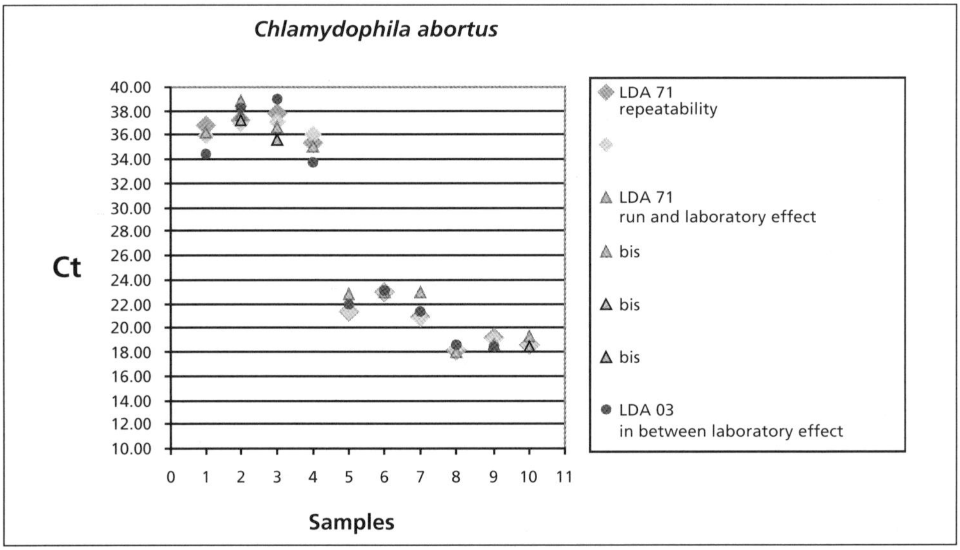

Fig. 1: Dispersion of the values of C_t in conditions of repeatability and reproducibility for the *C.abortus* target.

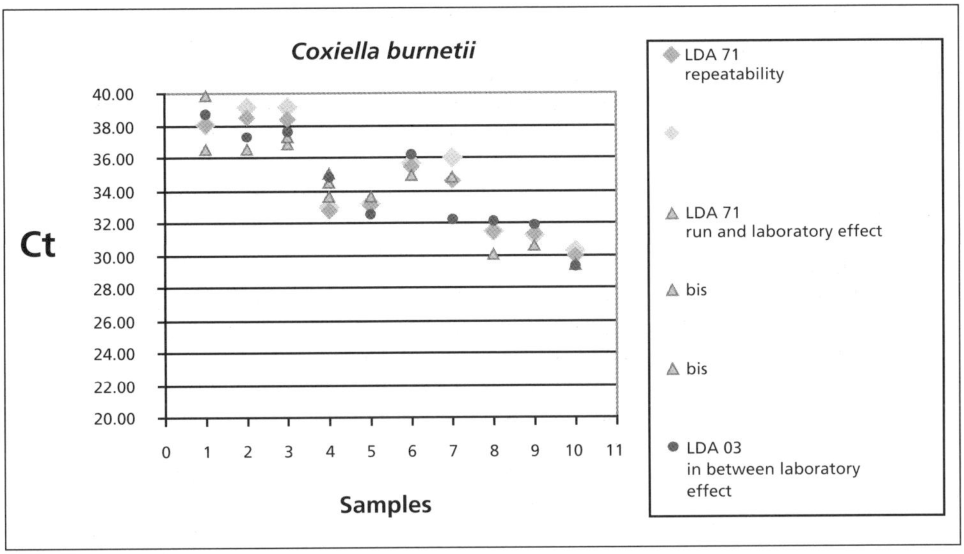

Fig. 2: Dispersion of the values of C_t in conditions of repeatability and reproducibility for the *C.burnetii* target.

Table 2: Coefficients of variation (percentage) for *C.abortus* target.

Samples	Positive low				Positive medium			Positive high		
	1	2	3	4	5	6	7	8	9	10
LDA 71 repeatability	1.67	0.50	1.51	1.50	0.07	0.40	0.48	0.27	0.15	0.08
LDA 71 run and laboratory effect	0.37	2.09	2.30	1.23	4.70	0.17	7.13	1.00	1.72	2.19
LDA 03 between laboratory effect	3.51	1.01	4.79	2.90	0.47	0.30	1.68	2.09	1.52	0.76

Table 3: Coefficients of variation (percentage) for *C.burnetii* target.

Samples	Positive low				Positive medium			Positive high		
	1	2	3	4	5	6	7	8	9	10
LDA 71 repeatability	0.17	1.24	1.40	0.45	0.28	0.46	2.77	0.07	0.29	0.77
LDA 71 run and laboratory effect	3.45	4.15	3.69	2.79	1.09	1.25	0.97	3.16	1.52	1.72
LDA 03 between laboratory effect	0.96	0.63	0.14	1.58	1.79	1.90	5.91	3.02	2.04	1.11

The accuracy of the PCR was checked on 10 positive samples of each valence and on 20 samples found negative with the two PCR of reference. These samples were analysed in duplicate in the same try. The positive samples in one or the other valence provided positive results, and the whole of the negative samples provided negative results for the valence *Chlamydophila abortus*. However, among the 20 negative samples with the two PCRs of reference, a certain number gave positive results for the valence *Coxiella burnetii*. Five samples thus gave a positive result for both duplicates (C_t ranging between 32.6 and 37.45), whereas four samples gave a positive signal only on one of the duplicates (higher C_t ranging between 38.04 and 39.46). As the specificity of the multiplex PCR was checked, and that in addition four samples found initially negative with the PCR of reference C.burnetii proved to be finally positive with two PCR in time real (internal multiplex PCR and commercial kit), the most probable assumption to explain these results seems to be the lack of sensitivity of the reference PCR. Moreover, the study of the relative threshold of detection consolidates this thesis.

Three positive samples for each valence and a doubly positive sample were selected. Five successive dilutions of 10 in 10 and framing the threshold of detection estimated for the multiplex PCR according to the value of C_t (detection threshold for $C_t = 40$) were analysed in duplicate for each sample with the multiplex PCR by the LDA71 and with the corresponding PCR of reference by the LDA03. The threshold of detection retained is the last dilution where a positive result is obtained for both duplicates. The thresholds of detection of each method are presented in Table 4. It is noted that the multiplex PCR has a lower threshold of detection than the PCR of reference with an improvement by a factor 10 for the *C.abortus* valence and by a factor 100 for the *C.burnetii* valence. Thus, the defect of accuracy of the multiplex PCR observed on negative samples with respect to the *C.burnetii* valence corresponds to a defect of initial sensitivity of the PCR of reference. Besides this difference seems to be accentuated at the time of the simultaneous presence of the two agents in the same sample. Even if one cannot conclude from only one result, it appears that the limit of detection of the multiplex PCR is not affected when the three PCR function simultaneously.

To compare the multiplex PCR with more traditional tools for diagnosis, we selected some placentas of bovines for which we had a positive result in Stamp's coloration, or associated with positive serologies in ELISA *C.burnetii* or *Chlamydophila sp*. The results obtained for the two target valences of the multiplex PCR are presented in Table 5. For the positive placentas in Stamp's coloration, only two gave a positive result in PCR with the *C.burnetii* valence, and only one of these two placentas is associated with a positive serology in ELISA *C.burnetii*. For the four placentas associated with the presence with sera antibodies directed against *C.burnetii*, only one gives a positive result for the *C.burnetii* valence in multiplex PCR. Lastly, for the three placentas associated with the presence with sera antibodies directed towards *Chlamydophila sp*., no positive result for the *C.abortus* valence was obtained with multiplex PCR. On the other hand, for one of the three placentas, a positive result was obtained for the *C.burnetii* valence.

Table 4: Comparison of the relative threshold of detection of the reference PCR and the multiplex PCR (last dilution positive for both duplicates).

	PCR of reference	Multiplex PCR
Samples positive for *Coxiella burnetii*	Pure Pure Pure	10^{-2} 10^{-2} 10^{-2}
Samples positive for *Chlamydophila abortus*	10^{-4} 10^{-4} 10^{-4}	10^{-5} 10^{-5} 10^{-5}
Sample positive for *C.burnetii* and *C.abortus*	10^{-3} for *C.burnetii* and *C.abortus*	10^{-6} for *C.burnetii* 10^{-5} for *C.abortus*

Table 5: Comparison of results obtained with Stamp's coloration, serology ELISA and multiplex PCR.

Spl. Nb	Stamp's coloration	ELISA *C.burnetii*	ELISA *Chlamydophila sp.*	Multiplex PCR	
				C.burnetii	*C.abortus*
1	**Presence**	**Positive**	Negative	**C_t 16.8**	Undetected
2	**Presence**	Negative	Negative	**C_t 18.1**	Undetected
3	**Presence**	Negative	Negative	Undetected	Undetected
4	**Presence**	Negative	Negative	Undetected	Undetected
5	Absence	**Positive**	Negative	Undetected	Undetected
6	Absence	**Positive**	Negative	Undetected	Undetected
7	Absence	**Positive**	Negative	Undetected	Undetected
8	Absence	**Positive**	Negative	Undetected	Undetected
9	Absence	Negative	**Positive**	Undetected	Undetected
10	Absence	Negative	**Positive**	Undetected	Undetected
11	Absence	Negative	**Positive**	**C_t 21.6**	Undetected

Discussion

The real time multiplex PCR developed with the LDA71 is today the only veterinary diagnosis tool allowing the simultaneous detection of *Coxiella burnetii* and *Chlamydophila abortus* in cattle in a broad range of samples: products of abortion, vaginal swabs and also milk. Thus, our method appears very suitable for the diagnosis of the abortions in cattle, while limiting the costs for the breeders.

Indeed, the commercial PCR currently on the market are targeting only one valence by test and they are sometimes not sensitive enough for samples low in bacteria such as milk. Moreover, for the diagnostic of infection to *Chlamydiae*, the selected target is rather *Chlamydophila species* than *C.abortus*.

Among the two species of *Chlamydiae* identified in the domestic ruminants (*C.abortus* and *C.pecorum*), we deliberately chose to target the species *C.abortus*, the only species implicated to date in abortive pathologies in cattle. Indeed, the frequent asymptomatic bearing of *C.pecorum* in faeces but also at the vaginal level [4,5] could lead, with a PCR not specifically directed to *C.abortus*, to a diagnosis in excess of the abortions with *Chlamydiae*.

In addition, we showed that the real-time multiplex PCR was very sensitive, even more than the PCR of reference. Furthermore, we validated it on samples not highly charged with bacteria, such as milk. Thus this protocol will also be able to find its place within the framework of monitoring herds while working, starting from a pool of milk.

The results obtained during the characterization, and comparison with the two individual PCR of reference being satisfactory allow our method to be validated. On the other hand, the comparative study of the PCR with the other techniques of diagnosis usually used (Stamp's coloration and ELISA) proved that it was not easy to establish a correlation between the various results of analysis. The only reliable technique to pose a diagnosis for the infections to *C.burnetii* and *C.abortus* seems to be the PCR: the serology by ELISA can highlight a bacterial passage former to the abortion and Stamp's coloration is neither very sensitive nor specific. The multiplex PCR is thus really a suitable tool and will undoubtedly bring a response to abortions which are not always explained.

By its originality as multiplex PCR and compared to the targets chosen, the method validated is innovative. Thus, the finality of this study would be the recovery of the multiplex PCR in the form of a commercial kit available soon (AES Laboratoires).

Acknowledgements

We thank Bryan Markey and Mary Gleeson from the Veterinary Faculty of Dublin, and the companies CEVA Santé Animale and ADIAGENE for their collaboration in this study by the gift of *Chlamydophila* strains.

References

1 Berri M, Laroucau K, Rodolakis A: The detection of *Coxiella burnetii* from ovine genital swabs, milk and fecal samples by the use of a single touchdown polymerase chain reaction. Vet Microbiol 2000;72(3-4):285-293.

2 Laroucau K, Souriau A, Rodolakis A: Improved sensitivity of PCR *Chlamydophila* using *pmp* genes. Veterinary Microbiology 2001;82:155-164.

3 Sheehy N, Markey B, Gleeson M, Quinn PJ: Differentiation of *Chlamydophila psittaci* and *C.pecorum* strains by species-specific PCR. Journal of Clinical Microbiology 1996;34(12):3175-3179.

4 DeGraves FJ, Gao D, Hehnen HR, Schlapp T, Kaltenboeck B: Quantitative detection of Chlamydophila psittaci and C.pecorum by high-sensitivity real-time PCR reveals high prevalence of vaginal infection in cattle. Journal of Clinical Microbiology 2003;41(4):1726-1729.

5 Jee JB, DeGraves FJ, Kim TY, Kaltenboeck B: High prevalence of natural *Chlamydophila* species infection in calves. Journal of Clinical Microbiology 2004;42(12):5664-5672.

Dr. C. Pelletier, Laboratoire Départemental d'Analyses de Saône-et-Loire, 267 rue des Epinoches, FR-71000 Mâcon, France.
E-mail : c.pelletier@cg71.fr.

Vannier P, Espeseth D (eds): New Diagnostic Technology: Applications in Animal Health and Biologics Controls. Dev Biol (Basel). Basel, Karger, 2006, vol 126, pp 227-236.

Development of a Real-Time, Differential RT-PCR TaqMan® Assay for Lyssavirus Genotypes 1, 5 and 6

P.R. Wakeley[1], N. Johnson[2], L.M. McElhinney[2], D. Marston[2], J. Sawyer[1], A.R. Fooks[2]

[1] Technology Transfer Unit, Biotechnology Department, Veterinary Laboratories Agency, Weybridge, Surrey, United Kingdom

[2] Rabies Research and Wildlife Zoonoses Group, Department of Virology, Veterinary Laboratories Agency, Weybridge, Surrey, United Kingdom

Key words: Rabies virus, TaqMan PCR, genotyping.

Abstract: A number of RT-PCR methods have been reported for the detection of rabies and rabies-related viruses. Here, a single, closed tube, non-nested RT-PCR TaqMan® assay to distinguish between Classical rabies virus and European bat lyssaviruses 1 and 2 in real time is described. The TaqMan® assay is rapid, sensitive, specific and allows for the genotyping of unknown isolates concomitant with the RT-PCR. It can be applied quantitatively and the use of an internal control enables the quality of the isolated template to be assessed. The efficiency and dynamic range of the PCR has been established using isolated viral RNA and cloned control template. Comparative performance of the TaqMan® assay against other diagnostic methodologies for the detection of rabies virus has been determined and the assay validated against a panel of archival samples and virus of unknown genotype from both Germany and the Sudan. Despite sequence heterogeneity between the different genotypes in the N-gene, a universal forward and reverse primer set have been designed allowing simplification of previously described assays. Rapid genotyping of two recent EBLV2 cases in the U.K. in Daubenton's bats and a recently imported human case of rabies has been performed using this test.

INTRODUCTION

Infection with rabies virus requires rapid diagnosis to ensure appropriate patient management. As infection with rabies virus is fatal in the absence of treatment, post-mortem confirmatory diagnosis allows for the timely administration of post-exposure treatment to all human contacts. Not only detection, but genotyping of the virus is essential for epidemiological purposes and to target control measures with respect to animal hosts.

Rabies and related viruses are in the genus *Lyssavirus* and are characterised as enveloped, single-stranded, negative-sense RNA viruses. Based on nucleotide sequences and the deduced amino acid similarities of the nucleoprotein, glycoprotein and phosphoprotein, the group has been separated into seven different genotypes. These include classical rabies virus (genotype 1), Lagos bat (genotype 2), Mokola virus (genotype 3), Duvenhage virus (genotype 4), European bat lyssavirus types 1 and 2 (genotypes 5 and 6 respectively) and the Australian bat lyssavirus (genotype 7). Infection with lyssavirus leads to clinically indistinguishable fatal disease in man. Genotype 1 is found world-wide with the exception of the United Kingdom, Ireland, an increasing number of Western European countries, Hawaii, New Zealand, Australia and the continent of Antarctica. Genotypes 2, 3 and 4 are restricted to Africa, whereas genotypes 5 and 6 (EBLV-1 and -2) are distributed throughout Europe.

Despite the U.K. being officially rabies-free, with only the occasional case of human rabies arising from contact with rabid animals while abroad, there has been a human death involving infection with EBLV after contact with an indigenous bat. Data from the first year of a longitudinal study combined with that of a 2003 pilot study, indicate that the prevalence of EBLV-2 seropositive Daubenton's bats in the U.K. is ~4.2%. However, due to the behaviour of these bats the risk of infection of members of the general public is low.

A number of reverse transcription polymerase chain reactions (RT-PCR) have been described for the detection and in some instances genotyping of lyssavirus. All these are to some extent dependent on the transfer of either first-strand cDNA or first-round PCR products (or both) between different tubes in the procedure with the inherent risk of contamination with positive control material. We have, therefore, developed a closed, single tube, real-time RT-PCR, incorporating TaqMan® probes to discriminate between three different genotypes of lyssavirus predominating in Europe (genotypes 1, 5 and 6). As all reactions occur in the same tube this reduces the possibility of cross-contamination. The assay compares very favourably with our previously published molecular method [1] and other routine diagnostic methods.

Materials and Methods

The lyssavirus RT-PCR primers and TaqMan® probes were designed following comparison of the nucleoprotein (N) gene sequences held in the VLA-Weybridge database and are described in Table 1. The upstream PCR primer (JW12) previously described [1] was paired with the novel primer N165-146 (positions relative to the Pasteur virus genome sequence). The genotype specific probes (LysGT1, 5 and 6) were designed following comparison of 557 partial N gene sequences from genotype 1 isolates, 35 genotype 5 isolates and 9 genotype 6 isolates. The probes were conjugated with different fluorophores and quenchers (LysGT1 FAM/TAMRA, LysGT5 HEX/Blackhole 1 and LysGT6 Cy5/Blackhole 2) to discriminate between amplicons in the PCR. The β-actin primers and TaqMan® probe used as a control in the assay were designed using a consensus sequence derived from the β-actin mRNA corresponding to exons 4 and 5 of mouse and inferred from the mRNA sequences of sheep, bat, cow, guinea pig, horse, human and rat β-actin genes. The fluorophore on the probe was ROX and the quencher Deep Dark Quencher 2. Total RNA from infected tissues was extracted using TRIzol following the manufacturer's instructions (Invitrogen). Both the reverse transcription reaction and the PCR occurred in a single tube with no requirement to open the tube once the RNA template was added. The RT-PCR TaqMan® reactions were performed in 50μl reaction volumes comprising 20.75μl of nuclease-free water, 5μl of 10x PCR buffer (Promega), 12μl of 25mM $MgCl_2$, 1μl of 10mM dNTPs, 1μl of each of the PCR primers (20μM each), 1μl of each of the TaqMan® probes (5μM each), 1μl of Triton X100 (10%v/v), 0.25μl of RNasin (20-40U/μl, Promega), 0.5μl of MMLV reverse transcriptase (200U/μl, Promega) and 0.5μl of Taq polymerase (5U/μl, Promega). To this mixture was added 1μl of total RNA at a concentration of 1μg/μl. The reactions were carried out in Thermo-Fast® 96 well PCR plates or Thermo-tube strips with Ultra Clear caps (ABgene) in a MX3000P Multiplex Quantitative PCR System (Stratagene). The RNA was reverse transcribed and amplified according to the following heating and cooling programme: one cycle of 42°C for 30 min and 94°C for two min followed by 45 cycles of 94°C for 30s, 55°C for 30s and 72°C for 20s. Positive control assays were run using RNA extracted from mouse neuroblastoma cell cultures infected with lyssavirus strains CVS (Challenge Virus Standard, genotype 1), RV1423 (genotype 5) and RV628 (genotype 6).

For each RT-PCR a critical threshold cycle number (Ct) was determined corresponding to the PCR cycle number at which the fluorescence of the reaction exceeded a value determined statistically to be higher than background by the software associated with the MX3000P (Stratagene).

Table 1: Oligonucleotide sequences of TaqMan probes and RT-PCR primers.

Name	Role	Length (nt)	Tm	Sequence	Position
JW12	RT/PCR	19	49°C	ATGTAACACCYCTACAATG	55-73
N165-146	PCR	20	53°C	GCAGGGTAYTTRTACTCATA	165-146
LysGT1	Probe	29	62°C	ACAAGATTGTATTCAAAGTCAATAATCAG	81-109
LysGT5	Probe	26	64°C	AACARGGTTGTTTTYAAGGTCCATAA	80-105
LysGT6	Probe	29	55°C	ACARAATTGTCTTCAARGTCCATAATCAG	81-109
β act intronic	PCR	22	63°C	CGATGAAGATCAAG/ATCATTGC	1051-1072
β act reverse	RT/PCR	17	62°C	AAGCATTTGCGGTGGAC	1204-1188
β-Actin	Probe	32	64°C	TCCACCTTCCAGCAGATGTGGATCAGCAAG	1128-1157

Primer and probe positions relative to the Pasteur virus genome Acc. No. M13215.
β-actin positions according to mouse β-actin gene sequence Acc. No. NM 007393.
The exon 4/5 boundary is indicated with (/). Y: Pyrimidine. R: Purine.

Results

The relative sensitivity of the TaqMan® assay, compared to our previously described hemi-nested RT-PCR (hnRT-PCR) agarose gel based method [1], was determined for each genotype specific TaqMan® RT-PCR using a dilution series of RNA extracted from mouse neuroblastoma cells infected with a candidate virus for each different genotype. For the TaqMan® assay the limit of detection was 1pg of total RNA added to the reactions whereas the limit of detection for the hnRT-PCR for genotypes 5 and 6 was 1ng and for genotype 1, 100pg of total RNA. The efficiencies of the genotype 1, 5 and 6 reactions were 92.1%, 98.8% and 99.0% respectively. The standard curve of the genotype 1 TaqMan® assay compared to the hnRT-PCR assay is shown in Figure 1. Using the genotype 1 amplicon generated using primers JW12 and JW6 previously described [1] the limit of detection of the PCR component of the TaqMan® assay was determined to approximate to 1 molecule.

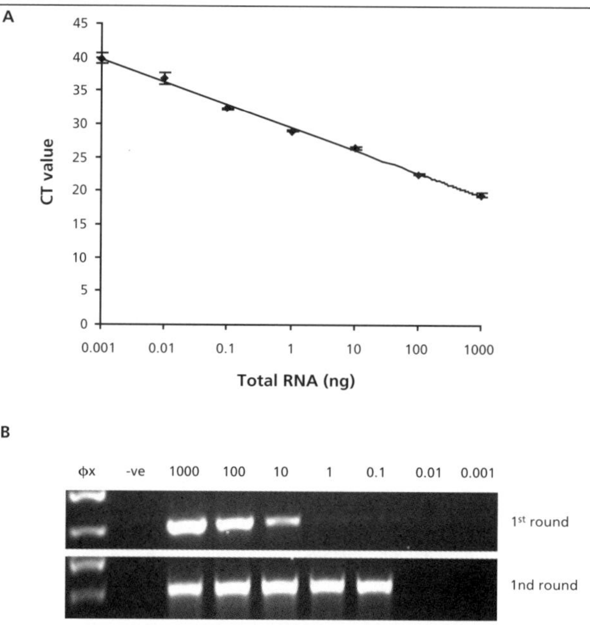

Fig. 1: Comparison of the single tube RT-PCR TaqMan assay for rabies virus (genotype 1) with the amplicon generated using the hemi-nested PCR method (12) using the same diluted RNA standards. A. Linear inverse relationship between RNA concentration and Ct values using the single tube RT-PCR TaqMan assay. The error bars show the standard deviation of the mean of three replicates. $r^2 = 0.994$. B. 1.8% agarose gel analysis of diluted total RNA expressed in nanograms following first and second round hemi-nested PCR. The negative control is uninfected mouse brain total RNA.

A total of 62 samples (40 designated as genotype 1, 15 as genotype 5, 6 as genotype 6 and 1 as genotype 7) were assayed using the RT-PCR TaqMan® assay and the result obtained was compared with the genotype derived by conventional sequencing methods [1]. All RNA templates were reverse-transcribed and amplified using primers JW12 and N165-146 to produce an amplicon of the expected size (92bp). The Ct values of the TaqMan® assay for each isolate, corresponding to all the probes in the reaction, are indicated in Table 2. Where no Ct is shown for a particular isolate and probe combination this indicates that for that genotype specific probe, the fluorescence readings failed to increase during the reaction. In all cases, where the fluorescence readings increased, only one probe hybridised to the template, indicating the genotype and that the virus isolate was not mixed. The Ct values were found to be quite variable due to the amount of template being added to the TaqMan® assay being standardised by reference to total RNA concentration rather than viral load. With the exception of RV48, all templates were readily genotyped (genotypes 1, 5 and 6). The RV48 amplicon only weakly hybridised with the genotype 1 probe. RNA from isolate RV634, an Australian bat lyssavirus isolate (genotype 7), was analysed. An amplicon of the correct size was produced in the RT-PCR TaqMan® assay (as determined by agarose gel electrophoresis) but none of the probes was hydrolysed during the reaction. The isolates tested and the divergence of each template in the region of the three different probes are indicated in Table 2.

Table 2: List of archival samples tested using the single tube Q RT-PCR TaqMan assay.

Isolate	Country of origin	Host species	Genotype	Ct values			% divergence		
				FAM	HEX	Cy5	Div GT1	Div GT5	Div GT6
8	Finland	Human	6			15.86	11.1	26.9	0
9	Germany	Bat	5		15.73		19.8	0	17.6
11	Germany	Bat	5		13.94		19.8	0	17.6
19	Denmark	Bat	5		16.46		19.8	0	17.6
29	Netherlands	Bat	6			12.12	15.3	26.9	0
31	Netherlands	Bat	5		14.12		19.8	0	17.6
32	Netherlands	Bat	5		13.62		19.8	0	17.6
33	Netherlands	Bat	5		15.67		19.8	0	17.6
37	Netherlands	Bat	5		12.95		19.8	0	17.6
38	Netherlands	Bat	5		16.57		19.8	0	17.6
44	USA	Bat	1	11.72			3.6	18.9	12.0
47	USA	Bat	1	16.19			7.4	24.8	16.6
48	USA	Bat	1	44.67			11.2	24.8	21.3
49	USA	Bat	1	10.95			7.2	18.9	21.3
56	USA	Fox	1	7.39			7.3	25.7	12.0
57	USA	Skunk	1	7.91			3.6	31.5	16.6
58	USA	Skunk	1	12.27			7.2	37.8	12.2
66	Poland	Bat	5		15.08		19.8	0	17.6
68	Peru	Dog	1	10.12			3.6	24.8	16.6
73	Belize	Dog	1	8.95			3.6	24.8	12.0
100	Morocco	?	1	20.27			3.6	31.5	16.6
114	Chile	Bat	1	15.36			7.4	18.9	12.0
144	Germany	Bat	5		29.13		19.8	0	17.6
193	Pakistan	Canine	1	9.25			3.6	31.5	16.6
202	Turkey	Canine	1	9.72			0	24.8	12.0
227	Nigeria	Canine	1	8.2			3.6	24.8	12.0
234	Russia	Canine	1	13.55			0	24.8	12.0
253	Russia	Wolf	1	12.31			3.5	25.7	7.8
264	Ukraine	Bat	5		24.78		15.3	26.9	0
266	France	Bat	5		8.07		24.9	0	17.6
303	Russia	Raccoon-dog	1	32.49			0	24.8	12.0
305	Georgia	Dog	1	10.27			0	24.8	12.0
307	Georgia	Cattle	1	8.15			0	24.8	12.0
313	Germany	Fox	1	9.08			0	24.8	12.0
318	Germany	Fox	1	25.89			3.5	19.4	7.8
338	China	Cattle	1	9.88			3.6	24.8	12.0
341	China	Dog	1	11.09			0	24.8	12.0
342	China	Cattle	1	11.27			3.6	24.8	12.0
345	Germany	Bat	5		23.85		19.8	0	17.6
349	Germany	Bat	5		24.77		19.8	0	17.6
350	Germany	Bat	5		24.06		19.8	0	17.6
399	Botswana	Jackal	1	10.46			11.6	18.9	21.6
437	Estonia	Raccoon-dog	1	12.58			0	24.8	12.0
447	Botswana	Dog	1	9.8			3.6	31.5	16.6
483	Botswana	Genet	1	13.55			11.6	18.9	21.6
484	Botswana	Cattle	1	12.62			11.6	18.9	21.6
498	S.Africa	Cat	1	8.25			3.6	24.8	16.6
519	?	Cattle	1	9.29			0	24.8	12.0
594	Switzerland	Bat	6			16.22	15.3	26.9	0
621	Switzerland	Bat	6			10.78	19.8	20.3	0
628	England	Bat	6			12.75	15.3	26.9	0
634	Australia	Bat	7				24.9	31.5	33.4
758	Tanzania	Canine	1	10.06			0	24.8	12
759	Tanzania	Canine	1	29.35			0	24.8	12
889	Czechoslovakia	Fox	1	13.95			3.7	19.9	12.5
960	Estonia	Raccoon-dog	1	13.7			NA	NA	NA
1124	Turkey	Fox	1	16.93			NA	NA	NA
1125	Turkey	Fox	1	19.08			NA	NA	NA
1176	Yugoslavia	Fox	1	13.92			NA	NA	NA
1177	Yugoslavia	Fox	1	17.65			3.6	31.5	16.6
1312	Poland	Badger	1	13.79			0	24.8	12.0
1333	Scotland	Human	6			13.05	19.8	26.9	3.8

No Ct value is recorded when the probes were not hydrolysed. Divergence was calculated using Clustal V from the Megalign multiple alignment package (DNASTAR Inc.).

A set of nine coded samples (brain homogenate suspensions), previously tested in a ring trial in Germany, kindly provided by Dr.Thomas Müller (Friedrich-Loeffler Institute, Federal Research Centre for Animal Health, Wusterhausen, Germany) were assayed blind for the presence of virus. Following testing, the samples were revealed to be seven mixtures of positive brain material from a single infected dog, diluted in a two-fold dilution series in uninfected fox brain material and samples of uninfected fox brain homogenate (Table 3). Diagnostic tests used included the fluorescent antibody test (FAT), rabies tissue culture inoculation test (RTCIT), the mouse inoculation test (MIT), all performed as previously described [2-4], the original first round RT-PCR as previously described [1] and the RT-PCR TaqMan® assay (Table 3). The β-actin control RT-PCR TaqMan® assay was included in a duplicate set of samples in addition to the lyssavirus-specific RT-PCR. Seven of the nine samples were shown to be positive for lyssavirus by the PCR-based assays and the MIT. The FAT detected all but the lowest dilution of rabies-infected material and the RTCIT failed to detect two of the positive samples. The β-actin positive control RT-PCR was positive with all samples analysed, did not cause an apparent overall change in the Ct values recorded for the rabies virus PCRs and indicated that there was less total RNA in the 1:8 dilution than the others (GT1 Ct = 18.01; β-actin Ct = 17.11). Using the RT-PCR TaqMan® method, only the genotype 1 probe hybridised with the amplicons produced, indicating that only classical rabies viral RNA was present in the sample.

Brain samples from various animals from the Sudan, kindly provided by Dr. Yahia Beik (Central Veterinary Research Laboratory, Khartoum, Sudan) suspected of being infected with rabies virus were tested using the RT-PCR TaqMan® assay and these results were compared with those obtained with the hnRT-PCR gel based assay [1]. The Ct values generated correlated with the results from the hnRT-PCR assay with higher Ct values corresponding to those samples with apparently less amplifiable material, determined by comparison of products generated in the hnRT-PCR assay analysed by agarose gel electrophoresis with Ct values, as shown in Figure 2. No band or Ct value was recorded for samples RV1343 or RV1344 with either the hemi-nested or the real-time RT-PCR TaqMan® assay. All the viruses identified were genotyped as classical rabies virus and the host species did not appear to affect the TaqMan® assay.

Table 3: Comparison of rabies diagnostic methods used on ring trial material.

Method	Result (+,- or Ct value) at indicated dilution							
	1:2	1:4	1:8	1:16	1:32	1:64	1:128	Negative control
FAT	+	+	+	+	+	+	-	-
RTCIT	+	+	-	+	+	+	-	-
MIT	+	+	+	+	+	+	+	-
hnRT-PCR	Weak +	+	+	+	+	+	+	-
Q-RT-PCR								
LysGT1 RT-PCR	16.55	16.08	18.01	12.35	13.22	11.78	15.33	No Ct
LysGT5 RT-PCR	No Ct	No Ct	No Ct	No Ct	No Ct	No Ct	No Ct	No Ct
LysGT6 RT-PCR	No Ct	No Ct	No Ct	No Ct	No Ct	No Ct	No Ct	No Ct
β-Actin RT-PCR	14.17	13.86	17.11	12.16	14.48	13.07	14.44	13.15

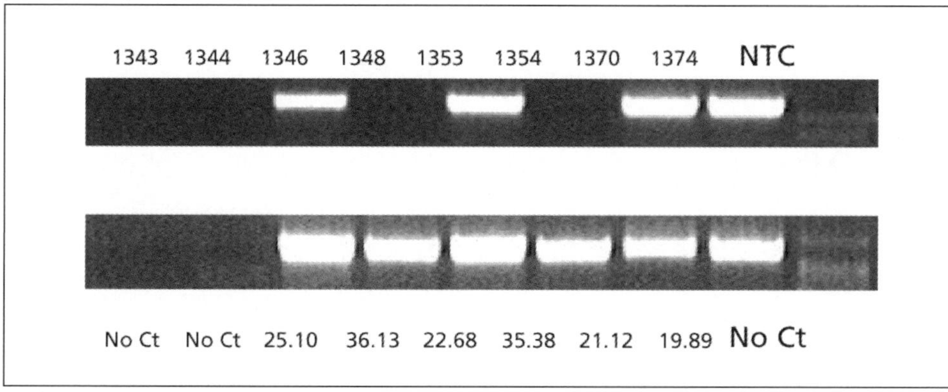

Fig. 2: Comparison of the conventional heminested RT-PCR (12) and the single tube Q RT-PCR methodology (following a 30 minute reverse transcription and 45 rounds of PCR) using 2µg of total RNA isolated directly from Sudanese brain samples from animals suspected to be infected with rabies. (A) Reverse transcription and first round PCR with JW12 and JW6 (dpl/e/m). (B) Second round PCR of products shown in panel A with JW12 and JW10 (dle2/me1/p). The corresponding Ct values are shown below. RV1343 (Goat), RV1344 (Cat), RV1346 (Dog), RV1348 (Goat), RV1353 (Donkey), RV1354 (Donkey), RV1370 (Cattle), RV1374 (Cattle).

The TaqMan® assay has been applied to the detection of lyssavirus from bat field samples. On September 25th 2004 as part of a routine passive surveillance of bats in Great Britain a brain sample from a juvenile female Daubenton's bat was submitted for testing using the fluorescent antibody test. This proved to be positive and the presence of lyssavirus was confirmed using the hnRT-PCR method [1] and the RT-PCR TaqMan® assay. In addition to confirming the presence of a lyssavirus the real-time RT-PCR TaqMan® assay suggested that the virus was a European bat lyssavirus type 2, i.e. lyssavirus genotype 6. A 405bp fragment of the N gene was sequenced and aligned with a panel of lyssavirus sequences held in the Weybridge database which confirmed that the virus was of genotype 6 [5]. The relative levels of viral RNA in different tissues was estimated for various tissues using calculations based on the method of Livak and Schmittgen [6]. No lyssavirus RNA was detected in heart, liver, salivary gland, tongue, wing, spleen or faeces. Viral RNA was found at the lowest detectable level in lung tissue and this was set as the level to which all other positive samples were compared (i.e. set at a nominal relative value of 1). Relative to the amount of virus found in the lung there was almost 1,600-fold more viral RNA found in brain tissue. EBLV-2 RNA was also detected in the intestine (4-fold), stomach (9-fold), bladder (19-fold) and thyroid (60-fold) relative to the levels found in the lung (Fig. 3). Equal amounts of total RNA from the different tissues was added to the assay. The RT-PCR TaqMan® assay has also been applied successfully to frozen bat material that had been stored for at least one year in this state [7]. Furthermore, the real time RT-PCR TaqMan® assay has also been used to confirm diagnosis of human infection by classical rabies virus [8]. Skin biopsy and saliva ante-mortem samples were submitted for testing using the hnRT-PCR and the RT-PCR TaqMan® assay and both proved to be positive. All other standard tests (FAT, RTCIT, MIT) used on post-mortem brain samples (cerebellum, medulla and hippocampus) confirmed the diagnosis, as did the RT-PCR tests on this material.

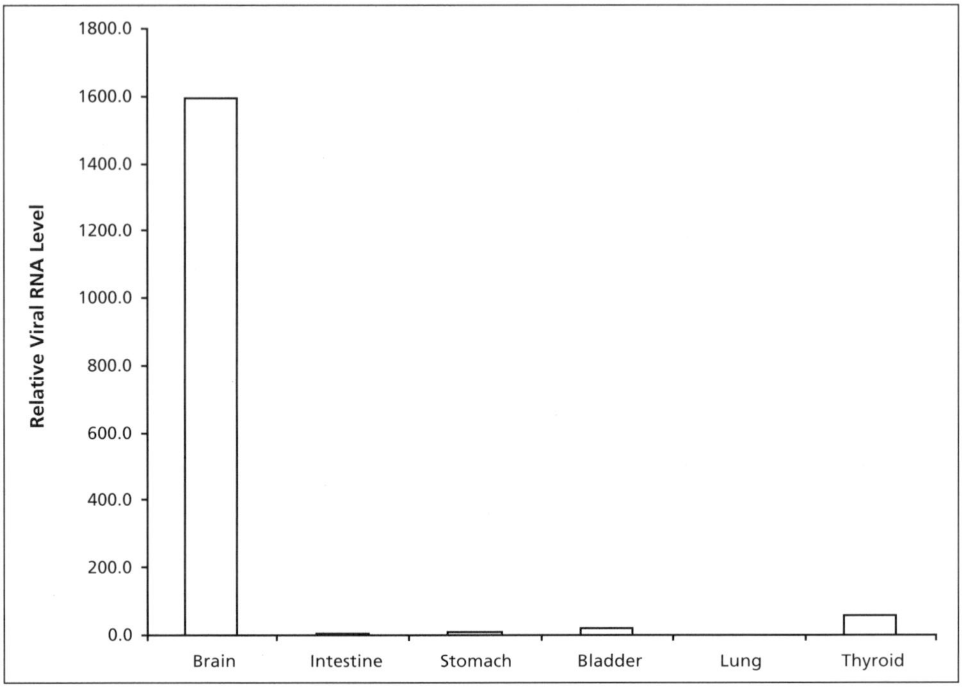

Fig. 3: Relative viral RNA levels compared to that found in the lung of an infected Daubenton's bat found in Staines, U.K. Values correspond to the $2^{\Delta\Delta Ct}$ as described in [6].

Discussion

A single tube RT-PCR TaqMan® method has been developed which addresses problems associated with a previously described hnRT-PCR assay [1] being its relative insensitivity and requirement for multiple transfers of material, with the concomitant opportunity for cross-contamination. The incorporation of TaqMan® technology in a real-time assay enables the lyssavirus template not only to be detected but also genotyped. The use of only two universal amplimers (JW12 and N165-146) plus a single probe for each genotype considerably simplifies the TaqMan® method developed for use on the Applied Biosystems LS50B [9] although it is acknowledged that the assay is only applicable to three different genotypes of the lyssavirus group. The β-actin RT-PCR TaqMan® assay allows the state of the starting material to be ascertained, which may be an important issue for the analysis of small quantities of material transported or stored under less than ideal circumstances. This control has not been used for normalisation purposes in the diagnostic assay and it is unlikely that it would be used as such due to the tendency of the primers to amplify genomic DNA in the absence of upstream RNase-free DNase treatment of template. However, with the correct enzymatic treatment, this assay could be applied quantitatively to estimate viral load.

The single tube RT-PCR TaqMan® assay performed well in comparison with the gel based RT-PCR assay with respect to sensitivity. A direct comparison of the hemi-nested assay with the TaqMan® assay revealed that a positive Ct value was still obtained when only 1pg of total RNA was added to the assay, whereas 100x more RNA was required to generate a visible band on an ethidium bromide stained agarose gel. In addition, using cloned classical rabies virus N-gene sequence, only one molecule was required to produce a positive result using the single tube TaqMan® assay, this being the theoretical minimum amount of template required to produce a positive Ct value. With the exception of RV48 (which only poorly hybridised with the LysGT1 probe) all archival samples corresponding to lyssavirus genotypes 1, 5 and 6 were readily detected and correctly genotyped. Positional/combinatorial effects of particular nucleotide substitutions in RV48 appear to preclude the effective genotyping of this isolate despite the amplimers binding and a PCR product being generated as determined by agarose gel electrophoresis. The ABLV isolate, RV634 was not detected using the TaqMan® assay despite an amplicon being produced, indicating that the PCR primers annealed to this template. As this assay was developed to test viruses likely to be submitted to European laboratories, an inability to detect ABLV was not deemed to be problematic and, as such, this assay will allow the rapid genotyping of suspect rabies samples in the absence of nucleotide sequencing.

In comparison to conventional virological methods to detect the virus used in the blind ring trial, the TaqMan® assay performed equivalently or in the case of the RTCIT was superior, and as such could work well as an adjunct to these OIE-prescribed assays. With respect to the detection of virus in samples from the Sudan, the host species from which the brain sample was taken did not apparently affect the detection or genotyping of the infecting virus. In addition, the Ct values obtained reflected the levels of template amplified in the conventional gel-based assay.

The single tube TaqMan® assay is considerably faster to perform than the hnRT-PCR assay with the process taking 2 hours and 11 minutes to perform including a 30 minute reverse transcription step. This can be reduced to a two-minute reverse transcription for experimental samples with known levels of template RNA (data not shown) reducing the overall time to perform the assay to 1 hour and 43 minutes. As the results are obtained in "real-time" it is possible to inform competent authorities as the reaction proceeds, more rapidly enabling appropriate decisions to be made with regard to patient management or disease control. The single tube TaqMan® assay has been introduced as a routine test in our laboratory and has been successfully used to detect and characterise virus from post-mortem field samples from Daubenton's bats found in the U.K. and from both ante- and post-mortem samples from a human case.

References

1 Heaton, PR, Johnstone P, McElhinney LM, Cowley R, O'Sullivan E, Whitby JE: Heminested PCR assay for the detection of six genotypes of rabies and rabies-related viruses. J Clin Microbiol 1997;35:2762-2766.

2 Dean DJ, Abelseth MK, Atanasiu P: The fluorescence antibody test: Laboratory techniques in rabies 4th edition, in Meslin FX, Kaplan MM, Koprowski H (eds): World Health Organization, Geneva, Switzerland, 1996, pp 88-93.

3 Webster WA, Casey GA: Virus isolation in neuroblastoma cell culture: Laboratory techniques in rabies 4th edition, in Meslin FX, Kaplan MM, Koprowski H (eds): World Health Organization, Geneva, Switzerland, 1996, pp 96-104.

4 Koprowski H: The mouse inoculation test: Laboratory techniques in rabies 4th edition, in Meslin FX, Kaplan MM, Koprowski H (eds): World Health Organization, Geneva, Switzerland, 1996, pp 80-87.

5 Fooks AR, McElhinney LM, Marston DA, Selden D, Joliffe TA, Wakeley PR, Johnson N, Brookes S: Identification of a European bat lyssavirus type 2 in Daubenton's bat found in Staines, Surrey, UK. Vet Rec 2004;155:434-435.

6 Livak KJ, Scmittgen TD: Analysis of relative gene expression data using real-time quantitative PCR and the 2(-Delta Delta C(T)) Method. Methods 2001;25:402-408.

7 Fooks AR, Selden D, Brookes S, Johnson N, Marston DA, Joliffe TA, Wakeley PR, McElhinney LM: Identification of a European bat lyssavirus type 2 in Daubenton's bat found in Lanchashire. Vet Rec 2004; 155:606-607.

8 Solomon T, Marston D, Mallewa M, Felton T, Shaw S, McElhinney LM, Das K, Mansfield K, Wainwright J, Kwong GN, Fooks AR: Paralytic rabies after a two week holiday in India. BMJ. 2005;331:501-503.

9 Black EM, Lowings JP, Smith J, Heaton PR, McElhinney LM: A rapid RT-PCR method to differentiate six established genotypes of rabies and rabies-related viruses using TaqMan™ technology. J Virol Methods 2002;105:25-35.

Dr. P.R. Wakeley, Technology Transfer Unit, Biotechnology, Veterinary Laboratories Agency – Weybridge, Addlestone, Surrey KT15 3NB, United Kingdom.
E-mail: p.wakeley@vla.defra.gsi.gov.uk

Vannier P, Espeseth D (eds): New Diagnostic Technology: Applications in Animal Health and Biologics Controls. Dev Biol (Basel). Basel, Karger, 2006, vol 126, p 237.

SESSION V

Examples of Application in Quality Control of Human and Veterinary Vaccines

Chairman: *Jacques Léchenet,* France

Vannier P, Espeseth D (eds): New Diagnostic Technology: Applications in Animal Health and Biologics Controls. Dev Biol (Basel). Basel, Karger, 2006, vol 126, pp 239-240.

Antigen Mass Evaluation Before Formulation of FMD Inactivated Vaccines

P. Dubourget[1], T.R. Doel[2]

[1] Mérial S.A.S., Lyon, France
[2] Merial Animal Health Ltd, Pirbright, U.K.

Introduction

In vitro measurement of the concentration of 146S particles (which comprise the immunogenic component of the foot-and-mouth disease [FMD] virion) has been originally described by Fayet et al [1] and refined by others such as Barteling et al [2] and Doel et al [3]. This remains a precise and invaluable procedure in FMD vaccine production and is the single most useful parameter for the formulation of effective FMD vaccine [4]. It has indeed raised the possibility of formulating vaccines that contain predetermined levels of antigen mass, rather than volume of antigen suspension, and made possible the aim of formulating vaccines with a given probability of passing a statutory potency test [5].

Material and Methods

Essentially, clarified tissue culture preparations of virus are centrifuged in sucrose density gradients and the gradients monitored at 254-260 nm by pumping the content of each tube through the flow cell of a u.v. photometer. The concentration of 146S component in each tissue culture preparation is proportional to the area of the absorbance peak above the absorbance background.

Gradients are prepared either by layering six different sucrose solutions or by means of a two-chamber device in which a concentrated solution of sucrose is continuously diluted with a weak solution. Gradients such as 5%-30% or 15%-45% (w/v in 0.04 M phosphate, 0.15 M NaCl or 0.05M tris-HCl, 0.15 M NaCl, pH 7.6) can be used.

1 to 2 ml of virus suspension is applied to each gradient, the precise amount being determined by weight. The centrifugation conditions must be adapted to the range of the gradient and to the rotor. For example, using a Beckman rotor SW41 at +4°C, 15%-45% gradients are centrifuged for 16 hours at 16,600 rpm. Using 5%-30% gradients, centrifuge conditions can be as short as 65 minutes at 40,000 rpm.

The area of the peak above the baseline can be measured in several ways (e.g. the shape of the peak is cut from a photocopy of the original chart record. The weights of the peak and an adjacent 7 x 7 cm square of paper allows the calculation of the peak area).

On a routine basis, the easiest way to convert peak area to µg/ml in the starting material is to use a standard curve established from a known solution measured in exactly the same conditions [2].

Results

The results obtained from the rapid determination of 146S concentration from the area of the virus peak on the spectrophotometer chart are used not only throughout the production process to monitor virus and antigen yields as well as formulation of the final product, but also for monitoring the stability of frozen antigen reserves (FMD Antigen Banks).

Discussion

It has been recognized for many years that the essential immunogenic component of FMD vaccines is the 146S particle of the virus. A number of procedures other than Sucrose Gradient Analysis have been directed towards its measurement, but most of them are at best non-specific because of their failure to discriminate between 146S and essentially non-protective 12S particles [6]. The 146S test is now a standard, internationally accepted technique.

Tissue culture preparation may be applied to a gradient usually without any previous treatment other than clarification. However, some authors routinely treat virus samples by ribonuclease (e.g. 10 μg/ml bovine pancreatic ribonuclease at 37°C for 20 to 60 mn) and treatment with trypsin may be necessary with some virus strains which have a tendency to aggregate.

The sucrose concentrations used to make gradients do not seem to be critical provided the centrifugations are adjusted appropriately.

Conclusion

The Sucrose Gradient Analysis is an accurate, sensitive method of quantifying 146S particles in virus preparations destined for vaccines. Although an ultracentrifuge and high sensitivity monochromatic u.v. photometer are essential and expensive pieces of equipment, the procedure is simple to carry out and requires little more than pipettes and centrifuge tubes for its continued use

References

1 Fayet MT, Fargeaud D, Louisot P, Stellmann C, Roumiantzeff M : Mesure physico-chimique des particules 140S du virus de la fièvre aphteuse. Ann Inst Pasteur 1971 ;121 :107-118.

2 Barteling SJ, Meloen RH : A simple method for the quantification of 140S particles of foot-and-mouth disease virus. Archiv fur die gesamte Virusforshung 1974;45:362.

3 Doel TR, Fletton BW, Staple RF: Further developments in the quantification of small RNA viruses by UV photometry of sucrose density gradients. Dev Biol Stand, 1981, vol 50, pp 209-221.

4 Doel TR: FMD vaccines, Virus Research 2003;91:81-99.

5 Garland AJM, Mowat GN, Fletton B: An evaluation of some methods of assay of foot-and-mouth disease antigen for vaccines. Dev Biol Stand, 1977, vol 35, pp 323-332.

6 Rweyemamu MM, Black L, Boge A, Thorne AC, Terry GM: The relationship between the 140S antigen dose in aqueous foot-and-mouth disease vaccines and the serum antibody response of cattle. Journal of Biological Standardization 1984;12:111-120.

Vannier P, Espeseth D (eds): New Diagnostic Technology: Applications in Animal Health and Biologics Controls. Dev Biol (Basel). Basel, Karger, 2006, vol 126, pp 241-250.

Development of an Inhibition ELISA Test for the Detection of Non-Capsid Polyprotein 3ABC in Viral Suspensions Destined for Inactivated Foot-and-Mouth Disease Vaccines

I.E. Bergmann[1], E. Neitzert[1], V. Malirat[1], R. de Mendonça Campos[1], M. Pulga[2], R. Muratovik[3], D. Quintino[4], J.C. Morgado[5], M. Oliveira[6], D. de Lucca Neto[7]

1 Pan American Foot-and-Mouth Disease Center, PAHO/WHO, Río de Janeiro, RJ, Brasil
2 Bayer S.A, São Paulo, SP, Brasil
3 Akzo Nobel Ltda – Intervet, Fortaleza CE, Brasil
4 Vallée S.A, Montes Claros, MG, Brasil
5 Merial Saúde Animal LTDA, Paulinia, SP, Brasil
6 Schering-Plough, São Paulo, SP, Brasil
7 Pfizer Ltda, São Paulo, SP, Brasil

Key words: Foot-and-mouth Disease (FMD), marker vaccines, vaccine purity to non-capsid proteins, serosurveillance.

Abstract: The use during the last decade of immuno-enzymatic tests based on the detection of antibodies to the non-capsid proteins (NCPs) of foot-and-mouth disease virus (FMDV) to assess viral circulation, irrespective of vaccination, supported the incorporation into the OIE code of the 'free from FMDV with vaccination' category and opened the way to a 'vaccination to live' policy. Eradication programmes in South America include systematic vaccination accompanied by large serosurveys through NCP antibody testing to ensure the absence of residual viral activity. For correct interpretation of serosurveys, a major prerequisite is that vaccines made of semi-purified preparations of inactivated virions do not contain levels of NCPs, which upon proper presentation conditions, could induce an antibody response under the conditions for field immunization. This work describes the development of an inhibition ELISA to detect NCP polyprotein 3ABC in viral suspensions destined for vaccine production as an in-process control during vaccine manufacture. Antibody responses against NCP 3ABC in vaccinated and revaccinated cattle, induced by vaccines with different purification processes and formulations, are discussed.

INTRODUCTION

Foot-and-mouth disease (FMD) is one of the most important diseases of domestic livestock. The agent, FMD virus (FMDV), consists of a non-enveloped, single-stranded positive sense RNA genome of approximately 8,500 bases, surrounded by four structural proteins to form an icosahedral capsid. The virus belongs to the *Aphthovirus* genus of the *Picornaviridae* family. The disease provokes a drastic negative impact on animal livestock, generating considerable losses to producers, to the animal industry and consequently to the affected countries. Many trade restrictions are placed on areas where the infection is present, with an imposition of national and international sanitary barriers to the movement of both animals and derived products.

Since 1987 a Continental Programme for the Eradication of FMD has been implemented in the Americas based on a resolution of the Inter-American Meeting at Ministerial Level on Animal Health [1]. The objectives of this programme included prevention of the disease in disease-free areas and eradication in endemic regions. Approaches to control and eradicate FMD where the disease is prevalent are generally based on quick detection and early response to an outbreak, high vaccine coverage, control of animal movement, slaughter of infected animals or herds and contacts, active epidemiological surveillance, and combinations of the above. An important characteristic to be considered when control/eradication measures are planned is the ability of the virus to establish sub-clinical infection, which can become persistent, the so-called carrier state [2], regardless of whether or not the animals were vaccinated before infection [3].

Considering that controversial points of view exist in the international community on the role of carrier animals in the epidemiology of the disease [4], particularly in vaccinated populations, and that over 250 million cattle are systematically vaccinated in South America for the control and eradication of FMD, the availability of methods for identification of asymptomatic viral circulation in a population, regardless of whether they have also been vaccinated, became of great importance. The need for such methods to accompany the progress of the eradication programmes is particularly important as eradication progresses and clinical symptoms diminish, as expected.

Novel approaches for the evaluation of antibodies to FMD non-capsid proteins (NCPs), misleadingly referred to as non-structural proteins, were developed and validated at the Pan American Foot-and-Mouth Disease Center, (PANAFTOSA) PAHO/WHO. These approaches proved valuable for assessing viral circulation in immunized populations [5-15], and supported the incorporation into the OIE code of the 'FMD-free with vaccination' category. The strategy is based on the detection of antibodies to the NCPs of FMDV as evidence of viral replication in animals exposed to live virus [15], through an immuno-enzymatic system [9] consisting of the use of an indirect ELISA to screen for antibodies against the non-capsid polyprotein 3ABC (NCPanaftosa screening test) [14,16], followed by confirmation of suspect or reactive samples through an enzyme-linked immuno-electro transfer blot test (EITB - NCPanaftosa confirmatory test) that detects antibodies against five NCPs (3A, 3B, 2C, 3D and 3ABC) [5,15,16]. Considering the large-scale application of the tests in South America where eradication programmes include systematic vaccination with vaccines from various manufacturers, a major pre-requisite to ensure that vaccination does not interfere with serosurveys, carried out to accompany the progress of eradication campaigns, is that vaccines made of semi-purified preparations of inactivated virions remove most NCPs, so that they do not induce antibodies against these proteins, under the conditions used for field immunization.

This paper describes the development of an inhibition ELISA to detect NCP polyprotein 3ABC in viral suspensions destined for vaccine production as an in-process control during vaccine manufacture. The significance of the antigen levels detected in the viral suspensions in terms of the induction of antibody responses against 3ABC in vaccinated and revaccinated cattle, which could interfere with the correct interpretation of serosurveys, is discussed.

Materials and Methods

Preparation of viral suspensions

FMD virus types O1 Campos, A24 Cruzeiro and C3 Indaial, were produced in either BHK-21 cells or the hamster embryo NIL-2 [17] cells lines. Culture supernatants were harvested, clarified and inactivated with 0.001 M binary ethyleneimine (BEI). Trivalent (O1 Campos, A24 Cruzeiro and C3 Indaial) inactivated viral suspensions containing about 30 μg/ml of FMDV antigen were prepared. Various concentration/purification approaches, individually or combined, as well as intermediate stages, were assessed (precipitation with polyethylene glycol, ultrafiltration, diafiltration, etc.), representing procedures conventionally used to manufacture vaccines in South America.

Preparation of a standard antigen containing 3ABC recombinant polyprotein

Expression and purification of 3ABC recombinant polyprotein was as described [15]. Identity of the antigen was determined by reactivity with specific antisera in a Western blot. A standard preparation of the purified reference antigen containing polyprotein 3ABC, quantified by a protein assay kit (Sigma Chemical Co., Saint Louis, MO), was established.

Detection of antibodies against 3ABC

Antibodies induced against 3ABC were evaluated by the NCPanaftosa screening test, performed as described [14].

Preparation of reference serum

The serum obtained from convalescent naturally infected cattle was used as a reference for all the tests. This reference serum used for the inhibition ELISA was derived from a pool of field sera from animals involved in field outbreaks known to have been infected with FMDV, collected at about 20 days after the episode. Serum was treated with chloroform, inactivated with BEI and innocuity-tested in susceptible cell cultures. The optimal dilution of this serum was predetermined by titration in the standard NCPanaftosa screening test. The selected working dilution (1:40) gave 100% reactivity at OD_{450} about 1.4, and 50 % reactivity at OD_{450} about 0.7.

Inhibition ELISA 3ABC

The inhibition ELISA 3ABC developed to establish the level of 3ABC in the viral suspensions consisted of the modification of the NCPanaftosa screening test described [14]. This method is based on the assumption that incubation of a given quantity of the reference positive serum with the sample containing 3ABC antigen will allow the formation of antibody-antigen complexes in a directly proportional manner. If this antibody-sample mixture is then incubated in an ELISA plate coated with 3ABC, only free antibodies will be able to react with the immobilized antigen on the plate. The antibodies against 3ABC in the reference serum as preformed serum-antigen complex cannot bind the immobilized antigen and are subsequently washed away. The antibodies left to react with the antigen can then be determined by the standard NCPanaftosa screening test procedure. The degree of residual reactivity (RR) in the binding of the reference sera is reciprocal to the concentration of antigen in the sample.

The final procedure adopted was as follows. In auxiliary plates, constant 80 μl aliquots of reference serum 1/20 were added to 80 μl of viral suspensions. The serum antigen mixture was incubated overnight at 37°C, to enable the formation of sera-antigen complexes, and 100 μl of the mixture further transferred to ELISA plates coated with 3ABC antigens. The NCPanaftosa screening test was performed as described [14]. The results were expressed as % RR: percentage of the OD_{450} value established for the sample to the mean OD_{450} recorded for quadruplicate tests in the reference serum in the absence of antigen containing 3ABC. An inhibition standard curve with known concentrations of the standard containing 3ABC was constructed for each plate and examination of the cumulative data from all test plates allowed the variation of the assay to be determined.

For the evaluation of 3ABC in the viral suspensions, serial dilutions in base 5, covering up to 1:125 were prepared. The level of 3ABC in the samples was estimated by reference to the curve obtained for the standard containing 3ABC through a regression line constructed by using the $\log_{10}$ of the concentration of 3ABC in the standard antigen and the values of % RR obtained for the viral suspension.

Vaccines and vaccination schemes

Vaccines were prepared from the different viral suspensions as water-in-oil single emulsion. Groups of about 16-20 cattle were vaccinated by the recommended route. Animals were boosted two or three times at 30-day intervals after initial vaccination and serum samples were taken at regular 15- or 30-day intervals for up to 90 or 180 days. All the cattle were shown to be sero-negative for FMDV before the start of the experiment and had never been vaccinated with FMD vaccine or exposed to FMD virus.

Results

Detection of the presence of 3ABC in viral suspensions by inhibition ELISA

Figure 1 shows the standard inhibition curve constructed with serial dilutions in base 5, covering from 1:2 to 1:156,250 of known concentrations of the standard antigen containing 3ABC, relating the $\log_{10}$ of the reciprocal dilution of the antigen preparation to the % RR of the reference serum in the inhibition ELISA 3ABC. The variation of each point was calculated by analysing the data obtained in 18 independent tests.

It can clearly be seen that the system can differentiate various antigen concentrations. The level of 3ABC was best estimated in a range corresponding to approximately 25 to 75% RR and corresponding to 7.7 to 192 ng/ml of the standard antigen containing 3ABC.

The variation in results is shown as bars (± 1 SD) for each of the dilutions in the standard curve, showing a coefficient of variation (CV) of the assay <15%.

Similar results were observed when viral suspensions were titrated in the inhibition ELISA, showing a dose-response curve with parallel regression lines and similar repeatability values as the ones in the standard curve, indicating that comparable antigenic determinants active in the 3ABC specific reaction are detected so that the standard curve could be used to estimate the relative level of 3ABC in the viral suspensions.

The overall consistency of the system was estimated in 48 monovalent viral suspensions, by determining for each one of them the volume expected to give 50% RR, as established by the standard curve, and performing the inhibition ELISA with the calculated volume to confirm the evaluation. The average RR value was 51% with a CV of 17%.

To determine whether the system was capable of discriminating the 3ABC levels in viral suspensions subjected to various concentration/purification procedures, eight preparations expected to have different levels of 3ABC were evaluated in the inhibition ELISA. To this end, monovalent viral suspensions were assembled in unique trivalent pools which were subjected to concentration and further sequential purification steps (Fig. 2).

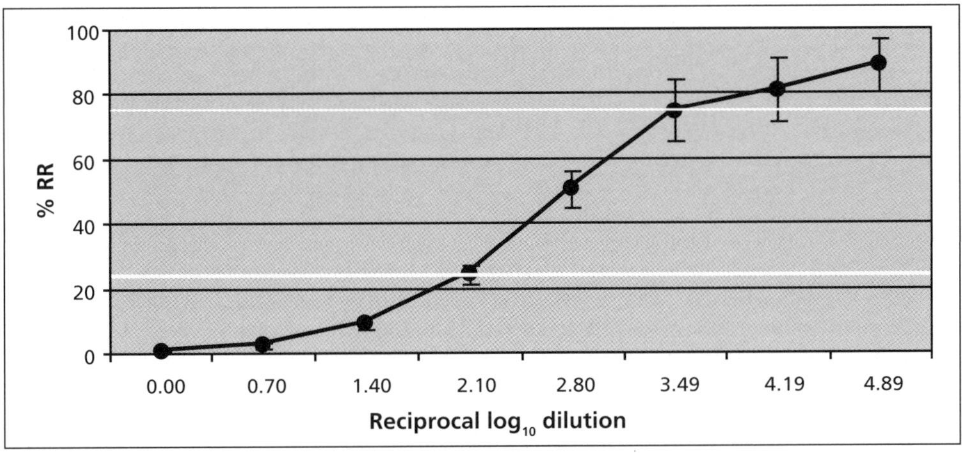

Fig. 1: **Standard inhibition curve for 3ABC in the inhibition ELISA.** The curve was constructed relating the log_{10} of the reciprocal dilution of the standard antigen containing known concentrations of 3ABC and the % RR, estimated from 18 independent assays. The bars correspond to the variation of results (1 ± SD Mean) for each of the dilutions. The horizontal lines limit the linear portion of the curve.

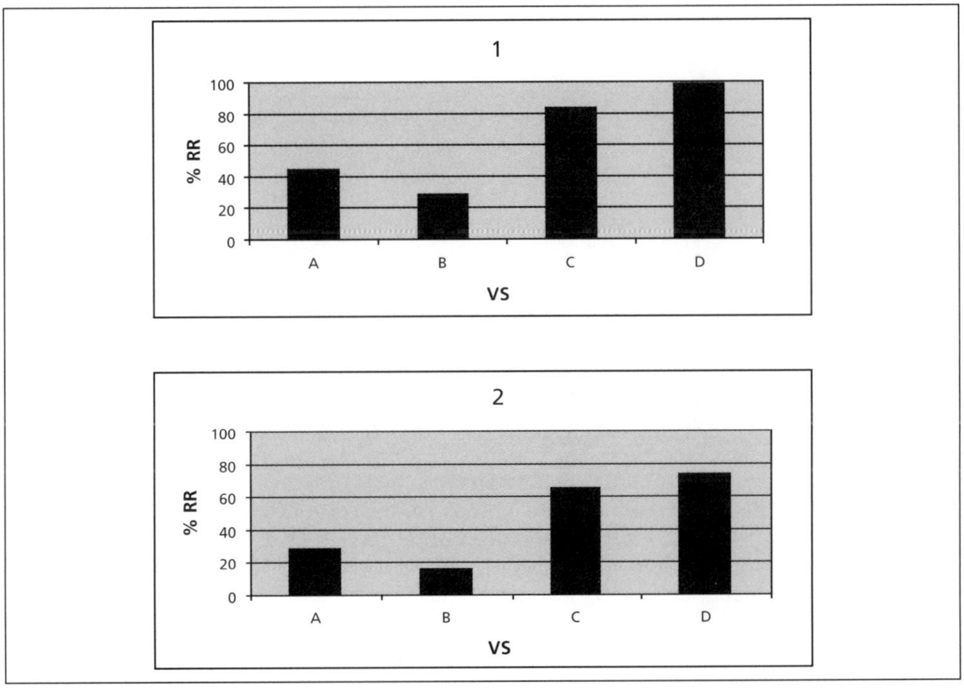

Fig. 2: **% RR in viral suspensions subjected to various concentration/purification conditions.** 1 and 2 represent viral suspensions (VS) subjected to different clarification/purification schemes. A. Trivalent clarified pool. B. Concentrated pool. C. Concentrated pool subjected to purification treatment. D. VS described in C subjected to a further purification procedure.

As can be seen, differences in the inhibitory responses were established for the various viral suspensions, consistent with the concentration/purification steps to which they were subjected. Clearly the % RR was proportional to the estimated degree of purification of the samples, with increasing values of % RR as the purification progresses. For schemes 1 and 2, decreased levels of 3ABC in the viral suspensions were discernible, also evident when additional treatments were included. For scheme 1 %RR ranged from 29 to 99 and from 16 to 74 for scheme 2. These % RR values indicate, for scheme 1, relative levels of 3ABC of 205 ng/ml in the concentrated suspension, already dropping to fewer than 5.2 ng/ml after the first purification treatment, and giving a purification factor of at least over 39-fold between these stages. For scheme 2, relative levels of 3ABC decreased from 433 ng/ml in the concentrated pool to 24 ng/ml and to 16 ng/ml after the first and second treatment, respectively, giving purification factors of about 27-fold.

Antibody responses to 3ABC induced by vaccination with vaccines prepared with various experimental viral suspensions

To establish to what extent the levels of NCP 3ABC in the viral suspensions accompanied an eventual induction of antibodies in vaccinated animals, antibody responses to 3ABC induced by vaccination with vaccines prepared with experimental viral suspensions were assessed in the NCPanaftosa screening test. Immunizations were under conditions that should induce an increased immune response such as several revaccination schemes as described in Materials and Methods. Results of antibody responses induced by the viral suspensions tested are shown in Figure 3.

Five different trivalent viral suspensions were assessed derived from three different pools (Fig. 3A). Pool 1 (the same as in Fig. 2-1) was used for vaccination in the concentrated form (VS 1B), after a further purification treatment (VS 1C) and after a second sequential treatment of the latter (VS 1D). As can be seen, the concentrated pool (VS 1B, 29 %RR) gave a considerable antibody response after the second vaccination. After subjecting this viral suspension to a further conventional treatment (VS 1C, 84 %RR) antibody responses were negligible, even when forcing the responses through two revaccinations. As expected, no antibody responses were observed after an additional treatment (VS 1D, 99 %RR). Also tested were VS 2 (not the same as the one mentioned in Fig. 2) and VS 3 which were experimental vaccines subjected to intermediate treatments leading to values in the inhibition ELISA 3ABC of 44%RR and 74%RR, respectively. As can be seen, VS 2 induced antibodies after first vaccination, while VS 3 gave antibody responses only after booster vaccination.

The results clearly indicate that, in principle, there is a tendency for a higher antibody response against 3ABC at lower %RR values. Moreover the relevance of other factors, such as purification schemes and presentation of the antigen is also evident. While VS 2 showing 44%RR already induced an antibody response after the first vaccination, VS 1B, with a lower value, 29%RR, induced a delayed response which became evident only after booster vaccination. Consequently, critical levels of NCP 3ABC will need to be established for each vaccine formulation. The results also indicate the relevance of including revaccination schemes when various suspensions and formulations are compared.

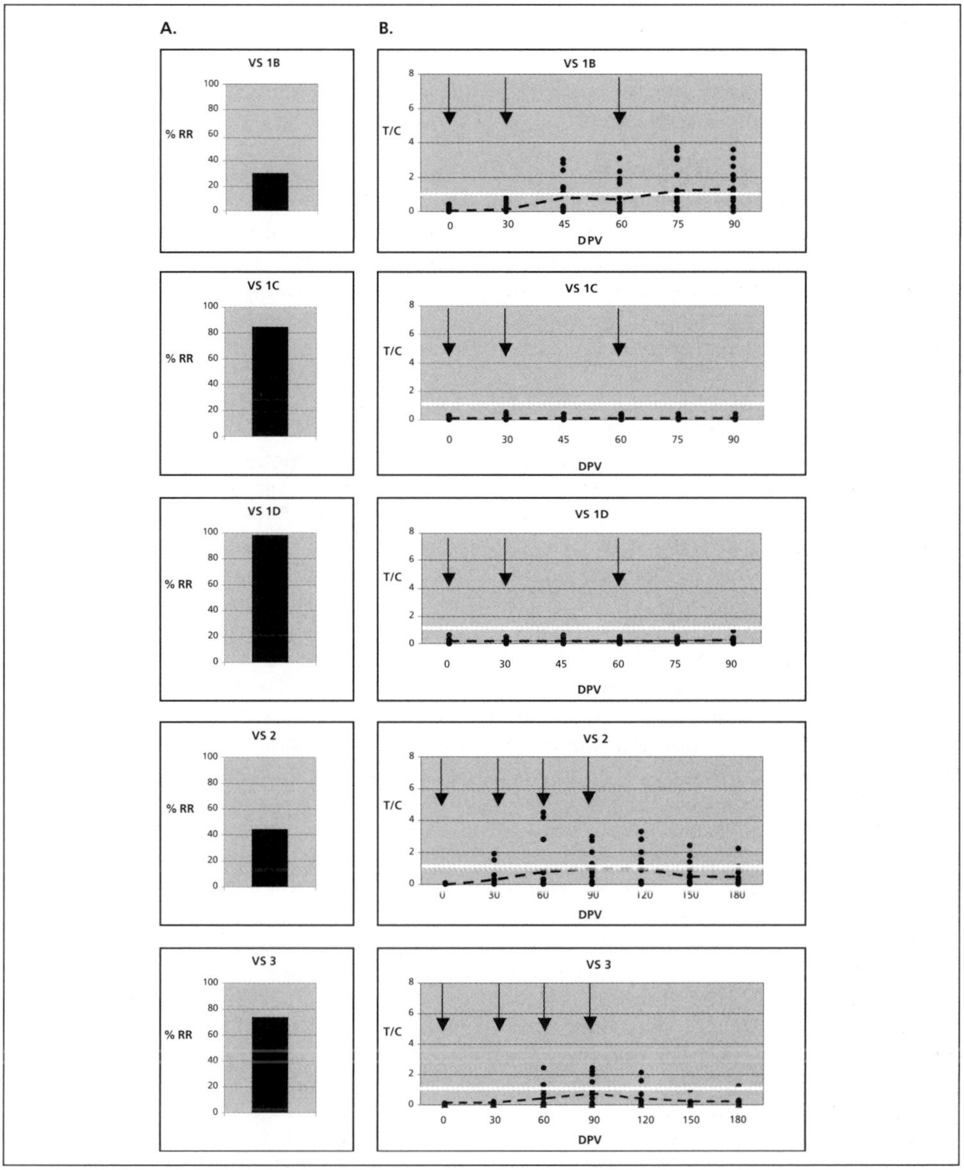

Fig. 3: **Antibody responses to 3ABC induced in animals vaccinated and revaccinated with vaccines prepared with experimental viral suspensions.**

A. %RR in the inhibition ELISA 3ABC for the viral suspensions used for vaccination. Vaccine formulations were different for VS 1, VS 2 and VS 3.

B. Antibody responses to 3ABC obtained in animals vaccinated and revaccinated with vaccines prepared with the viral suspensions following the immunization schemes described in Materials and Methods were evaluated by the NCPanaftosa screening test. Antibody levels are expressed as T/C (absorbance of the sample/absorbance of the cut-off control serum) as described in [14]. Cut-off value is T/C = 1 indicated by the solid line. The dotted line indicates the average value of T/C for all animals on the day post-vaccination (DPV) indicated. Arrows indicate the vaccination schemes.

Discussion and Conclusions

Sero-surveys based on the detection of antibodies against NCPs is of great relevance for accompanying the progress of eradication campaigns in South America, and for supporting the certification of areas free of FMDV by the OIE. In areas under systematic vaccination, it is critical to guarantee that the vaccines made of semi-purified FMD virions do not induce antibodies against those proteins which act as serological probes in the immuno-enzymatic tests used in sero-surveys.

It is unlikely that any purification process can completely remove NCPs, and there is currently no method for predicting the immunogenicity of any proteins remaining. Therefore demonstrating that the vaccine does not induce significant levels of NCP in immunized animals currently represents the only means of supporting information on the purity related to NCP.

Although various experimental and field models have demonstrated essentially no vaccine-induced NCP antibody interference with the PANAFTOSA's immuno-enzymatic system used in South America, there has been a higher background of NCP antibody reactivity with negative test values in vaccinated regions without FMDV for at least the past five years when compared with regions without vaccination [11]. Understanding that this is a critical issue and being aware that modern vaccines may use alternative procedures for antigen concentration and purification, the development of a tool to evaluate the levels of NCP 3ABC in the viral suspensions that may interfere with correct interpretations of sero-surveys was of utmost importance.

The inhibition ELISA 3ABC proved to be an appropriate tool for assessing the levels of 3ABC in viral suspensions destined for vaccines. To visualize the effect of minimal levels of 3ABC in the induction of antibodies, forced vaccination conditions were applied in our studies. This approach supports the choice of the most appropriate purification scheme during vaccine production. The results presented in this work indicate that for each manufacturing procedure the critical levels of NCPs that avoid interference in sero-surveys will need to be established. This is somehow expected, taking into account that the induction of antibodies is highly dependent on various factors including adjuvants, immuno-stimulants, scheme of virus concentration/purification, and inherent variations of individual animal responses, etc. Additionally this approach can be easily adapted to continuous on-line testing.

Studies need to be conducted to define correlations between %RR and induction of antibodies under the standard vaccination conditions used in the field, for each production process. Once these correlations have been evaluated, the measurement of NCPs/dose of viral suspensions through the inhibition ELISA 3ABC opens the way for its application for batch control, allowing for a well-timed decision-making policy before final vaccine formulation.

Results presented in this work indicate that with the use of conventional procedures for concentration/purification, vaccines can be produced with purity regarding NCPs that should not induce antibody responses that interfere with sero-surveys.

To minimize the effect of vaccination in the interpretation of surveys, National Reference Laboratories in South America are establishing procedures for the systematic control of vaccine batches through the evaluation of antibody induction to NCP at 30 days post-vaccination, together with background experience showing that revaccinations do not induce antibody responses. To implement guidelines to respond to the quality required to avoid NCP interference in the sero-surveys, the Veterinary Authorities need to take into account, among others, conditions such as the time elapsing between the last vaccination and collection of samples and duration of immune responses. In this context, it can be clearly seen that short live responses were frequently observed (results herein and data not shown).

A deeper understanding of the overall behaviour of vaccine purity in terms of the induction of NCP antibodies, under the conditions used for field immunization and taking into account sero-survey designs, is highly beneficial for implementing regulatory requirements that are scientifically sound and feasible.

Acknowledgements

The valuable and excellent technical work of Mr. Carlos A. de Senna Costa, Luis Claudio Amparo dos Santos, Alberto Tadeu Botelho de Freire, Airton Gaudenci, is greatly appreciated.

References

1 Organización Panamericana de la Salud (OPS): Plan de acción: Programa Hemisférico de Erradicación de la Fiebre Aftosa en América del Sur. Reunión del Comité Hemisférico de Erradicación de la Fiebre Aftosa, 6-7 de junio, Washington, DC, OPS, Washington DC, 1988, p 32.

2 Van Bekkum JG, Frenkel HS, Frederiks HHJ, Frenkel S: Observations on the carrier state of cattle exposed to foot-and-mouth disease virus. Tijdschr Diergeneeskd 1959;84:1159-1164.

3 Sutmoller P, McVicar JW, Cottral GE: The epizootiological importance of foot-and-mouth disease carriers. I. Experimentally produced foot-and-mouth disease carriers in susceptible and immune cattle. Arch. Gesamte Virusforsch 1968;23:227-235.

4 Sutmoller P, Barteling SS, Casas Olascoaga R, Sumption KJ: Control and eradication of foot-and-mouth disease. Virus Research 2003;91:101-144.

5 Bergmann IE, Augé de Mello P, Neitzert E, Beck E, Gomes I: Diagnosis of persistent aphthovirus infection and its differentiation from vaccination response in cattle by use of enzyme-linked immunoelectrotransfer blot analysis with bioengineered nonstructural viral antigens. Am J Vet Res 1993;54:825-831.

6 Bergmann IE, Malirat V: Performance of a rapid enzyme-linked immunoelectrotransfer blot assay for detection of cattle exposed to foot-and-mouth disease virus. Bull Panam Foot-and-Mouth Dis Center 1995;61:40-44.

7 Bergmann IE, Malirat V, Dias LE, Dilandro R: Identification of foot-and-mouth disease virus-free regions by use of a standardized enzyme-linked immunoelectrotransfer blot assay. Am J Vet Res 1996;57:972-974.

8 Bergmann IE, Astudillo V, Malirat V, Neitzert E: Serodiagnostic strategy for estimation of foot-and-mouth disease viral activity through highly sensitive immunoassays using bioengineered nonstructural proteins. The Vet Quart 1998;20:S6-S9.

9 Bergmann IE, Malirat V, Neitzert E, Beck E, Panizzutti N, Sánchez C, Falczuk A: Improvement of a serodiagnostic strategy for foot-and-mouth disease virus surveillance in cattle under systematic vaccination: a combined system of an indirect ELISA 3ABC with an enzyme-linked immunoelectrotransfer blot assay. Arch Virol 2000;145:473-489.

10 Bergmann IE, Malirat V, Neitzert E, Correa Melo E: Validation of the I-ELISA/EITB system for use in foot-and-mouth disease surveillance: overview of the South American experience, in Dodet B, Vicari M (eds): Foot-and-Mouth Disease: Control Strategies. Editions Scientifiques et Médicales. Elsevier SAS, 2003, pp 361-370.

11 Bergmann IE, Malirat V, Neitzert E, Correa Melo E: Vaccines and companion diagnostic tests for Foot-and-Mouth Disease Virus. An overview of the experience in South America, in Brown F, Roth J (eds): Vaccines for OIE list A and Emerging Animal Diseases. Dev Biol. Basel, Karger, 2003, vol 114, pp 59-65.

12 Bergmann IE, Neitzert E, Malirat V, Ortiz S, Colling A, Sánchez C, Correa Melo E: Rapid serological profiling by enzyme linked immunosorbent assay and its use as an epidemiological indicator of foot-and-mouth disease viral activity. Archives of virology 2003;148:891-901.

13 Bergmann IE, Malirat V, Neitzert E, Correa Melo E: Vaccination: Foot-and-Mouth Disease Experience in South America, in Schudel A, Lombard M (eds): Control of Infectious Animal Diseases by Vaccination. Dev Biol. Basel, Karger, 2004, vol 119, pp 273-282.

14 Malirat V, Neitzert E, Bergmann IE, Maradei E, Beck E: Detection of cattle exposed to foot-and-mouth disease virus by means of an indirect ELISA test using bioengineered nonstructural polyprotein 3ABC. The Vet Quart 1998;20:S24-S26.

15 Neitzert E, Beck E, Augé de Mello P, Gomes I, Bergmann IE: Expression of the Aphthovirus RNA polymerase gene in Escherichia coli and its use together with other bioengineered nonstructural antigens in detection of late persistent infection. Virology 1991;184:799-804.

16 Manual of Standards for Diagnostic Tests and Vaccines: Chapter 2.1 Foot-and-Mouth Disease. Office International des Epizooties, Paris, 2000.

17 Favre H, Mougeot H, Fargeaud D, Brun A, Dubouclard C, Fontaine J: Utilisation d'une lignée cellulaire adaptée à la culture en suspension pour la préparation des vaccins anti-aphteux. Bull Off Int Epiz 1975;81:1201-1229

Dr. I.E. Bergmann, Pan American Foot-and-Mouth Disease Center, PO Box 589, Av. Pres. Kennedy 7778 Duque de Caxias, Río de Janeiro, RJ, Brasil.
E-mail: ibergman@panaftosa.ops-oms.org

Vannier P, Espeseth D (eds): New Diagnostic Technology: Applications in Animal Health and Biologics Controls. Dev Biol (Basel). Basel, Karger, 2006, vol 126, pp 251-252.

Correlation Between Results of a Candidate in vitro Assay and Established in vivo Assays for Potency Determination of Inactivated Newcastle Disease Vaccines

I. Claassen[1], *C. Milne*[2], *R. Maas*[1]

[1] CIDC-Lelystad The Netherlands
[2] European Directorate for the Quality of Medicines, Strasbourg, France

Introduction

Quality control of both veterinary and human vaccines often involves the use of experimental animal studies to prove potency and safety. A number of alternative assays that reduce or replace animal testing have been developed over the past few years. Wherever possible and appropriate these assays are eventually included in European Pharmacopoeia (Ph.Eur) monographs.

The development, validation and finally the acceptance of alternative assays for quality control of biologicals is a long and costly process. Manufacturers may develop assays that can be used for a specific product but that are not always applicable to a wider range of similar vaccines made by other companies. This is however necessary for inclusion into specific vaccine monographs. Another important consideration in the development of a standardised alternative assay is the availability of specific reagents over time and the need and availability of a standardised references preparations.

In this paper we will describe the process of development, validation and implementation of a specific alternative assay for quality control. In this particular case, choices were made early in the process that proved to be essential for the success of the project. We will place this in the broader perspective of development of alternatives for animal tests for quality control of biologicals.

If combined with experience from other laboratories that have been involved in similar projects this could lead to guidance based on "best practices".

Results

Over the past eight years an assay to quantify Newcastle disease virus (NDV) antigens has been developed, validated, presented to Group 15V of the Ph. Eur., and once again validated in two ring trials. It was found to be fit for the purpose and a proposal has been made to include the test in the relevant monograph of the Ph.Eur.

A quantification assay for the Haemagglutinin-Neuraminidase (HN) protein of (NDV) was developed at CIDC-Lelystad as a candidate in vitro potency test for inactivated Newcastle disease (ND) vaccines. In studies performed at CIDC-Lelystad, a high correlation was demonstrated between the results of this candidate in vitro potency assay and the results of the serological potency assay (Ph. Eur. monograph 0870; test A). Furthermore, a high correlation between the serological data (Haemagglutination Inhibition-antibody titres) and clinical protection after challenge was demonstrated.

The potential of this alternative assay was realized and a project was started with the goal of having the test included in the monograph for inactivated ND vaccines. In-house validation experiments were performed and critical reagents such as monoclonal antibodies and reference preparations were produced in large quantities. Hybridoma cell lines were produced as seed lots for future production and proof was generated that the cell lines were stable producers to guarantee future needs. It was demonstrated that the assay could be easily transferred to other laboratories. A dossier containing this information and the scientific rationale to support the assay was submitted to Group 15V of the Ph. Eur.

The group concluded that further proof of the suitability of the assay was still needed. Therefore correlation between in vivo and in vitro potency assays was confirmed in a collaborative pre-validation study. This study was organized in close collaboration with the European Directorate for the Quality of Medicines (EDQM). In the pre-validation study three Official Medicines Control Laboratories (OMCLs) determined both the NDV-HN antigen content and the in vivo potency (vaccination-serology and vaccination-challenge) of six commercial vaccine batches. The conclusion of the pre-validation study was that a large-scale collaborative study should be organised under the aegis of the EDQM Biological Standardisation Programme to validate the in vitro method and the suitability of the reference preparation.

In brief, 14 laboratories (eight OMCLs and six vaccine manufacturers) determined the NDV-HN antigen content of nine different vaccines in three independent tests (BSP055). The vaccine batches were produced by five different manufacturers and represent a quantitative range of NDV antigen content. Importantly, one vaccine batch with insufficient potency and one poultry vaccine not containing NDV were also included. Statistical evaluation of the results indicated that the antigen content could be determined with high precision. A good repeatability as well as reproducibility was demonstrated. Furthermore, all laboratories found a similar ranking of the vaccines, based on the antigen content. Comparison of the antigen content and the in vivo potency of a series of vaccines with relatively low potencies indicated that a threshold relative antigen level of 7.0 antigen units per dose would discriminate between vaccine batches with sufficient and insufficient potency. An in vitro assay with this threshold level for antigen content did not result in any false positive results and only a limited number of false negative results in the BSP055 study. It was concluded that the in vitro measurement of the antigen content of inactivated ND-vaccines with the proposed method is a reliable alternative potency assay. As a result, the critical reagents and reference preparation have been adopted by the Ph Eur Commission as BRPs and Group 15V has proposed that the method be included as a new method in monograph 0870 on inactivated ND-vaccines.

Dr. I. Claassen, CIDC-Lelystad, Dept. of Bacteriology & TSE, P.O. Box 2004, NL-8203 AA Lelystad, The Netherlands.
E-mail: ivo.claassen@wur.nl

Vannier P, Espeseth D (eds): New Diagnostic Technology: Applications in Animal Health and Biologics Controls. Dev Biol (Basel). Basel, Karger, 2006, vol 126, pp 253-259.

Risk Assessment in New and Conventional Vaccines

D. Dory, R. Gravier, A. Jestin

Viral Genetics and Biosafety Unit, French Food Safety Agency (Afssa), Ploufragan, France

Abstract: New technology, and in particular new molecular biology tools, have opened up access to new generations of vaccines or control tests. DNA vaccines are of particular interest since only the gene is injected and, as a consequence, specific risks are associated with this kind of vaccination. Some strategies are presented here to improve DNA vaccine efficacy, which can in some cases result in the reduction of the quantity of plasmid needed and therefore limits the risk. Specific tools to study the distribution of plasmids inside the body as well as tools to detect plasmid insertion inside the host genome have been developed. A final part of this paper will present briefly the new tools which have been or can be developed to detect contaminants in vaccines.

INTRODUCTION

Vaccines can be classified as conventional and new vaccines. Conventional vaccines are directly derived from the pathogen, as for example from viruses, responsible for the disease. These viruses can be attenuated, inactivated, or protein fractions can be extracted and purified.

New vaccines on the other hand are obtained after molecular biology modifications, as for example insertion or deletion of genes in live viruses, insertion of genes in viral vectors, in plasmids for DNA vaccines or in baculoviruses for the production of sub-unit recombinant vaccines.

DNA vaccines are of particular interest since only the gene is injected. Plasmids encoding the protein of interest are injected inside the body where the protein will be synthesized. The immunity induced against the endogenously produced vaccine results in antigen presentation through both the major MHC I and II pathways, resulting in a broad spectrum of humoral and cellular immune responses, similar to what one would expect from a natural viral infection. Furthermore, there are also specific risks different from those encountered with other kind of vaccines. The improvements of the vaccine efficacy can help to limit the risk by reducing the quantity of plasmids injected. Tools have been developed to detect plasmids in blood and tissues, and more importantly to detect integration events inside the genomic DNA.

DNA Vaccine Improvements

Strategies are being tested to enhance vaccine efficiencies and progresses have been made. As an illustration, the case of DNA vaccination against Pseudorabies virus infection in swine can be used. In this model, the base of the DNA vaccine is the association of three plasmids encoding PRV-gB, gC and gD (Fig. 1). A first way of investigation targets the amplification of the level of transcription of the encoded protein (Fig. 1). For this purpose, a new generation of non-replicating plasmid derived from the Sindbis virus (pSIN) amplifying the level of gene transcription of the proteins of interest has been used [1,2]. A low dose of DNA, i.e. 13 μg of each PRV gB-, gC- and gD-pSIN corresponding to 25 times fewer plasmids than the quantity commonly used, can induce production of neutralizing antibodies as high as that obtained in animals receiving a high dose of DNA, i.e. 340 μg of plasmids (Fig. 2) and induce similar clinical protection against an experimental infection [2]. The fact that a similar clinical protection is achieved by a small quantity of plasmids is important in terms of safety, as the quantity of foreign DNA injected can be considerably reduced. The addition of plasmids encoding the cytokine GM-CSF also helped to improve the DNA-vaccine efficacy (Fig. 1), as the body-weight loss was more attenuated compared to the vaccine without pGM-CSF [3]. CpG-containing oligodeoxynucleotides (Fig. 1), which contain sequences recognised by the TLR-9 receptor, improved the PRV-specific humoral and cellular immune responses and provided better clinical protection against lethal PRV-infection [4]. Heterologous prime-boost immunizations consisting of single DNA priming followed by booster immunization with replication-incompetent recombinant viruses (Fig. 1) have also been shown to improve the vaccination efficacy strongly. Animals primed with Sindbis virus-derived plasmids encoding viral glycoproteins gC and gD (gC- and gD-pSIN) and subsequently booster immunized with Orf virus (ORFV; Parapoxvirus) recombinants expressing gC and gD (D1701-VrVgC and -VrVgD) were protected against lethal challenge infection with no signs of PRV-specific clinical symptoms and excellent body weight loss attenuation. In this last case, the DNA vaccine and the recombinant viruses have to be taken into account to assess the risk of this kind of vaccination.

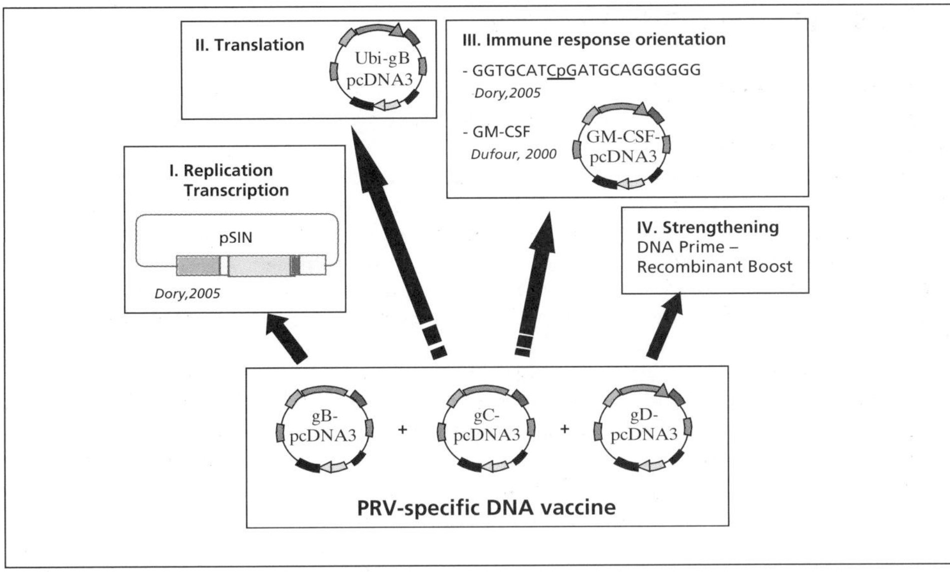

Fig. 1: Improvements of PRV-specific DNA vaccine at Afssa.

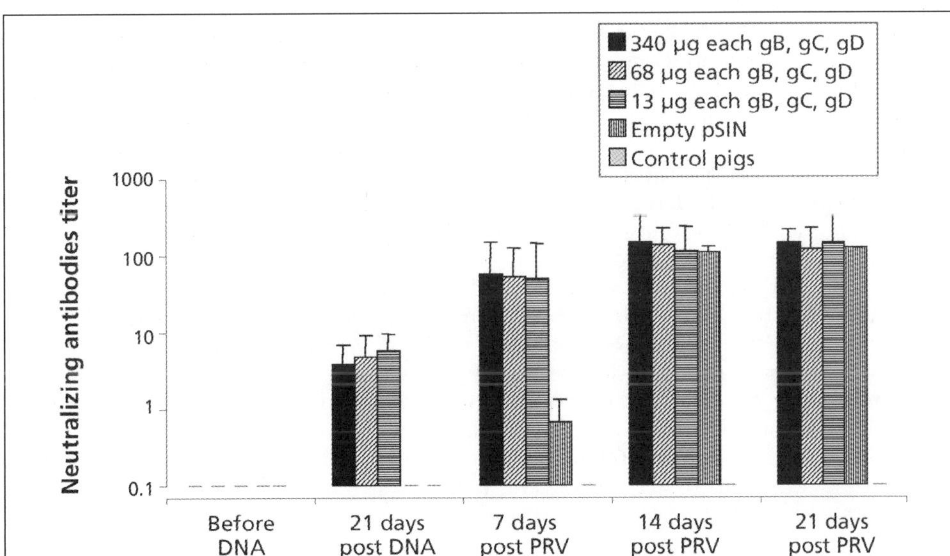

Fig. 2: Virus neutralizing antibody levels in serum before DNA immunization, 21 days after DNA immunization by a single injection with various quantities of each PrV gB-, gC- and gD-pSIN or empty pSIN, 7, 14 and 21 days after viral challenge with 2 x $10^{6.3}$ $TCID_{50}$ of the NiA3 virulent PrV strain via nasal instillation. A non-immunized and non-challenged control group was assessed in parallel. The seroneutralization assay was performed on a swine kidney cell line. The mean titres, including standard error bars, for each group of pigs are expressed as the highest serum dilution inhibiting the cytopathogenic effect in 50% of PRV-infected PK15 cell line cultures.

Biosafety of DNA Vaccines

Antigen-expressing plasmids are amplified in bacterial cultures, making the production of DNA vaccines easy, rapid and economical. Bacterial production of plasmid DNA also avoids the risk of potential contamination with the viruses/proteins present in eukaryotic cell lines used to produce conventional vaccines. The first risk due to the nature of the vaccine is to elicit the production of anti-DNA antibodies. The level of these antibodies is normally low after vaccination in healthy organisms, but can increase dramatically in individuals with auto-immune disease [5]. Another important issue is the incidence of plasmids inside the body. Techniques based on PCR or real-time PCR have been developed to detect and quantify plasmids in blood or tissues. For example, plasmids encoding PRV glycoproteins gB, gC and gD were injected by intravenous or intramuscular routes and were researched in swine sera by real time PCR at various times following injection (Fig. 3). After intravenous injection, the quantity of plasmid in serum was maximal at 2 min, and then diminished rapidly, becoming undetectable at 10 min. After intramuscular injection, plasmids started to be detected at high levels between seven and 13 min. About 3000 copies/ml of blood were needed to detect signals (Fig. 3). Plasmids can also be detected in different tissues in mice following intramuscular DNA injection [6]. One hour and a half after DNA injection, plasmids were detected in all the organs analysed (liver, brain, kidney, spleen, heart, testis, lung and muscle). Eight hours after injection, plasmids were undetected in the majority of the tissues, except for brain, lung and muscle. Thereafter, only the muscle was positive until day 14. One month after injection plasmids were undetected in all the tissues tested. The limit of detection of the technique was of 1000 copies of plasmid/mg of tissue, meaning that traces (fewer than 1000 copies/mg) of plasmids can still be present within the tissue without being detected.

With the classical PCR techniques, it is not possible to distinguish between plasmids which are present within or associated with the tissue and plasmids which are integrated in the host genome. It is important to be able to detect integration events since integration can be responsible for insertional mutagenesis (activation of an oncogen or functional deletion of a tumour suppressor gene) or of transmission of foreign DNA to the descent. Techniques based on the separation of high molecular weight genomic DNA from extrachromosomal plasmid digested by restriction enzymes [7] or not [8] by gel electrophoresis were useful for detecting plasmids associated with genomic DNA, but were not sufficient for researching insertion events. For this purpose a technique called Repeat-Anchored Integration Capture PCR has been developed which, after a two step nested PCR, permits the amplification of a DNA fragment composed of a part of the integrated plasmid and a part of the genomic DNA [9]. The detection limit of this technique was one copy of plasmid per 150,000 diploid cells. As an example, the authors identified four random integration events in mice muscles after intramuscular DNA injection followed by electroporation.

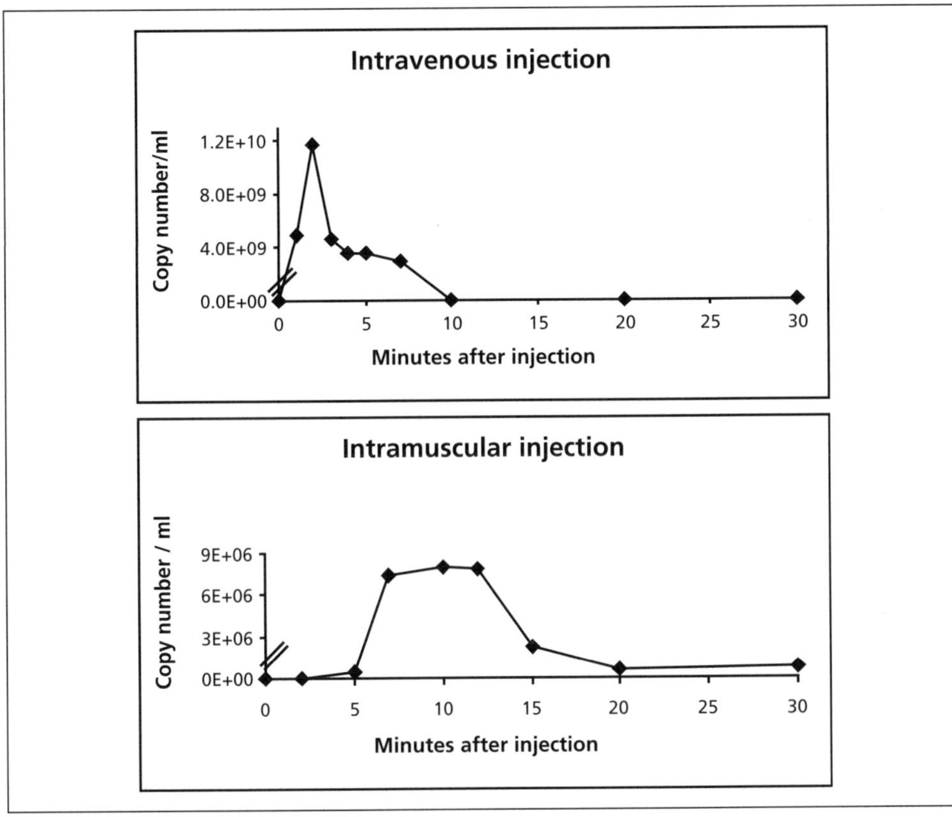

Fig. 3: Measurement of PRV gB encoding plasmids in the blood of pigs following intravenous and intramuscular injections.
The quantity of plasmids was determined by real time PCR.
Limits of detection: 3000 copies/ml serum

Molecular Biology Based Tools for the Control of Vaccines

Molecular biology not only enables new vaccines to be developed, but can also be used to develop new tools to control new and conventional vaccines. PCR-based techniques have been developed for the detection of contaminants in vaccines. For example, pestiviruses [10], leucosis virus [11] or avian reoviruses [12] have been researched in vaccines by this technique. It is also possible to research several viruses at a time by multiplexing PCR [13].

We can also imagine adapting newly developed techniques to research several pathogens simultaneously. DNA macro-array with a limited number of spotted oligonucleotides corresponding to different pathogens can potentially be developed to research contaminants in vaccines. For this purpose it is necessary to perform first a PCR or RT-PCR to amplify contaminant fragments. Alternatively, the amplicons can be added to microspheres with specific physical characteristics. Each kind of microsphere is adsorbed with one type of specific oligonucleotide corresponding to one pathogen. The hybridization is analysed by flow cytometry [14].

Conclusion

Molecular biology offers the possibility of developing new vaccines. Among these, DNA vaccines are of particular interest. There is no risk of contamination of the plasmid during the production process. The risk with this kind of vaccination is due to the DNA nature of the vaccine. In fact, the plasmids can potentially be integrated into the host genome. Techniques have been developed to detect plasmids in the blood and tissues. The Repeat-Anchored Integration Capture PCR is able to detect and to characterize integration events.

For the conventional and new vaccines, techniques can be adapted from molecular biology tools to research contaminants. These tests have the advantage of offering flexibility, multiplexing and high throughput.

Acknowledgements

Research workers from the Viral Genetics and Biosafety Unit (Afssa, Ploufragan, France) are acknowledged for their help: Véronique Béven, Yannick Blanchard, Yann Moalic, Béatrice Grasland and Christophe Loizel. Collaboration with Hanns-Joachim Rziha (Friedrich-Loeffler-Institute, Tübingen, Germany), Fernando Rodriguez (Cresa, Barcelona, Spain) and Michel Laurentie (Afssa, Fougères, France) were greatly appreciated. The authors are grateful to the staff of the Healthy Pigs Production and Experimentation Section (Afssa, Ploufragan, France) headed by Roland Cariolet for the expert manipulation of pigs. Some studies were supported by grant QLRT-1999-459 from the European Community.

References

1 Dufour V, De Boisseson C: Use of a Sindbis virus DNA-based expression vector for induction of protective immunity against pseudorabies virus in pigs. Vet Immunol Immunopathol 2003;93:125-134.

2 Dory D, Torché AM, Béven V, Blanchard P, Loizel C, Cariolet R, Jestin A: Effective protection of pigs against lethal Pseudorabies virus infection after a single injection of low-dose Sindbis-derived plasmids encoding PrV gB, gC and gD glycoproteins. Vaccine 2005;23:3483-3491.

3 Dufour V, Chevallier S, Cariolet R, Somasundaram S, Lefèvre F, Jestin A, Albina E: Induction of porcine cytokine mRNA expression after DNA immunization and pseudorabies virus infection. J Interferon Cytokine Res 2000;20:885-890.

4 Dory D, Béven V, Torché AM, Bougeard S, Cariolet R, Jestin A: CpG motif in ATCGAT hexamer improves DNA-vaccine efficiency against lethal Pseudorabies virus infection in pigs. Vaccine 2005;23:4532-4540.

5 Klinman DM, Takeno M, Ichino M, Gu M, Yamshchikov G, Mor G, Conover J: DNA vaccines: safety and efficacy issues. Springer Semin Immunopathol 1997;19:245-256.

6 Kim BM, Lee DS, Choi JH, Kim CY, Son M, Suh YS, Baek KH, Park KS, Sung YC, Kim WB: In vivo kinetics and biodistribution of a HIV-1 DNA vaccine after administration in mice. Arch Pharm Res 2003;26:493-498.

7 Nichols WW, Ledwith BJ, Manam SV, Troilo PJ: Potential DNA vaccine integration into host cell genome. Annals of the New York Academy of Sciences 1995;772:30-39.

8 Ledwith BJ, Manam S, Troilo PJ, Barnum AB, Pauley CJ, Griffiths TG, Harper LB, Beare CM, Bagdon WJ, Nichols WW: Plasmid DNA vaccines: Investigation of integration into host cellular DNA following intramuscular injection in mice. Intervirology 2000;43:258-272.

9 Wang Z, Troilo PJ, Wang X, Griffiths TG, Pacchione SJ, Barnum AB, Harper LB, Pauley CJ, Niu Z, Denisova L, Follmer TT, Rizzuto G, Ciliberto G, Fattori E, Monica NL, Manam S, Ledwith BJ: Detection of integration of plasmid DNA into host genomic DNA following intramuscular injection and electroporation. Gene Ther 2004;11:711-721.

10 Studer E, Bertoni G, Candrian U: Detection and characterization of pestivirus contaminations in human live viral vaccines. Biologicals 2002;30:289-296.

11 Hauptli D, Bruckner L, Ottiger HP: Use of reverse transcriptase polymerase chain reaction for detection of vaccine contamination by avian leukosis virus. J Virol Methods 1997;66:71-81.

12 Bruhn S, Bruckner L, Ottiger HP: Application of RT-PCR for the detection of avian reovirus contamination in avian viral vaccines. J Virol Methods 2005;123:179-186.

13 Liu HJ, Lee LH, Shih WL, Lin MY, Liao MH: Detection of infectious bronchitis virus by multiplex polymerase chain reaction and sequence analysis. J Virol Methods 2003;109:31-37.

14 Dunbar SA: Applications of Luminex xMAP technology for rapid, high-throughput multiplexed nucleic acid detection. Clin Chim Acta 2006;363:71-82.

Dr. D. Dory, Viral Genetics and Biosafety Unit, Afssa, BP 53, FR-22440 Ploufragan, France.
E-mail : d.dory@ploufragan.afssa.fr

Vannier P, Espeseth D (eds): New Diagnostic Technology: Applications in Animal Health and Biologics Controls. Dev Biol (Basel). Basel, Karger, 2006, vol 126, pp 261-270.

DNA Vaccines: Safety Aspect Assessment and Regulation

T.D.E. Medjitna, C. Stadler, L. Bruckner, C. Griot, H.-P. Ottiger

Institute of Virology and Immunoprophylaxis, Mittelhäusern, Switzerland

Abstract: For licensing purposes, besides the immunogenic aspects, deoxyribonucleic acid (DNA) vaccines present safety considerations that must be critically assessed during preclinical or/and clinical safety studies. The major concerns with regard to safety are integration of the plasmid DNA into the host genome, adverse immunopathological effects, the formation of anti-DNA antibodies resulting in auto-immune disease and the use of novel molecular adjuvants. Moreover, for veterinary vaccines intended to be used in husbandry animals, food safety aspects will become an important issue. All new vaccine candidates should therefore be thoroughly tested in target animals, keeping in mind that for food producing animals, the products will be consumed. Finally, a further safety aspect of interest concerns the possible spread of genetic material to the environment, by the potential transformation of the environmental microflora with only a few copies of complete or fragmented plasmid.

These are issues that need to be considered in the final scientific decisions underpinning the registration of vaccines. Thus, to establish criteria for guidance and regulations for industry and licensing authorities, a project has been initiated to assess such risks of plasmid DNA vaccinations. Major emphasis will be placed on aspects such as the biodistribution of plasmid in vaccinated animals. This paper is intended as a contribution to the debate on the use of biotechnology in the future and should facilitate further discussions on the various safety aspects of DNA-based immunisations.

Introduction

Vaccines are an important tool in the control and prevention of a number of viral, bacterial and parasitic diseases of humans, domestic livestock and pet animals, improving both the health and welfare of animals and also in protecting public health.

Nucleic acid-mediated or DNA vaccines are a new way for inducing an immune response in animals or humans for protective or therapeutic purposes. This method of immunization involves the direct introduction of naked DNA into the muscle or epidermis of the vaccinated subject by direct injection or particle bombardment. Thus, the use of these vaccines has potentially wide-ranging applications, from vaccinations against a variety of infectious diseases to use in cancer therapy.

Nucleic Acid Immunization

A DNA vaccine is composed of a plasmid carrying an inserted gene sequence encoding one or more antigen(s) of interest, such as proteins or glycoproteins which may be in vivo expressed by the cellular machinery, when taken up by cells of the subject to be vaccinated or inoculated, initiating then a specific immune response, strong enough to protect against disease. Besides the gene(s) encoding one or more immunogenic proteins, in most cases, DNA vaccine vectors contain genetic elements which are of importance to optimize the expression and immunogenicity, such as eukaryotic promoter sequences necessary to start gene transcription in mammalian cells, multiple cloning sites for the insertion of heterologous genes, polyadenylation (polyA) signals for terminating and stabilization of mRNA transcript, selectable markers and origin of replication sequence for providing high plasmid copies in bacteria [1].

The concept of DNA vaccination was first described in the early 1990s by Wolff et al [2], who delivered a reporter gene inserted into a naked plasmid DNA directly into mice muscle, resulting in the expression of encoded proteins in the muscle cells. Although the exact process by which transcription occurs in this system has not been completely elucidated, it seems that muscle injection is more effective at inducing an immune response than other routes or tissues that have been evaluated. The effectiveness of naked DNA in provoking immunity opened new perspectives, potentially to achieve immune responses against cancer and a wide variety of viral, bacterial and parasitic pathogens [3-9]. Moreover, the simplicity of the technology and its efficacy in protecting animals (companion and husbandry animals), including neonatal animals in the presence of maternal antibodies, have led to the implementation of intense researches on the development of DNA vaccines in the veterinary field [10].

There are several advantages of DNA-based immunizations over the use of conventional live-attenuated or inactivated vaccines (see Table 1). A unique advantage of DNA vaccines is the ability of the protein to be expressed in vivo by the host cell which in turn can be manipulated through recombinant engineering to mimic the conformational and post-translational form produced during a natural infection. This method of immunization uses the production of antigen in vivo for the induction of specific cellular and humoral immunity. Some of the expressing cells display the protein on the cell surface, leading to a cell-mediated immune response (mobilization of the T lymphocytes). The protein can also be secreted or released from the cells where it is taken up and processed by antigen-processing cells, leading to a humoral immune response (stimulation of the production of antibodies specifically directed against the immunogen). Beside the advantage of eliciting both humoral and cell-mediated responses [1] as compared to primarily humoral responses of conventional (inactivated) vaccines, their use is not affected by pre-existing antibodies (i.e. protection of neonatal animals in the presence of maternal antibodies). Furthermore, in contrast to modified live vaccines, no replication, no reversion to virulence or shedding of infectious pathogens can occur. The ease of production of plasmid vectors and their ability to encode more than one immunogen also offers the possibility to design multivalent or even combined vaccines, expressing different antigens including sequence motifs (cytokines) as molecular adjuvants.

Table 1: Issues for consideration in DNA vaccination.

Advantages	Problems
Simple principle (delivery of gene(s) for desired pathogen/antigen)	No experience with potential long-term safety aspects
Easy production (generic manufacturing)	Public perception and consumer acceptance
Long-term stability of final product (heat stability)	Unexpected and undesirable consequences of the persistence expression of a foreign antigen (risk of adverse immune effects by induction of immunological tolerance, anti-DNA antibodies, anaphylaxis, hyperimmunity, tumour development)
More than one gene can be targeted simultaneously (priming in mixed immunizations)	Risk of integration of DNA sequences into host chromosomes and/or germ-line transmission (insertion of an active oncogene, insertional activation of a host proto-oncogene, insertional deactivation of a tumour-suppressor gene)
No reversion to virulence or live virus shedding	Risk of spread of genes to the environment, antibiotic resistance markers
Efficacy in the presence of pre-existing antibodies (protection of neonatal animals in the presence of maternal antibodies; foetal immunization)	Local reactogenicity and systemic toxicity (injection site tissue, persistence of vector in somatic cells)
Designer immune response: Incorporation of molecular adjuvants (cytokines/costimulatory molecules) able to stimulate desired immune response not induced by wild-type disease	Biodistribution (fate of plasmid in vaccinated animals is not known)
Immune profile (induction of cellular and humoral immunity)	Development of regulatory requirements needed (harmonized acceptance/rejection criteria)

Field Application of Nucleic Acid Vaccines

Until recently, no veterinary or human DNA vaccines had been licensed. In humans a number of clinical trials with DNA vaccines against diseases such as acquired immune deficiency syndrome, hepatitis and malaria have been conducted [11,12]. Unfortunately, they usually failed to demonstrate in human volunteers, during clinical studies, a potent immune response as in experimental animals.

On the other hand, in 2005, the first DNA vaccine efficient in protecting horses against West Nile virus has been approved. Shortly after, a second DNA vaccine for aquacultured fish received marketing authorization. This vaccine is intended to protect against Infectious Haematopoietic Necrosis Virus (IHNV) infections responsible of serious losses for the salmon breeding industry in case of outbreaks. Among the many experimental nucleic acid-based vaccines tested in various animal species, other DNA vaccines against viral diseases (i.e. in fish) have given promising results in inducing rapid and long-lasting protection. It is very likely that in due course the number of other nucleic acid vaccines entering the market will increase.

Quality and Safety Issues

In recent years, quality assurance and Good Manufacturing Practice (GMP) have become increasingly important in the pharmaceutical industry. The need to maintain control over all aspects of GMP cannot be over-emphasized. In addition to those quality considerations common to all veterinary vaccines, DNA vaccines raise a number of specific concerns that must be addressed. Many of the studies that are required will be similar to those for live viral vaccines.

As with other veterinary vaccines, three aspects must be addressed when it comes to safety: the vaccinated animals, the environment and the consumer. Because of the novelty of the DNA-based immunization technology, however, there are a number of theoretical safety concerns that need to be considered. These include (i) the fate of the plasmid in the vaccinated animals; (ii) the risk of the integration of DNA sequences into host genome or germline, and (iii) the risk of inducing anti-DNA antibodies. Additional concerns regarding DNA vaccines safety issues are summarized in Table 1.

At the preclinical safety stage the potential integration of the whole or fragmented plasmid sequences into the host genome after administration must be addressed. One of the theoretical consequences could be the insertional mutagenesis in inadequate positions of the genome inducing genomic instability and disturbing some regulatory genes, such as the possible activation of proto-oncogenes or inactivation of tumour suppressor genes which could then give rise to malignant cell development. This may occur by insertion of promoter/enhancer sequences or by specific-motif methylation responsible for cellular gene silencing. So far, chromosomal integration has not been reported [13], but further studies should still be implemented, since integration probability may differ depending on tissue type, method of delivery or adjuvant formulations with increased cellular uptake. The fate and bio-distribution of plasmid DNA must be tested in a wide range of tissues and body fluids at different time intervals.

Since the beginning of the use of plasmid DNA as a vaccine, there have been concerns about whether repeated DNA vaccinations could generate anti-DNA responses or the induction of systemic auto-immune disorder. Auto-immunity could not be demonstrated despite the fact that circulating anti-DNA antibodies have been identified after exposure to bacterial DNA or administration of naked plasmids. Induction of anti-DNA antibodies to mammalian DNA seems to need the association of proteins, or Freund's adjuvant to the administered DNA [14]. Moreover, the immune system becomes exposed to DNA on several occasions such as the administration of conventional bacterial or viral vaccines and, to date, no incidents related to anti-DNA auto-antibodies have been reported. Consequently, on this basis, the likelihood that anti-DNA antibodies induce pathogenic effects following the administration of a highly purified DNA vaccine appear to be very low. This was the reason that FDA will not require specific preclinical studies to be performed to assess this issue [15]. It is important to keep in mind that other undesired immuno-pathological reactions could theoretically occur as a consequence of long-term expression of a foreign antigen and the use of molecular adjuvants/co-stimulatory molecules such as cytokines which may modulate a deviated reaction. Thus, an accurate assessment of such additional risks should be performed, identifying possible further causes of adverse reactions. Other safety considerations are specifically related to the environment, particularly the possible spread of genetic material with only a few copies of complete or fragmented plasmids. For food-producing animals, all new vaccine candidates should be thoroughly tested in target animals, bearing in mind that their products will be consumed. A comprehensive review of safety aspects related to the DNA vaccination of food-producing animals has been given by Holm [16].

Regulatory Issues of Veterinary DNA Vaccines

To provide advice and guidance to manufacturers and vaccine developers, regulatory documents have been established and published by national and international administrative organisations, such as the European Medicines Evaluation Agency (EMEA), the Food and Drug Administration (FDA) in the United States of America and the World Health Organization (WHO) (see Table 2).

These guidelines provide useful information and recommendations to assist scientists and industry in the development of such products, give assistance to define tests to be carried out and help to prepare data for the submission of a dossier. The purposes are to enhance and optimize the chances of gaining regulatory pre-approval for clinical trials or approval for market licensing. The recommendations primarily focus on the manufacturing quality process of the products as well as on the preclinical and clinical efficacy and safety issues that need to be addressed. Some guidance documents are general and provide global recommendations and principles that could be applied to all kinds of medicines, while others were established more specifically for DNA vaccines and gene therapeutic products. In general, although regulatory authorities ensure that gene products are effective and safe by drafting strict safety requirements, the public perception does not always seem favourable to such new approaches. As a result of the current debate about the genetically modified organisms (GMO) in food production, the potential relationship between GMO and DNA vaccination could become a sensitive issue. This is mainly due to a general distrust of everything which is related to gene technology. It is therefore important to have a clear regulatory strategy as well as to keep the public clearly informed.

In Switzerland, the legal basis which rules over all aspects of gene technology is laid down in the constitution. Switzerland is probably among the few countries where this is mentioned in its constitution. Indeed, two articles, Art. 119 and Art. 120, for human and non-human field application respectively, highlight the intention to protect people and the environment "against the abuse of gene technology". The next level for specific gene technology regulations are in the laws and ordinances (see Table 2). The following fields are covered: medicine, agriculture and the environment. Activities with gene transfer products (i.e. trial or application for marketing authorization) are more specifically regulated by the federal law on "Therapeutic Products" (SR 812.21) and the ordinance on "Clinical Trials of Therapeutics" (SR 812.214.2).

Several institutions are involved in the regulatory process. The Swiss authority for therapeutic products (Swissmedic) is responsible for approving pre-clinical and clinical trials of gene therapy products and DNA vaccines and for marketing authorisations. For the veterinary vaccines, the Swiss Federal Veterinary Office (BVET-OVF) is the competent authority. The dossiers are reviewed also by the Federal Office for the Environment (BAFV-OFEV, the Federal Office for Public Health (BAG-OFSP) as well as by the Swiss Expert Committee for Biosafety (EFBS-CFSB), which is an independent expert advisory committee in the area of biotechnology, gene technology, environment and health. Furthermore, approval also requires the consent of an ethics commission.

Table 2: Guidance related to quality, safety and efficacy evaluation of DNA-based vaccines and Swiss regulatory framework in the gene technology field.

Institution	Adoption	Reference	Guidance document
EMEA	1995	CPMP/ICH/302/95	Preclinical Safety Evaluation of Biotechnology-Derived Pharmaceuticals.
	1997	CPMP/SWP/465/95	Note for Guidance on Preclinical Pharmacological and Toxicological Testing of Vaccines.
	2001	CPMP/BWP/3088/99	Note for Guidance on the Quality, Preclinical and Clinical Aspects of Gene Transfer Medicinal Products.
	2001	CVMP/IWP/07/98	Note for guidance: DNA Vaccines non-Amplifiable in Eukariotic Cells for Veterinary Use.
	2004	CPMP/VEG/17/03/2004v5	Guidelines on adjuvants in vaccines
FDA	1996	CBER/FDA/96	Points to Consider on Plasmid DNA Vaccines for Preventive Infectious Disease Indications
	2005	CBER/FDA/05	Guidance for Industry: consideration for plasmid DNA vaccines for infectious disease indication.
WHO	1998	-	Guidelines for assuring the quality of DNA vaccines.
Swiss legislation			
Medicine	2000	SR 812.21	Federal Law on Therapeutic Products
	2001	SR 812.214.2	Ordinance on clinical trials of therapeutic products
	1996	SR 818.111	Federal Decree on the control of transplants
	1996	SR 818.111.3	Ordinance on the control of transplants
	1999	SR 814.912	Ordinance on the contained use of organisms (Containment Ordinance)
	1999	SR 832.321	Ordinance on the protection of workers from risks related to exposure to biological agents at work (SAMV)
Environment & Agriculture	2003	SR 814.91	Federal Law relating to Non-human Gene Technology (Gene Technology Law)
	1983-1997	SR 814.01	Federal law on the protection of the environment
	1999	SR 814.911	Ordinance on the release of organisms into the environment
	2003	SR 832.321	Ordinance on the Transboundary Movements of Genetically Modified Organisms
	1998	SR 910.1	Federal law on agriculture
	1998	SR 916.151	Ordinance on seeds
	2001	SR 916.20	Ordinance on plant protection
	2005	SR 916.307.11	Ordinance on GMO feed list
Food	1992	SR 817.0	Federal law on food
	1996	SR 817.021.35	Ordinance concerning the authorisation procedure related to food containing GMO, to food additives containing GMO and auxiliary substances with GMO

NEW PROJECT: SAFETY ASPECTS ASSESSMENT OF DNA VACCINES

Most of the studies assessing potential risks of DNA vaccination have been performed in mice from which conclusions were extrapolated to other species, including humans. Few researchers have addressed the safety issues in livestock animals or poultry. Despite the fact that DNA vaccination in livestock shares similar safety issues as in humans, others are unique in the intended target species.

Thus, a pilot project was initiated to assess the various potential risks of DNA-based immunisation in farm animals. The ultimate purposes of this project are not only to establish experimental protocols and methods for the evaluation of these risks, but also to gain experience with safety issues related to gene transfer immunisation, permitting the establishment of specific criteria for guidance and regulation.

A major emphasis was placed on aspects such as uptake, bio-distribution, persistence and secretion of naked DNA following administration. These are issues concerning food safety aspects and the possible spread of genetic material to the environment if the plasmids are present in body fluids. Other issues are integration and the risk of germ-line transmission. For this reason and because in ovo vaccinations are a commonly applied delivery method, egg-laying hens seem to be a suitable model for such evaluation.

As a first approach, a preliminary study focusing on the stability and excretion of naked DNA plasmids in vitro was conducted in the framework of this project. This approach is related to safety issues that concern the potential transformation of environmental microflora. Thus, the stability was studied by exposing a plasmid DNA to nuclease degradation of different body fluids (urine, saliva, blood serum, milk, egg yolk and albumen). Samples from various species were spiked in vitro with a standard amount of NpCI plasmid expressing the nucleocapsid protein of canine distemper virus [17]. Results of the stability of the plasmid are summarized in Table 3. The presence of the plasmid DNA after spiking and incubation at 37°C for intervals ranging from five seconds to 24 hours indicated that DNA degradation was species-dependent, especially in the matrices of urine and saliva. Specific amplification products could be detected in urine (with the exception of dog) over the whole experimental period of 24 hours, while in saliva, fragments were obtained no more than after 30 minutes of incubation. Considering these results, even if the PCR is a sensitive method, we have to keep in mind that it only detects fragments but gives no evidence of whether plasmid DNA preserves its functional integrity. To test for this, DNA preparations from spiked urine were subsequently transformed and cloned into E. coli to demonstrate the ability of replication after incubation at 37°C. Using the 10 Gold Ultracompetent Cells Cloning System (Stratagene, CA, USA), transformants were obtained from DNA preparations from different animal urine (see Table 4). Surprisingly, spiked swine urine still contained NpCI DNA after an incubation time of 24 hours at 37°C, which was able to replicate in bacteria. This could therefore be an indication that a transformation event might also occur naturally if plasmid DNA is excreted in urine. In mice, detection in urine of 3.2% of the whole radiolabelled λ–phage DNA injected into the abdominal cavity, has been reported [18]. The genetic transformation of the microflora of soil and groundwater with free DNA is also possible and has been demonstrated in several studies [19,20]. In this respect, another aspect would be the spread of antibiotic-resistant genes present in many constructs used for gene transfer immunisation.

Table 3: In vitro stability of naked plasmid DNA in body fluids and hens' eggs.

		Maximum incubation time for plasmid detection	
Samples	**Species**	**1% Agarose gel**	**PCR**
Urine	Cattle	1 h	24 h
	Swine	3 h	24 h
	Dog	5 sec	3 h
	Guinea pig	1 h	24 h
	Monkey	3 h	24 h
Saliva	Cattle	5 sec	30 min
	Goat	15 min	30 min
	Dog	1 min	5 min
	Sheep	2 min	30 min
	Human	30 min	30 min
Blood serum	Cattle	1 h	1 h
	Swine	1 h	24h
	Dog	1 h	1 h
Milk	Dairy cow	1 h	1 h
Egg yolk	Chicken	1 h	24 h
Egg albumen	Chicken	10 min	1 h

Several body fluids sampled from various species were spiked with 100ng of NpCI plasmid, designed to express the nucleocapsid protein of the canine distemper virus, and incubated at 37°C for intervals from five seconds to 24 hours. Phenol/chloroform/isoamylalcohol extraction and DNA purification were performed after the addition of TES buffer (Tris-EDTA-SDS) to stop the nucleases activity. The whole plasmid or just fragments were detected by loading on to a 1% agarose gel or by performing PCR analysis.

Table 4: Transformation into *E.coli* after incubation in urine.

	Incubation time (37°)	**CFU/100ng NpCI**
Dog	5 sec	8
Guinea pig	30 min	8×10^3
Monkey	3 h	3×10^3
Swine	24 h	40

Transformation was performed with XL-10 Gold Ultracompetent Cells (Stratagene cloning system, CA, U.S.A.) by heat shock according to the manufacturer's instructions. Transformants containing NpCI plasmid were selected on LB amp_{50} plates. Transformation rates are expressed as the number of Colony Forming Units (cfu) per 100ng NpCI plasmid DNA initially added to urine sample (iNpCI). The different time steps indicate the maximum incubation interval tested. Small-scale preparations of NpCI were prepared using the Promega Wizard Plus Minipreps DNA purification system (Promega, Madison, U.S.A.) and the integrity was checked by restriction analysis with *XhoI* or *EcoRI*.

As a first in vivo excretion experiment, interest was focused on egg-laying hens to check whether a DNA translocation occurred either during the vitellus and albumen production, or in the germ-line after intramuscular inoculation. Preliminary data show no traces of NpCI plasmid detectable by PCR within 21 days post-inoculation. Because only a small number of animals were used, these data need to be confirmed.

These experiments give a first impression of the stability and in vivo excretion of naked plasmid DNA. Further studies will be needed to assess accurately the secretion of DNA vaccines via other routes (i.e. milk, saliva, urine, faeces etc.).

Conclusion

New technologies have given access to a new generation of vaccines. This new type of vaccines with their molecular adjuvants and delivery system have different action mechanisms and pharmacological properties than traditional products. Although great advances have been accomplished, the whole concept is still in its early stages. Little is still known about the fate of the DNA molecules within and outside the body. More emphasis should therefore be placed on bio-distribution studies. These and other issues today challenge the commercialization of DNA vaccines including, but not limited to, public perception and associations with other recombinant technologies. Regulatory authorities are continually challenged with the task of establishing standards for veterinary vaccines that ensure appropriate levels of purity, safety and efficacy, yet are not so demanding that they inhibit product availability. A reasonable risk-benefit assessment should be provided, taking into consideration animal welfare, consumer health and the environment. Finally, there is a need for well-defined science-based and harmonized regulatory requirements specific to veterinary DNA vaccines in order to provide suitable solutions.

Acknowledgements

Parts of this work has been funded by the Swiss Federal Veterinary Office.

References

1 Gurunathan S, Klinman DM, Seder RA: DNA vaccines: immunology, application and optimization. Ann Rev Immunol 2000;18:927-974.

2 Wolff JA, Malone RW, Williams P, Chong W, Acsadi G, Jani A, Felgner PL: Direct gene transfer into mouse muscle in vivo. Science. 1990;247:1465-1468.

3 Tang DC, DeVit M, Johnston SA: Genetic immunization is a simple method for eliciting an immune response. Nature 1992;356:152-154.

4 Ulmer JG, Donnelly JJ, Parker SE, Rhodes PL, Felgner PL, Dwarki VJ, Gromkowski SH, Deck RR, DeWitt CM, Friedman A, Hawe LA, Leander KR, Martinez D, Perry HC, Shiver JW, Montgomery DL, Liu MA: Heterologous protection against influenza by injection of DNA encoding a viral protein. Science 1993;259:1745-1749.

5 Davis HL, McCluskie MJ: DNA vaccines for viral diseases. Microbes Infect 1999;1:1-7.

6 Strugnell RA, Drew D, Merciaca J, Dinatale S, Firez N, Dunstun SJ, Simmons CP, Vadolas J: DNA vaccines for bacterial infections. Immunol. Cell Biol 1997;75:364-369.

7 Rothel JS, Waterkeyn JG, Strugnell RA, Wood PR, Seow HF, Vadolas J, Lightowlers MW: Nucleic acid vaccination in sheep: use in combination with a conventional adjuvanted vaccine against *Taenia ovis*. Immunol Cell Biol 1997;75:41-46.

8 Wang R, Doolan DL, Le TP, Hedstrom RC, Coonan KM, Charotenvit Y, Jones TR, Hobart P, Margalith M, Ng J, Weiss WR, Sedegah M, de Taisne C, Norman JA, Hoffman SL: Induction of antigen-specific cytotoxic T lymphocytes in human by a malaria DNA vaccine. Science 1998;282:476-480.

9 Benton PA, Kennedy RC: DNA vaccine strategies for the treatment of cancer, in Koprowski H, Weiner DB (eds): DNA Vaccination / Genetic Vaccination. Berlin, Heidelberger, Springer-Verlag, 1998, pp 1-20.

10 Babiuk LA, Lewis PJ, Van Drunen Little-van den Hurk S, Tikoo S, Liang X: Nucleic acid vaccines: veterinary applications. Curr Top Microbial Immunol 1998;226:90-106.

11 Hasan UA, Abai AM, Harper DR, Wren BW, Morrow WJ: Nucleic acid immunization: concepts and techniques associated with third generation vaccines. J Immunol methods 1999;229:1-22.

12 Smith HA: Regulation and review of DNA vaccine products, in Brown F, Cichutek K, Robertson J (eds): Development and Clinical Progress of DNA Vaccines. Dev Biol. Basel, Karger, 2000, vol 104, pp 57-62.

13 Ledwith BJ, Manam S, Troilo PJ, Barnum AB, Pauley CJ, Griffith TG, Harper LB, Beare CM, Bagdon WJ, Nichols WW: Plasmid DNA vaccines: Investigation of integration into host cellular DNA following intramuscular injection in mice. Intervirology 2000;43:258-272.

14 Gilkeson GS, Ruiz P, Howell D, Lefkowith JB, Pisetsky DS: Induction of immune-mediated glomerulonephritis in normal mice immunized with bacterial DNA. Clin Immunol Immunolpathol 1993;68:283-292.

15 FDA: Guidance for industry: Consideration for plasmid DNA vaccines for infectious diseases indication. 2005

16 Holm A: DNA vaccines for food-producing animals. Technical review and discussion of safety issues. Ministry of Food, Agriculture and Fisheries, Danish Veterinary Institut, 2003.

17 Griot C, Moser C, Cherpillod P, Bruckner L, Wittek R, Zurbriggen A, Zurbriggen R: Early DNA vaccination of puppies against canine distemper in the presence of maternally derived immunity. Vaccine 2004;22:650-654.

18 Botezatu I, Serdxuc O, Potapova G, Shelepov V, Alechina R, Molyaka Y, Ananev V, Bazin I, Garin A, Narimanov M, Knysh V, Melkonyan H, Umansky S, Lichtenstein A: Genetic analysis of DNA excreted in urine: a new approach for detecting specific genomic DNA sequences from cells dying in an organism. Clin Chem 2000;46:1078-1084.

19 Lorenz MG, Wackernagel W: Bacterial gene transfer by natural genetic transformation in the environment. Microbiol Rev 1994;58:563-602.

20 Sikorski J, Graupner S, Lorenz MG, Wackernagel W: Natural genetic transformation of Pseudomonas stutzeri in a non-sterile soil. Microbiology 1998;144:569-576.

Dr. T.D.E. Medjitna, Institute of Virology and Immunoprophylaxis (IVI), Sensemattstrasse 293, CH-3147 Mittelhäusern, Switzerland.
E-mail: djamel.medjitna@ivi.admin.ch

Vannier P, Espeseth D (eds): New Diagnostic Technology: Applications in Animal Health and Biologics Controls. Dev Biol (Basel). Basel, Karger, 2006, vol 126, pp 271-281.

Real-Time Quantitative PCR for Marek's Disease Vaccine Virus in Feather Samples: Applications and Opportunities

S. Baigent[1], V. Nair[1], R. Currie[2]

[1] Viral Oncogenesis Group, Institute for Animal Health, Compton, Berkshire, U.K.
[2] Fort Dodge Animal Health, Southampton, U.K.

Abstract: Marek's disease, an economically-important lymphoid neoplasm of chickens, is controlled by vaccination with CVI988 strain of Marek's disease herpesvirus. Sub-optimal vaccinal protection can have multiple causes. Accurate quantification of CVI988 in vaccinated chickens will assist in understanding the causes of these vaccine breaks. We developed, optimised and validated a real-time PCR assay for quantification of CVI988 vaccine virus (in terms of CVI988 genomes per 10,000 cells) in the feather tips, a rich source of viral DNA which can easily be sampled in a non-invasive manner. CVI988 load in feathers was predictive of CVI988 load in spleen, so is anticipated to be a good predictor of protection. The optimal age of chicks for feather collection is between two to five weeks, and feathers are preferentially taken from the axillary tract. This PCR test is now used to monitor vaccine virus levels in commercial chicks. For each flock under test, fifty birds are feather sampled for q-PCR. We describe how the feather CVI988 genome profile can show the flock's response to vaccination, and the likelihood of vaccine breaks. Furthermore, the q-PCR test can be applied to researching optimal timing and vaccine delivery routes, and optimal vaccination regimes for different breeds of chick.

Introduction

Marek's disease (MD) is a lymphoproliferative disease of chickens, caused by virulent serotype 1 strains of Marek's disease herpesvirus (MDV-1). The pathogenesis of MD has recently been reviewed in detail [1,2]. Three-to-four weeks post infection with virulent MDV, gross lymphoid tumours develop in the visceral organs and peripheral nerves. The resulting morbidity and mortality leads to decreased productivity and major economic losses. The virus also replicates in feather tissues, and virus shed with skin and feather debris from infected birds is the source of infection for their flock-mates [3].

Vaccination against MD effectively protects against disease and mortality, and the total annual worldwide vaccinations number over 20 billion. Cell-associated live vaccine virus is administered either to embryonated eggs, or to neonatal chicks using automated vaccinating machines. CVI988 (Rispens) strain (an avirulent MDV-1) was introduced in the mid-1990s [4,5] and is the most effective MD vaccine. However, vaccine breaks and excessive MD losses in a vaccinated flock do occur and can have several causes. In young chickens, the effects of vaccine are severely compromised by homologous maternal antibodies which delay vaccine virus replication and establishment of immunity. Suppression of cell-mediated immunocompetence, by environmental stresses or concurrent bacterial and viral infections, can decrease vaccine efficacy. Virulent strains can break through vaccinal protection in chicks which are infected at an early age before the establishment of full vaccinal immunity. However, even in correctly vaccinated and fully immunocompetent chicks, hypervirulent MDV strains can cause disease. The final important contributory factor to lowered protection is the reduction of effective vaccine activity due to inappropriate storage, handling and administration, which can result in chickens not receiving the full dose. This is an area where problems could and should easily be reduced, and yet its importance is often overlooked by those preparing and administering the vaccine.

Without testing for the presence and level of vaccine virus, it is difficult to determine the main cause of a vaccine break in a particular flock. Traditionally, CVI988 is detected by time-consuming virus isolation from blood. With this in mind, we developed, optimised and validated a rapid, sensitive and reproducible real-time PCR for the quantification of CVI988 in tissue samples from commercial vaccinated chicks.

Materials and Methods

Tissue samples from CVI988-vaccinated experimental and commercial chickens

All chickens were vaccinated at one day of age with one commercial dose of monovalent CVI988 vaccine (Fort Dodge Animal Health). Specified pathogen-free chickens, of either experimental or commercial breeds, were housed in a controlled environment at IAH-Compton, and tissue samples were taken on various days post-vaccination (dpv). A variety of flocks in the commercial environment, in France and Spain, were sampled. Feather samples were obtained from 50 birds in each flock. For feather sampling, six pinfeathers (short, newly growing feathers with plenty of pulp) were plucked and the proximal 1 cm of the shaft (the feather tip) was cut off and stored at 4°C short-term or -20°C long-term.

Preparation of tissue DNA

DNA was prepared from peripheral blood lymphocytes and spleen lymphocytes using phenol-chloroform extraction. DNA was prepared from feather tips using the Nucleospin Tissue Kit (Macherey-Nagel).

Real-time PCR for absolute quantification of the MDV genome

The method has previously been described in detail [6]. Briefly we optimised and validated a duplex real-time PCR which measures, in a single reaction, both an MDV gene (meq) and a chicken gene (*ovotransferrin, ovo*). The *ovo* reaction serves as an internal standard allowing normalisation of the number of host cells per reaction, confirming the presence of PCR reagents and amplifiable DNA, and the absence of PCR inhibitors. Standard curves are prepared for each reaction, enabling direct conversion of q-PCR Ct values to a number of copies of the *meq* or *ovo* gene [6]. The standard curve for the *meq* gene is prepared using pCVI988-BAC10 (a bacterial artificial chromosome clone of the complete CVI988 genome, [7]). The standard curve for the *ovo* gene is prepared using pGEMT-ovo plasmid (1 kb fragment of *ovo* gene cloned into the Promega pGEM-T Easy vector). Test samples are used at 50 ng DNA per reaction. Fluorescent probe ('Taqman') real-time PCR is used to amplify and detect the reaction products. For a given test sample, MDV genome copies per 10,000 cells is calculated from the Ct values of the *meq* and *ovo* reactions from that sample using the standard curves.

Adaptation of the method for use with feather DNA samples

Melanin pigment, present in extracts of feather tips particularly from brown chickens, is inhibitory in conventional PCR reactions but inhibition can be overcome by incubation of DNA samples with bovine serum albumin (BSA) prior to PCR [8]. We investigated the possibility of including BSA as a component of the q-PCR reaction itself, to overcome the melanin-induced inhibition. We examined which of the major feather tracts (cervical, humeral, spinal, axillary and femoral) was preferable for feather sampling to detect CVI988 in terms of ease of sampling, amount and quality of DNA obtained, and level of MDV detectable. We also examined over what age range of chicks it was possible to sample feather tips for preparation of good quality DNA.

Results

Choice of feather tissue for sampling

A study of CVI988 replication kinetics in experimental birds compared various tissues [9]. In peripheral blood lymphocytes the virus did not replicate to high levels. In spleen, there was an early peak of replication, before a decline to lower levels. In feather tips, detection of CVI988 was delayed compared with the other tissues but, from 14 dpv, virus load reached very high levels (peaking at almost 10^6 genome copies per 10,000 cells), which were maintained long-term. Furthermore, from 10 dpv, CVI988 load in feathers was predictive of CVI988 load in spleen in individual birds. Feathers were selected as the tissue of choice for developing the commercial test.

Optimisation of feather sampling and q-PCR

The optimal age of chicks for feather collection was two to five weeks, when there were many growing 'pinfeathers' with soft shafts and rich cellular pulp for DNA extraction [6]. Feathers plucked from the axillary feather tract were found to have the greatest and most uniform virus load [6]. It was possible to overcome the melanin-associated inhibition of PCR by inclusion of BSA in each reaction, 10 μg BSA being optimal (Fig. 1). BSA is now added to q-PCR reactions for all feather samples, whether from brown or from white birds.

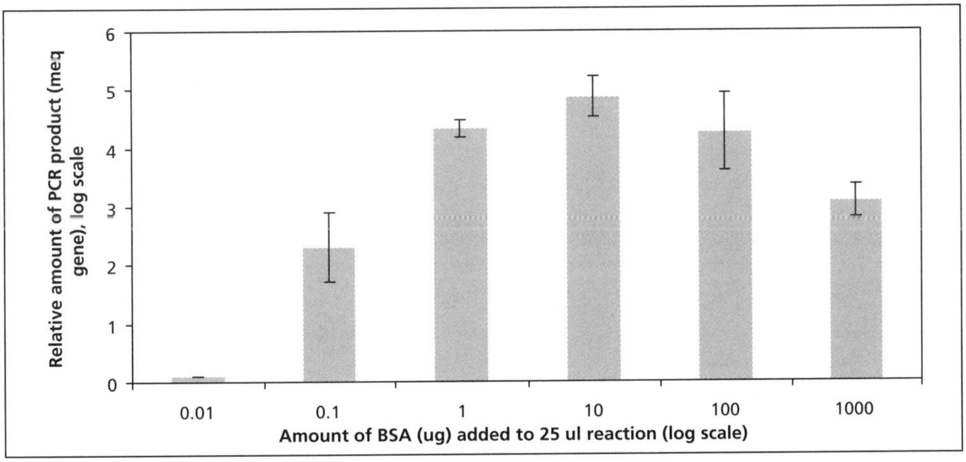

Fig. 1: **Effect of bovine serum albumin on PCR amplification in feather DNA samples from brown chickens.** MDV q-PCR was performed using 50 ng of brown feather DNA template per reaction and a range of concentrations of bovine serum albumin (BSA). The relative amount of MDV (*meq*) PCR product is plotted (mean +/- the standard error of the mean for triplicate samples) against concentration of BSA.

Setting a baseline for scoring samples as CVI988-positive

We set a 'baseline' CVI988 genome copy number, below which samples are scored negative and above which samples are scored positive, for CVI988. The baseline is set such that all samples which are truly negative, or samples which have an unreliably low level of CVI988 detection, will be scored negative. The baseline value is determined using a theoretical sample which has a *meq* Ct readout value of 38 and an *ovo* Ct readout value equal to the maximum value obtained for any of the test samples, and will therefore vary slightly between q-PCR runs.

Commercial feather q-PCR test on CVI988-vaccinated chicks

Figure 2 shows the replication kinetics of CVI988 in feather tips of two commercial breeds and one experimental breed of chicken, vaccinated with one dose of monovalent Fort Dodge CVI988 vaccine at one day of age. The replication kinetics in the commercial layer breed closely mirrored that in the experimental breed (although the virus load was lower in the layers) with an early peak and relatively high CVI988 load being maintained thereafter. In the commercial broiler breed, the peak of CVI988 virus load was delayed until 21 dpv, after which it plummeted to very low levels. Thus, in some commercial breeds of chicken, the replication curve is skewed and the peak of replication delayed and, if a single sampling time is to be used to determine the flock's vaccinal status, it is important that an appropriate sampling time is chosen.

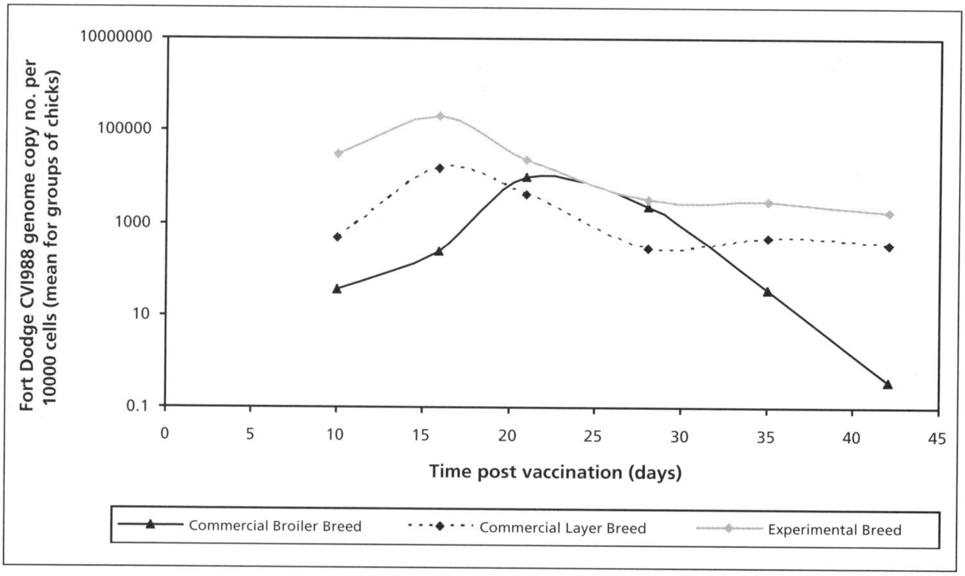

Fig. 2: **Replication kinetics of CVI988 in feather tips of different breeds of chicken.** The feather q-PCR test was used to examine kinetics of replication of Fort Dodge monovalent CVI988 vaccine in an experimental breed of chicken, a commercial broiler breed and a commercial layer breed. All chickens were vaccinated at one day of age with one commercial dose of vaccine, and kept under experimental conditions. Each point represents the mean CVI988 genome copy no. value for groups of five chickens.

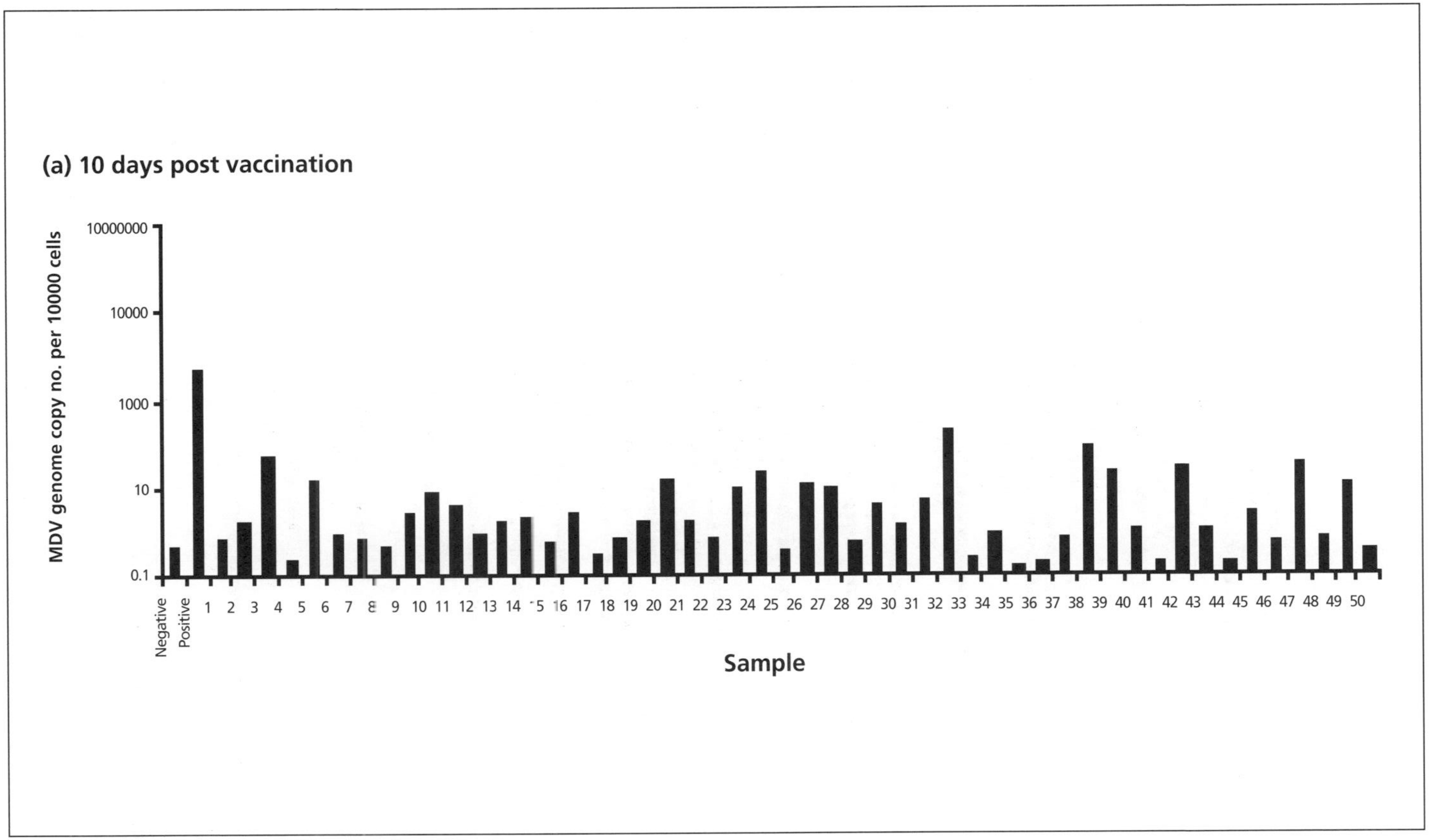

Fig. 3: **Use of the feather q-PCR test to monitor a flock's response to CVI988 vaccination.** A commercial flock vaccinated at one day of age with Fort Dodge monovalent CVI988 vaccine was feather sampled at 10, 17 and 21 dpv for q-PCR. Data from 45 or 50 chicks are shown for each time: since chicks were randomly selected from a large flock, the data are not repeated measures from the same 50 chicks. Negative control: feather DNA from a non-vaccinated experimental chicken. Positive control: feather DNA from a CVI988-vaccinated experimental chicken. The baseline, above which samples are scored CVI988-positive, was 1.3 copies of the virus genome, calculated as described in the text. The data for each group are summarised in the chart below Figure 3c.

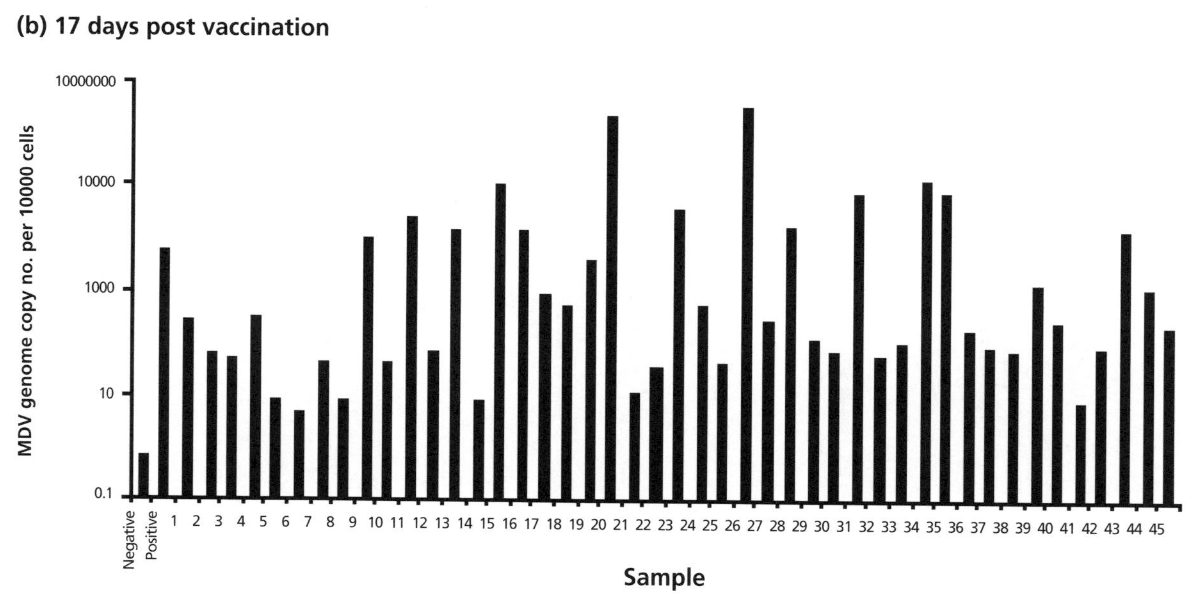

Fig. 3: **Use of the feather q-PCR test to monitor a flock's response to CVI988 vaccination.** A commercial flock vaccinated at one day of age with Fort Dodge monovalent CVI988 vaccine was feather sampled at 10, 17 and 21 dpv for q-PCR. Data from 45 or 50 chicks are shown for each time: since chicks were randomly selected from a large flock, the data are not repeated measures from the same 50 chicks. Negative control: feather DNA from a non-vaccinated experimental chicken. Positive control: feather DNA from a CVI988-vaccinated experimental chicken. The baseline, above which samples are scored CVI988-positive, was 1.3 copies of the virus genome, calculated as described in the text. The data for each group are summarised in the chart below Figure 3c.

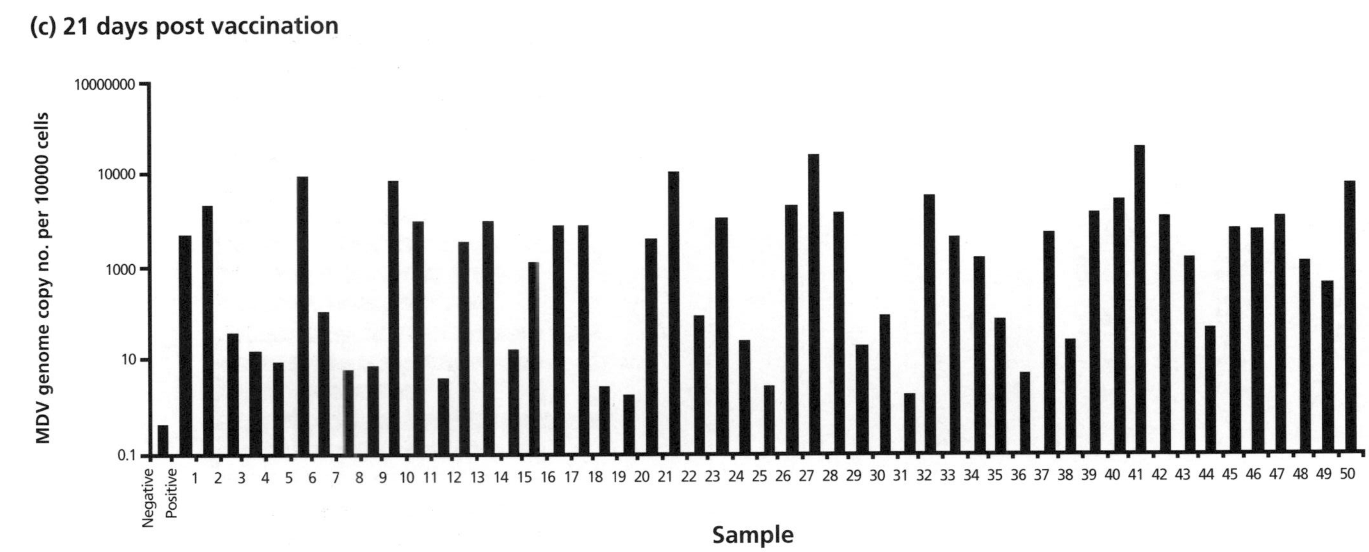

Days post vaccination	Range of CVI988 genome copy no.	% CVI988-positive samples
10	0.3 - 209	58%
17	4 - 3124829	100%
21	2 - 411966	100%

Fig. 3: **Use of the feather q-PCR test to monitor a flock's response to CVI988 vaccination.** A commercial flock vaccinated at one day of age with Fort Dodge monovalent CVI988 vaccine was feather sampled at 10, 17 and 21 dpv for q-PCR. Data from 45 or 50 chicks are shown for each time: since chicks were randomly selected from a large flock, the data are not repeated measures from the same 50 chicks. Negative control: feather DNA from a non-vaccinated experimental chicken. Positive control: feather DNA from a CVI988-vaccinated experimental chicken. The baseline, above which samples are scored CVI988-positive, was 1.3 copies of the virus genome, calculated as described in the text. The data for each group are summarised in the chart below Figure 3c.

Figure 3 shows feather q-PCR data from a commercial flock sampled at three time-points post vaccination (10, 17 and 21 days). Clearly a sampling time of 10 dpv is too early, because little or no CVI988 is detected in the majority of chicks at this time, so this would give a false impression of the flock's vaccinal status. In all test samples at 17 dpv and 21 dpv, the calculated CVI988 genome copy no. per 10,000 cells was above that of the baseline, showing that CVI988 DNA is present in each. However, CVI988 genome load was highly variable between chickens within the flock.

Thus time-points between 17 and 21 dpv are considered appropriate for sampling since, between these times, CVI988 levels should be sufficient in flocks showing delayed CVI988 replication, as well as in flocks showing an earlier peak of replication. Furthermore, the onset of feathering can often be delayed until around 14 days of age in commercial flocks (H. le Galludec, personal communication) so that sampling before 14 days is not possible.

The PCR test is now applied to feather samples taken from chicks in the commercial environment (PCT application WO 2004/035821) to confirm the successful vaccination of flocks. A number of 'blind' trials of the feather q-PCR test have generated some interesting findings which confirm useful application of the test in a commercial environment. For example, an exceptionally high and uniform CVI988 virus load in one flock correlated with the fact that this flock received two CVI988 vaccine doses (instead of one dose) on the day of hatch (Fig. 4). In another test, we found very low levels of CVI988 in all chicks (data not shown), and the customer later reported that there had been problems with the automatic vaccinator at the time this flock was vaccinated.

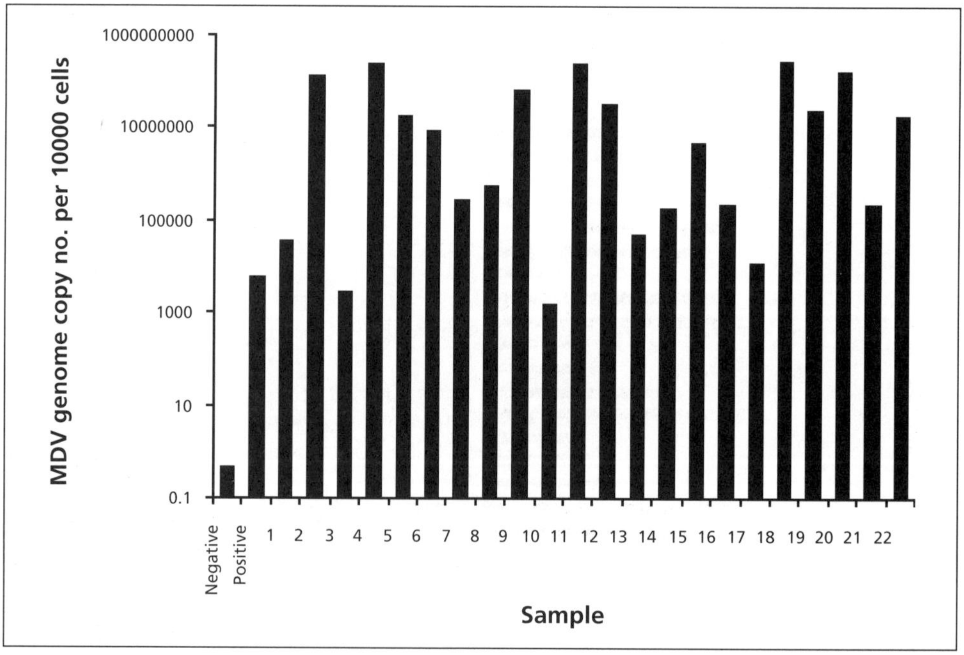

Fig. 4: **Effect of two vaccinations on CVI988 virus genome load in feather tips.** The data shown are from a commercial flock which received two doses of CVI988 vaccine (instead of a single dose) on the day of hatch. This correlates with the exceptionally high and uniform CVI988 virus load in the flock.

Discussion

We developed, validated and optimised a real-time quantitative duplex PCR assay for absolute quantification of the genome of CVI988 in feather tips [6]. In addition to being an excellent source of MDV DNA [6,9], feather tips can be easily and regularly sampled non-invasively, and they are easy to store and transport.

The real-time PCR assay was equally as sensitive as virus isolation (data not shown) but was more reproducible and considerably less time-consuming. Compared with conventional PCR to detect the CVI988 genome, real-time PCR was considerably more sensitive: 20-fold less input DNA was required, and the CVI988 genome could be detected, at low levels, in many samples which were negative by conventional PCR [9]. Furthermore, the quantitative nature of our real-time PCR, and the ability to normalise to a given number of host cells, allows direct and meaningful comparison among samples, both within a flock and between flocks.

For each flock subjected to the commercial feather q-PCR test, a CVI988 genome load profile is obtained. This profile should be interpreted as giving an overall picture of the flock's vaccinal status: if the majority of the flock is adequately protected, it should not matter if a minor percentage is not fully protected. It is presently unknown what level of CVI988 should be classed as 'positive' in terms of vaccinal protection against challenge. We can predict that vaccine efficacy will correlate with virus load and persistence in the lymphoid tissues and feather tips. However, at present, we can only speculate how differences in CVI988 replication kinetics (timing of the peak, virus load at the peak, decay following the peak) may affect vaccinal protection. We know that CVI988 load in feathers is predictive of CVI988 load in spleen (the predominant site of immune responses), in individual birds [9], so it is anticipated to be a good predictor of protection. A delay in the peak of CVI988 replication in feathers is therefore likely to reflect a delay in the onset of immunity which could be a disadvantage in the commercial environment, when chicks may be exposed to virulent strains in the first week of life. The level of CVI988 at the peak of replication is also likely to be significant since greater CVI988 replication in the spleen during the cytolytic phase should better stimulate immune responses [10]. The importance of maintenance of high levels of latent vaccine virus for protection is not clear.

Forthcoming experiments will address how replication kinetics and peak load of CVI988 in feathers correlates with protection against challenge by virulent MDV. Then it should be possible to understand fully how the feather CVI988 genome load profile in any given flock relates to the level of protection of that flock against virulent MD.

CVI988 genome load was found to be highly variable among birds from the same flock. It is quite possible that those chicks with a low level of CVI988 did not receive the correct dose of vaccine, perhaps because they were not pushed correctly against the needle on the automatic vaccinator. However, there is also variability within a commercial flock in terms of genetic background, maternal antibody levels and infection by other pathogens, which can all alter, delay or suppress replication of vaccine virus.

Using the q-PCR alone, we cannot exclude the possibility that those birds showing very high virus load are co-infected with a virulent field strain of MDV. Although this PCR is specific for serotype-1 MDV, it cannot distinguish between CVI988 and virulent strains. Since MD vaccines do not prevent super-infection, replication and shedding of virulent challenge virus, both CVI988 and challenge virus could be present in an individual bird. To confirm that the virus which is quantified by feather q-PCR is only CVI988, it is necessary to perform another test. CVI988 and virulent strains differ in

132-bp repeat genomic region (CVI988 has multiple copies of the repeats, while virulent strains have predominantly two repeats), and can be distinguished using a conventional PCR to amplify this region [11,12]. Flock feather DNA samples are pooled and subjected to 132-bp repeat PCR to confirm the absence of virulent MDV in the flock. The ultimate goal, a real-time q-PCR to distinguish CVI988 and virulent MDV strains, is not easy to attain since genomic differences are limited. The repetitive nature of the 132-bp repeat genomic region precludes its use in quantitative PCR. Differences in the *meq* gene and the ICP4 gene are not consistent between CVI988 and virulent strains. The pp38 gene shows a consistent single nucleotide difference between CVI988 and virulent strains, but it is not easy to design q-PCR primers to distinguish such a small difference.

Once the relationship between feather virus load and vaccinal protection is understood, we should be able to understand what kinetic profile of CVI988 replication is associated with maximum protection. The feather q-PCR test provides the opportunity to investigate the effects of type of vaccine, vaccine dose, timing of vaccination and vaccine delivery route on vaccine virus load in the feathers. This should ultimately enable tailoring of vaccination strategies to optimize protection in different breeds of chicken.

Conclusions

We developed, optimised and validated a real-time PCR assay for absolute quantification of CVI988 MD vaccine virus in feather tips. The test is now used commercially to monitor a flock's response to vaccination and to identify the likelihood of vaccine breaks. The test can also be applied to researching optimal vaccination strategies. The immediate priority is to investigate precisely how CVI988 replication kinetics and peak load in the feather tips correlates with protection, in challenge studies using virulent strains.

Acknowledgements

We are grateful to the staff of Fort Dodge Animal Health in Spain (Dr. X. Vila, Dr. M. Balasch, Ms. M. Toldra and Ms. C. Ramirez) and France (Dr. H. Le Galludec) for collection and processing of feather samples from commercial birds. We also thank staff from the Institute for Animal Health, Mrs. L. Smith and Mr. K. Howes for sampling of experimental chickens, Drs. J. Young, P. Kaiser and C. Butter for useful discussion on quantitative real-time PCR.

References

1 Calnek BW: Pathogenesis of Marek's disease virus infection, in Hirai K (ed): Current Topics in Microbiology and Immunology. Berlin, Springer, 2001, pp 25-55.

2 Baigent SJ, Davison TF: Marek's disease virus: Biology and life cycle, in Davison F, Nair V (eds): Marek's disease, an evolving problem. Oxford, Elsevier, 2004, pp 62-77.

3 Calnek BW, Adldinger HK, Kahn DE: Feather follicle epithelium: a source of enveloped and infectious cell-free herpesvirus from Marek's disease. Avian Dis 1970;14:219-233.

4 Rispens BH, Van Vloten J, Mastenbroek N, Maas HJL, Schat KA: Control of Marek's disease in the Netherlands. I. Isolation of an avirulent Marek's disease virus (strain CVI 988) and its use in laboratory vaccination trials. Avian Dis 1972;16:108-125.

5 de Boer GF, Groenendal JE, Boerrigter HM, Kok GL, Pol JM: Protective efficacy of Marek's disease virus (MDV) CVI-988 CEF65 clone C against challenge infection with three very virulent MDV strains. Avian Dis 1986;30:276-283.

6 Baigent SJ, Petherbridge LJ, Howes K, Smith LP, Currie RJW, Nair, VK: Absolute quantification of Marek's disease virus genome copy number in chicken feather and lymphocyte samples using real-time PCR. J Virol Methods 2005;23:53-64.

7 Petherbridge L, Howes K, Baigent SJ, Sacco MA, Evans S, Osterrieder N, Nair V: Replication-competent bacterial artificial chromosomes of Marek's disease virus: novel tools for generation of molecularly defined herpesvirus vaccines. J Virol 2003;77:8712-8718.

8 Handberg KJ, Nielsen OL, Jorgensen PH: The use of serotype 1- and serotype 3-specific polymerase chain reaction for the detection of Marek's disease virus in chickens. Avian Pathol 2001;30:243-249.

9 Baigent SJ, Smith LP, Currie RJW, Nair VK: Replication kinetics of Marek's disease vaccine virus in feathers and lymphoid tissues using PCR and virus isolation. J Gen Virol 2005;86:2989-2998.

10 Gimeno IM, Witter RL, Hunt HD, Reddy SM, Reed WM: Biocharacteristics shared by highly protective vaccines against Marek's disease. Avian Pathol 2004;33:59-68.

11 Becker Y, Asher Y, Tabor E, Davidson I, Malkinson M, Weisman Y: Polymerase chain reaction for differentiation between pathogenic and non-pathogenic serotype 1 Marek's disease viruses (MDV) and vaccine viruses of MDV-serotypes 2 and 3. J Virol Methods 1992;40:307-322.

12 Silva RF: Differentiation of pathogenic and non-pathogenic serotype 1 Marek's disease viruses (MDVs) by the polymerase chain reaction amplification of the tandem direct repeats within the MDV genome. Avian Dis 1992;36:521-528.

Dr. S. Baigent, Institute for Animal Health, Compton Laboratory, Compton, Newbury, Berkshire, RG20 7NN, U.K.
E-mail: sue.baigent@bbsrc.ac.uk.

Vannier P, Espeseth D (eds): New Diagnostic Technology: Applications in Animal Health and Biologics Controls. Dev Biol (Basel). Basel, Karger, 2006, vol 126, pp 283-290.

Detection Breadth and Limits for Potential Adventitious/Endogenous Contaminants in Biopharmaceutical Processes: A Reality Check for Innovative Methods

P. Duncan, L. McKerral, S. Feng, P.K. Tsai

Merck Research Laboratories, West Point, PA, USA

Abstract: In this presentation we address the "gold standards" among diagnostic tests used to detect some potential adventitious and endogenous contaminants in human biologics and vaccines, reflecting on the breadth as well as limits of detection of these assays. This perspective may help developers of innovative diagnostics to address the unique needs of this highly regulated industry.

INTRODUCTION

The cell substrates, viral seeds, and routine production cultures used to make pharmaceutical-grade proteins, nucleic acids, vectors, and viruses, are subjected to a systematic screening process for potential microbial and viral contaminants [for review see 1]. The tests suggested or defined in the various regulatory literature provide reasonable assurance of the purity of the cultures. However, the long culture times of many of the methods preclude timely corrective actions in response to results during processing. Innovative tools are being developed to detect potential microbial and viral contaminants. These tools can add value in the biopharmaceutical industry if they provide reliable results earlier, in a cost-effective manner, and with sufficient sensitivity and breadth of detection. While innovative methods may have a place for in-process testing, it is also conceivable that they might replace terminal (release) tests if equivalence or superiority to existing methods can be demonstrated. This presentation addresses the analytical sensitivity and susceptibility of several key tests for cell and virus banks and their cultures with hopes of beginning a rational comparison with innovative methods.

Breadth of the Screening Tests

Bacteria and fungi

Tests for bacteria and fungi are usually conducted using compendial sterility tests and/or plate-based bioburden tests [2,3]. Limitations of the methods are well known: some organisms may not grow well on these two media formulations and/or at the specified incubation temperatures [4]. The ability of these methods to detect potential contaminants in the sample matrix is generally demonstrated using isolates relevant to the production environment, and/or recommended controls. Contaminants observed in sterility and bioburden testing at Merck (Table 1) are essentially consistent with those listed by Cundell [4]. Knowledge of the relevant contaminants can help determine the necessary breadth of detection for innovative methods.

Recent guidance has reaffirmed that the bacteria and fungi isolated from true contamination events in biological processes should be identified using genetic methods [5]. Therefore, innovative methods that do not directly recover the contaminant in a form suitable for identification should be supplemented with methods capable of doing so. Similarly, it is also informative to know whether cells or genomes detected by innovative methods represent viable organisms.

Mycoplasma

Noteworthy among the bacteria not-detectable in the sterility test media are the mycoplasma (mycobacteria as well, but it is not addressed in this paper), for which a compendial test is available [2]. Limitations of the mycoplasma test are not unlike those of the sterility test, with perhaps the added annoyance of occasional inconclusive DNA stain in cell cultures [6]. While the mycoplasma are represented by over 100 species, only six of these species (*M. fermentans, M. hyorhinis, M. orale, M. arginini, A. laidlawii, M. hominis*, in order of decreasing frequency) accounted for 94% of the mycoplasma contaminations in 79 PCR-positive cell culture samples [7]

Table 1: Bacteria and fungi recovered in sterility and bioburden tests 2002-2005. A survey of sterility and bioburden tests.

Proteobacteria	**High GC Gram positive**
Klebsiella pneumoniae	*Brevibacterium diminuta*
Methylobacterium sp.	*Corynebacterium aquaticum, afermentans*
Paracoccus sp.	*Kocuria varians*
Ralstonia pickettii	*Micrococcus luteus, lylae*
Stenotrophomonas maltophila	*Propionibacterium sp*
	Rhodococcus fascians
Low GC Gram positive	**Fungi - Ascomycota**
Bacillus cereus, circulans, pumilus	*Aspergillus*
Enterococcus sp.	*Penicillium*
Paenibacillus glucanoyticus	*Stachybotry*
Staphylococcus epidermidis, hominis,	*Trichphyton*
pasteuri, saccharolyticus, warneri,sp	*Verticillium*

Innovative methods have addressed the panel of potential contaminants differently, in some cases using a single set of primers sufficient to cover a limited [8,9] or potentially broader range of species [10], in another case using multiplex primers targeting >50 mycoplasma species coupled with limited culture to enhance recovery [11].

Viruses

The classical screens for adventitious viruses and retroviruses include inoculation of susceptible cell cultures as well as animals and embryonated eggs. Inoculated cell cultures are monitored for viral-induced cytopathology, and tested for endpoints such as haemadsorption, specific immunofluorescence, reverse transcriptase, or focus formation. Embryonated eggs are monitored for viability, haemagglutination, and specific pathology in tissues of the embryo or embryonic membranes. Suckling and adult mice, as well as other animal species if used, are monitored for illness or death. These methods are not completely harmonized among service laboratories, or even among the descriptions in the regulatory literature (for instance, compare descriptions of in vivo testing in section 5.2.3 and 2.6.16 of the European Pharmacopoeia) [2]. Specific method descriptions are not described in a key reference on viral safety [12]. The method variations, and limited range of validation experience for the existing methods, make it difficult to demonstrate the equivalence or superiority of innovative methods. A summary of the breadth of coverage of the viral screen tests, compiled from a review of several diagnostic virology texts [13-18], is provided in Table 2, as an introduction to the breadth of detection of the current methods.

Analytical Sensitivity of the Screening Tests

This brief discussion of sensitivity provides some perspective for developers of innovative methods on the relevant levels to meet or exceed so that a new method can be considered a potential replacement for some component of the existing screen tests. Analytical sensitivity is determined in part by the method itself (i.e. no less than one replicating unit per vessel) and in part by the volume tested – in some cases specified by regulations or regulatory guidance.

Bacteria and fungi

The direct inoculation "sterility" test applied to mammalian cell banks can detect no fewer than one replication-competent bacterium or fungus or spore in the vessel(s) that are inoculated, and per ICHQ5D, at least 1% of the cell bank vials (or minimum of two vials) are to be tested [19]. The theoretical limit of detection (LOD) in this system, then, is one replicating unit per 1% of the cell bank containers (minimum two), or 1 unit/2 mL in a bank with 200 vials filled with 1 mL/vial. Similarly, if a bioburden test designed to detect contamination in a large-scale recombinant protein production culture examines only a minimal volume, say 100 mcl, on an agar plate, then the theoretical LOD is 10 colony-forming units per mL. Testing a larger volume increases sensitivity: if 10 mL are tested in each of two broth media, then the theoretical LOD = one replicating unit per 10 mL. Practical LODs would be considerably higher than these theoretical LODs, based on the probability of detecting rare events distributed in large volumes. In fact, inoculation of positive controls is routinely conducted at levels of 10-100 replicating units per system to achieve recovery most of the time.

Table 2: Selected relevant virus families and their detection by the classical methods.*

Family	Significance	Viruses of particular concern	Tests in routine cell cultures	Tests in embryonated eggs (via allantoic and yolk sac routes)	Tests in mice (via intracranial IC and intraperitoneal IP routes)	Tests in guinea pigs, rabbits	Antibody production in vivo for specific viruses
Pox-	Generally non-lethal disease except for Smallpox	Vaccinia, Monkeypox, Variola, Cowpox	Most, but recovery variable for variola	Vaccinia and variola best by CAM route	Suckling IC, except variola	rabbits (except variola); guinea pigs for vaccinia	Ectromelia
Herpes-	Oncogenic (EBV)	HSV-1,2; VZV; CMV; EBV; HHV-6,7,8; IBR	HSV, VZV, CMV, some others; but not HHV-6,7,8, EBV	Pseudorabies, HSV; not VZV	Suckling IC/IP for HSV; not VZV		MCMV, Mouse thymic virus
Paramyxo-	Respiratory	RSV, HMPV, Measles, Mumps, PIV 1-4, Newcastle	PIV, mumps, measles by hemadsorption; measles, RSV, some PIV by syncytia formation; but wild measles may grow slowly	Mumps, Newcastle; less sensitive for PIV; RSV not recovered			Sendai (PI-1), PVM, SV5
Orthomyxo-	Respiratory	Influenza A, B, C	In primary MK cells with HAd detection, MDCK with trypsin; poor otherwise	Avian influenza; influenza viruses A,B;			
Picorna	Enteric; respiratory	Polio-, Rhino-, Echo-, Entero-, Coxsackie- A and B; Hep A	Most except CoxA, Hep A; some rhinoviruses using specific detection		Suckling IC/IP for Cox A, B		GDVII
Adeno-	Oncogenic; enteric; respiratory;	Adenovirus	most are recoverable in routine tests	Only for epidemic keratoconjuctivitis in allantois			Mouse Adenovirus
Reo-	Includes enterics, arboviruses	Rotavirus, BTV, Colorado tick fever, REO-3	Reoviruses; BT+IFA for BTV; rotaviruses with trypsin	BTV	Suckling mice by IC/IP for arboviruses	Arboviruses in guinea pigs	REO-3, EDIM
Toga-	Includes arboviruses	Rubella, Sindbis; Equine encephalitis EEV, WEV, VEV	Alphavirses & rubella, especially in MK or RK cells	EEV, WEV, VEV	Suckling mice by IC; most but not all alphavirses	Arboviruses in guinea pigs	
Arena-	Includes arboviruses	LCMV, Junin, Lassa, Machupo	Recoverable in Vero, but may require specific detection	LCMV replicates via CAM route, but without lesions	Suckling mice by IC/IP; adult for LCM	Arboviruses in guinea pigs	LCM
Bunya-	Arboviruses	Bunyamwera, Hantaan, Rift Valley Fever, La Crosse	Arboviruses, esp in MK cells; some require specific detection	Rift Valley Fever recovered weakly by CAM route	Suckling mice by IC/IP for arboviruses	Arboviruses in guinea pigs	Hantaan
Rhabdo-	Includes arboviruses	Rabies, VSV	VSV; rabies in BHK-21 & BT with specific detection	VSV by CAM route	Suckling IC for rabies and other arboviruses	Arboviruses in guinea pigs	
Flavi-	Arboviruses	Hep C, Yellow fever, Dengue, WNV, St Louis encephalitis, BVDV	Arboviruses esp in MK cells; BT + IFA for BVDV; not Hep C	Weak recovery of some by CAM route	Suckling mice by IC/IP for arboviruses	Arboviruses in guinea pigs	
Filo-	Hemorrhagic disease	Marburg, Ebola	Recoverable in Vero but with specific detection			Marburg, Ebola	
Corona-	Enteric respiratory	Coronavivrus, SARS, TGE	TGE in PPK or PT-1+ IFA; recovery of others generally poor				MHV; RCV/SDAV
Parvo-	Enteric	B19, RA-1; also AAV, MMV, BPV, PPV	BPV, PPV with specific detection; MMV in 324K cells				MMV, Toolan's H1, KRV
Polyoma-	Oncogenic	JC, BK, SV40; BPyV	SV40 in HEK; BK with HA or IFA; others not recovered well				Mouse K virus, polyomavirus
Retro-	Oncogenic	HIV 1&2; HTLV 1&2; mammalian type C, BLV, BIV; monkey B	HIV in lymphatic-derived cell lines; HTLV in PBMC cocultures with specific detection				
Arteri-	Mouse colonies & possible emerging category	Lactate Deh'ase elev. Virus LDV					LDV
Astro-	Enteric	Astrovirus					
Calici-	Enteric; respiratory	Norwalk, Hep E					
Circo-	Possible relationship to human pathogens	TTV, Porcine circovirus, Chicken anemia virus					
Hepadna-	Oncogenic (HepB)	HepB					
Papilloma-	Oncogenic	HPV					

* Darkly shading indicates unproven or unsuitable combinations; light shading suggests possible, but perhaps not method of choice, or not without specific cell lines and detection strategies.

Mycoplasma

The LOD for mycoplasma depends on which part of the test is considered. Sensitivity for a cell bank from which one vial is tested, per ICHQ5D [19], can be no greater than the total volume represented by the test (i.e. one replicating unit per container with a volume of 1 mL, from the example above). The theoretical LOD for a culture is calculated differently, however, because now the denominator is not the single vial represented in the test, but rather the total volume of culture inoculated into the test. Per EP2.6.7 [2], the theoretical LOD of the test for cultivable species can be assessed as follows: broth – one replicating unit per 10 mL for both the aerobic and micro-aerophilic conditions (minimum of one replicating unit/tube; maximum inoculation volume of 10% of 100 mL volume of tube; one tube per aeration condition); direct plating on to agar – one replicating unit per 0.2 mL (0.2 mL/plate, one plate per aeration condition); cell culture and nucleic acid stain – one viable unit/2 mL (one mL/flask, two flasks/sample). Note that these estimates assume that the single mycoplasma unit originally inoculated into the test systems at the LOD replicates rapidly enough in the broth, or on agar, or in cell culture, to be readily distinguished when examined at the corresponding endpoint. Again, practical limits would be defined mathematically or by frequency of recovery in spiking studies.

Viruses

Similarly, analytical sensitivity of in vivo and in vitro screen tests for adventitious viruses is determined largely by the arbitrary choice of the volume of material to test. The theoretical LOD for an adventitious virus is one replication-competent unit per vessel or amount tested. For mammalian cell banks, at least 10^7 cells or cell equivalents are tested in each indicator cell system, each route of egg inoculation, and each set of suckling and adult mice, per EP5.2.3 [2]. For vaccine production cultures, the theoretical LOD would be one replication-competent virus in 500 dose-equivalent volume or 50 mL, per EP2.6.16 [2]. Limits (or rather minimum amounts to test) are not clearly defined in regulatory literature for therapeutic protein production cultures, to our knowledge.

PCR-based tests are now routinely used as part of a screen for specific viral agents. Merck has used these methods to supplement the general screen for viruses that might not otherwise be detectable, in addition to some other relevant viruses which might be detected in other systems. The sensitivity of PCR methods are limited by the mass of nucleic acid that the PCR reaction can accommodate without compromising recovery – for instance around 1 mcg of cellular genomic DNA per reaction, or roughly 10^5 cell equivalents. Typically no effort is made to examine10^7 cell equivalents fully, nor does it appear to be expected. In our experience with a continuous cell line and vaccine seeds made in it, the analytical sensitivity of a set of commercially available PCR tests for a broad range of DNA and RNA viruses ranged from 1-10,000 cell equivalents per viral genome equivalent detected in our sample matrix (Table 3). Recoveries were calculated based on the amount of nucleic acid tested or cells extracted. For comparison, a recovery of 100 viral sequence copies in a reaction with one microgram of diploid human DNA would be ~1500 cell equivalents per viral copy detected.

Viral copies based on PCR are not necessarily the same as infectious viruses detected by infectivity assay, or pathogenic doses in in vivo systems, complicating a direct comparison. Converting viral genome copies to infectious or pathogenic equivalents is speculative, requiring an assumption about the ratio of particle or genomes per infectious virion for wild viruses, and the pathogenic dose in the various animal or egg systems. A PCR recovery of 100 copies of a virus with particle to infectivity ratio of 1 or 100 for example, in a test of one microgram of total human diploid DNA, translates to a sensitivity of roughly 10^3 or 10^5 cell equivalents per infectious virus, respectively. Higher genome (or particle) to infectivity ratios effectively increase the calculated sensitivity (calculated cell equivalents per infectious virus), but do not change the actual cell equivalents tested.

Table 3: Analytical sensitivity of a set of commercially available PCR tests in the presence of test article nucleic acid representing 10^4–10^5 cells. Results are shown for spike recovery, rounded to one significant figure. Cell equivalents per infectious or pathogenic unit would depend on the number of viral genomes per infectious or pathogenic unit.

	Cell equivalents/viral genome equivalent detected		
Virus	**Lot 1**	**Lot 2**	**Lot 3**
Bovine immunodeficiency virus	8000	8000	8000
Bovine leukemia virus	8000	8000	8000
Bovine herpesvirus I and IV	5000	10000	2000
Simian spumavirus	2000	2000	2000
Human immunodeficiency virus I and II	400	1000	400
Adeno-associated virus	500	1000	200
Bovine polyomavirus	500	1000	200
Epstein-Barr virus	500	1000	200
Human herpesvirus type 8	500	1000	200
Human polyomavirus JC & BK	500	1000	200
Human T cell lymphotrophic virus I and II	500	1000	200
Porcine parvovirus	500	1000	200
Human herpesvirus type 6	50	1000	200
Human herpesvirus type 7	50	1000	20
Murine minute virus	50	1000	20
Bovine/porcine circovirus	500	100	200
Enterovirus *	100	90	80
Measles	100	90	80
Hepatitis B virus	50	100	20
Human cytomegalovirus	50	100	20
Parvovirus B19	5	100	20
Human papillomavirus	9	10	40
Hepatitis C virus	10	1	8
Bovine rotavirus	10	2	0.4

* Enterovirus results reported as cell equivalents per TCID50-equivalent detected.

Opportunities for Novel Detection Methods in Biopharmaceutical Processes

Regulatory precedent essentially establishes the standard by which innovative methods are judged for their suitability for terminal (release) testing. The new methods need to have a relevant scope and analytical sensitivity, and be cost/time effective to be attractive to a manufacturer. As an alternative when novel methods in themselves are not considered sufficient, their novel detection systems might be useful as alternative endpoints when combined with existing culture-based assays.

"In-process" testing is perhaps more amenable to creativity and innovation, because of less regulatory precedent. In-process microbial monitoring could add value by enabling near real-time process response, or risk reduction to further downstream processing. False positives in an innovative test for adventitious contaminants would counteract potential benefits of the test by occupying resources in investigations. False negatives related to inadequate breadth of detection or analytical sensitivity would limit their viability. The benefit of an innovative method needs to be weighed against the probability and/or impact of occurrence of that which the test is designed to detect. For instance, if mycoplasma and viruses have not been detected in a well-controlled manufacturing process, but bacterial or fungal contaminations have occurred, then a risk/benefit assessment may favour in-process testing for bacterial/fungal contaminations over mycoplasma or viral contaminations, assuming the normal scope of terminal release testing.

Finally, innovative methods may be necessary for emerging potential contaminants that may not be detected in existing screens, or for processes for which the turn-around time or sampling size of the existing methods are unsuitable. The new methods can leverage the increasing availability of genomes, bio-informatic tools, and detection technologies.

Conclusions

In conclusion, methods discussed above have a finite, but not entirely known, breadth of detection. This makes it hard to establish equivalency of innovative methods to existing methods. Detection limits for most of the classical screening methods reflect replication and/or pathogenicity, consistent with regulatory interest in product safety. The amount of material (i.e. cell equivalents) to test by PCR methods for specific adventitious agents has not been defined in the regulatory literature, nor has the desired sensitivity for any specific agents by PCR been defined. Thus far, there appears to be no expectation that PCR testing be conducted on amounts of material comparable to the existing in vitro and in vivo tests. Finally, there is less regulatory precedent for in-process testing, for which innovative methods may have a significant advantage over culture-based methods.

References

1 Schiff LJ: Review: Production, characterization, and testing of banked mammalian cell substrates used to produce biological products. In vitro Cell Dev Biol – Anim 2005;41:65-70.

2 European Pharmacopoeia 5th Edition, 2005.

3 United States Pharmacopeia–National Formulary (USP28–NF23), 2005.

4 Cundell A: Review of the media selection and incubation conditions for the compendial sterility and microbial limit tests. Pharm Forum 2002;28:2034-2041.

5 Food and Drug Administration (FDA) Center for Biologics Evaluation and Research. Aseptic processing guidance, Sept 2004.

6 Hu M, Buck C, Jacobs D, Paulino G, Khouri H: Application of PCR for detection and identification of mycoplasma contamination in virus stocks. In Vitro Cell Dev Biol -Anim 1995;31:710-715.

7 Uphoff CC, Drexler HG: Comparative PCR analysis for detection of mycoplasma infections in continuous cell lines. In vitro Cell Dev Biol -Anim 2002; 38:79-85.

8 Wong-Lee JG, Lovett M: Rapid and sensitive PCR method for identification of mycoplasma species in tissue culture, in Persing DH, Smith TF, Tenover FC, While TJ (eds): Diagnostic Molecular Microbiology Principles and Applications. Washington, DC, American Society for Microbiology, 1993, pp 257-260.

9 Eldering JA, Felten C, Veilleux CA, Potts BJ: Development of a PCR method for mycoplasma testing of Chinese hamster ovary cell cultures used in the manufacture of recombinant therapeutic proteins. Biologicals 2004;32(4):183-193.

10 McAuliffe L, Ellis RJ, Ayling RD, Nicholas RAJ: Differentiation of mycoplasma species by 16S ribosomal DNA PCR and denaturing gradient gel electrophoresis fingerprinting. J Clin Microbiol 2003;41: 4844-4847.

11 Chang A, Prud'homme I, Armstrong A, Hsieh WT, Presente E, Kerr A, Dean A, Dao MT, Onions D: Using Q-PCR for MycoplasmaTesting of Biopharmaceutical Products: The "HyMy" Assay. Presented at Well Characterized Biopharmceutical Products WCBP2005, Washington DC, Jan 2005.

12 ICH Harmonized Tripartite Guideline. Viral safety evaluation of biotechnology products derived from cell lines of human or animal origin, 1997.

13 Beveridge WIB, Burnet FM: The cultivation of viruses and rickettsiae in the chick embryo. Med Res Council 1946, Special Report Series No. 256.

14 Fields BN, Knipe DM, Howley PM (eds): Fields Virology 3rd Ed. Philadelphia, Lippincott Raven, 1995.

15 Hsiung GD, CKY Fong, ML Landry (eds): Hsiung's Diagnostic Virology. 4th Ed. New Haven, Yale University Press, 1994.

16 Merten OW: Virus contaminations of cell cultures – A biotechnological view. Cytotechnology 2002;39:91-116.

17 Murray PR (ed): Manual of Clinical Microbiology 7th Edition. Washington DC, ASM Press, 1999.

18 Richman DD, Whitley RJ, Hayden FG (eds): Clincal Virology 2nd Ed. Washington DC, ASM Press, 2002.

19 ICH Harmonized Tripartite Guideline. Derivation and characterization of cell substrates used for production of biotechnological/biological products, 1997.

Dr. P. Duncan, Merck Research Laboratories, WP17-101, Sumneytown Pike, West Point, PA 19486-0004, USA.
E-mail: paul_duncan@merck.com

Vannier P, Espeseth D (eds): New Diagnostic Technology: Applications in Animal Health and Biologics Controls. Dev Biol (Basel). Basel, Karger, 2006, vol 126, pp 291-292.

Bovine Virus Diarrhoea Virus in Bovine Serum Used for Vaccine Manufacturing

W. Hesselink, B. Makoschey

Intervet International BV, Boxmeer, The Netherlands

Abstract: Most cell culture-based manufacturing processes for animal and human vaccines use bovine foetal serum (FCS). Potentially, FCS can be contaminated with a number of viruses that can infect the bovine foetus transplacentally. Contaminations with BVDV are more frequent than with other agents that infect cattle because of (i) the high prevalence of BVDV in the main countries of origin for FCS, and (ii) the capacity of this virus to cause persistent infections of the bovine foetus. Due to the occurrence of BVDV strains with A non-cytopathic phenotype, contamination with BVDV may escape detection by classical means. Nowadays FCS is therefore treated to inactivate viral contaminants, including BVDV, before it is used in cell culture. Treatment is mainly by gamma-irradiation. Nevertheless, after a case of BVDV-2 outbreaks due to the use of a BVDV-2 contaminated commercial vaccine in cattle, two EMEA guidelines with special attention for BVDV were developed for the use of bovine serum in the manufacture of veterinary and human immunological medicinal products (EMEA/CVMP/743/00 and EMEA/BWP/1793/02). This is remarkable, as it was shown that the virus contaminating the vaccine was not derived from the batch of FCS used to produce the vaccine. The two guidelines give instructions on how to test FCS for contamination with BVDV. As a manufacturer of veterinary vaccines, we have chosen a testing procedure that is very sensitive and allows discrimination of infectious and non-infectious BVDV nucleic acids. After giving the non-irradiated serum three passages in tissue culture, BVDV infection is detected both by immunofluorescent staining and by Reverse Transcription – Polymerase Chain Reaction (RT-PCR). The numbers of BVDV RNA molecules are calculated and compared between the serum sample and the third passage supernatant sample. An increase in the calculated copy number after the cell culture passages indicates that the RNA originates from infectious BVDV. The results of seven foetal calf serum batches tested will be presented. Four batches were found positive for BVDV. Based on the results of the RT-PCR, two of them were classified as BVDV type 1, one batch contained BVDV type 2 and the fourth batch was contaminated both with BVDV type 1 and type 2.

Only three of the seven batches were found to be free of infectious BVDV before inactivation treatment. One of them contained a very high number of BVDV RNA copies, but these transpired to be non-infectious. This particular batch was selected to test whether the non-infectious BVDV antigen present in the serum batch was sufficient or not to induce sero-conversion in cattle. An experimental vaccine batch was prepared containing 1 ml of FCS per dose of 2 ml and alusalt as adjuvant. Five heifers, sero-negative and free of BVDV, were vaccinated twice at a four-week intervals. Blood samples were taken at the times of vaccination and four weeks after the second vaccination, and tested for BVDV specific antibodies. No such antibodies were detected in any of the vaccinated animals at any time. These results demonstrate that the BVDV in the respective batch of serum, despite the high levels, was indeed non-infectious and did not result in sero-conversion.

The supply of batches completely free of BVDV is insufficient to satisfy the demands of vaccine manufacturers. As has been shown, adequate tests to detect BVDV and to discriminate between infectious and non-infectious BVDV have been developed and are now available. However, given the adequacy of the gamma-irradiation treatment, we would like to question the justification of the strong emphasis put upon testing bovine serum batches for BVDV in both EMEA guidelines.

Dr. W. Hesselink, Intervet International BV, P.O. Box 31, NL-5830 AA Boxmeer, The Netherlands.
E-mail: wim.hesselink@intervet.com

Vannier P, Espeseth D (eds): New Diagnostic Technology: Applications in Animal Health and Biologics Controls. Dev Biol (Basel). Basel, Karger, 2006, vol 126, pp 293-299.

Isolation of BVDV from Bovine Serum by EMEA/CVMP and 9 CFR: A Comparison

J.W. Black

American BioResearch Laboratory, Sevierville, TN, USA

Abstract: Data derived from both experience and validation exercises on the behaviour and properties of BVDV is presented as applied to recovery of the virus from bovine serum used in the industrial production of animal biologicals. Licensed products both in the EU and the U.S. must use ingredients shown to be free of this agent and both regulatory agencies have published regulations governing this topic (EU: EMEA/CVMP and US: 9 CFR). The two systems differ in some specifics covering cell cultures used, timing of testing within the assays and the use of multiple approaches to the isolation of this virus. It was found that selected cell lines instead of primary cells could be used for BVDV isolation. FA tests, however, needed to be performed weekly for maximum sensitivity instead of once at the end of the test. BVDV interference assay (comparative titration) was found to be an unreliable test for the detection of non-CPE BVDV in cell culture. BVDV neutralization assays, while desirable from an informational standpoint, were determined to be unnecessary in serum that was to be adequately irradiated. Ultracentrifuge concentration of BVDV from serum was found to be a satisfactory method of increasing the sensitivity of assays for this agent, although quantification of the virus was found unnecessary in serum to be irradiated. Recommendations for the harmonization of 9 CFR and EMEA/CVMP into one assay are given.

Introduction

Bovine Viral Diarrhoea Virus (BVDV), or Bovine Pestivirus, is a moderate-sized, enveloped, single-stranded RNA virus belonging to the genus *Pestivirus*. It may be found in multiple animal species but most especially bovines. Since bovine serum, in some form, is almost universally used as a media supplement for the growth of cell cultures and since many different species of cells may be infected with this virus, it becomes a major goal of the QC laboratory to avoid this agent in the production of licensed biologicals.

BVDV is the most commonly encountered viral contaminant in bovine serum. Commercial batches of foetal bovine serum (FBS) are frequently positive for infectious BVDV where large pools of adult or calf serum are not. The origin of the BVDV found in bovine serum is almost entirely from animals that are persistently infected. BVDV may be found exhibiting varying degrees of cytopathic effect (CPE) in cell culture ranging from complete 4+ CPE [biotype I (btI)] to no discernible CPE at all [biotype III (btIII)]. BVDV isolated from bovine serum is of the btIII, or non-CPE group, and is capable of producing persistent infections in cell culture that are inapparent and difficult to detect by standard methods. For purposes of this discussion, direct fluorescent antibody (FA) techniques will be used for detecting acute or persistent BVDV infections in cell cultures.

Methods for detecting and demonstrating BVDV in bovine serum are described in the EU and U.S. regulations found in EMEA/CVMP/743/00 (4.3.2) (4.3.3) (4.3.4) and US 9 CFR 113.53c (113.47). These regulations are intended for producers of animal biologicals and the suppliers of bovine serum used in their manufacture [1,2]. These assays are similar but not identical and differ somewhat in the approach taken to demonstrate this agent. The American BioResearch Laboratory (ABR) has performed many assays entirely compliant with both systems over the years and conducted sensitivity comparisons and validation of the various methods. The main differences between the systems will be detailed in the following section (Materials and Methods).

As the global market in animal biologicals grows, it becomes imperative to devise one harmonized extraneous virus assay that satisfies both the EU and the U.S. for the movement of regulated goods across international boundaries. Detection of BVDV in bovine serum is sufficiently difficult that significant contaminations may be encountered in materials thought to be free of this agent. After evaluating the various methods and becoming familiar with this virus over the years, we at ABR will disclose in this presentation our experiences with BVDV in many types of bovine sera and give our recommendations for structuring one harmonized assay for BVDV detection in bovine serum that should be accepted by both the EU and the U.S. [3].

Materials and Methods

Cell cultures used for BVDV assays were various primary bovine kidney and spleen cultures (1BK, 1Bsp), Madin-Darby Bovine Kidney (MDBK) cell line, Embryonic Bovine Lung (EBL) cell line and Bovine Turbinate (BT) cell line. MDBK cells were used between passages 110 and 225, EBL and BT cells between passages 12 and 35, and 1BK and 1Bsp cells between passages 1 and 5. The cell culture medium used consisted of EMEM + 20 mM HEPES + NEAA + 5 mM bicarbonate + 2X pyruvate and glutamine pH 7.5. Stock cultures were grown in donor goat serum 5-10% and Ciprofloxicin 10 μg/ml. Cultures under assay used test article serum only for growth.

FA techniques were performed by the method of ABR (contact laboratory for details) and specific BVDV FA conjugate was obtained from commercial stocks at ABR. Validation documents for the various methods are obtainable from ABR as a dossier on request from the author.

A summary of requirements for EMEA/CVMP/740/00 in force as of the last quarter of 2005 for the isolation of BVDV from bovine serum mandates that suitable primary bovine cells, below passage 6, are grown in the presence of test article (TA) serum for three passages, performing a specific assay for BVDV at the end of each passage (EMEA/CVMP 5.2.5 Extraneous viruses) [1]. In addition, an unspecified requirement for quantification of any BVDV that may be present in the TA is mandated (4.3.2). ABR performs this quantification by ultracentrifuge concentrating the sample at 100,000 x G for 60 minutes and titrating the pellet. A further CVMP requirement (4.3.4) for BVD Comparative Titration (CT) is performed by titrating a CPE (btI) BVDV strain in TA-conditioned cells and control cells in an effort to detect interfering agents present in the test article. An assay for BVDV neutralizing antibody is mandated by CVMP in an attempt to determine the amount of BVDV that can theoretically be masked from detection in the sample and to demonstrate that this quantity cannot exceed the titretitre of BVDV shown to be inactivated by irradiation (4.3.3).

BVDV detection in bovine serum by the method of US 9 CFR 113.53c (113.47) was done by growing a suitable cell line of bovine origin (EBL, BT) in the presence of TA serum for not fewer than 21 days (two subcultures) and performing a specific test (FA) for BVDV on these cells at the terminal end of the assay only. BVDV quantification, serum neutralization and BVDV comparative titration assays are not required by 9 CFR.

Results

Not all species' cells were found to be sensitive to BVDV infection. Cell cultures from bovine, porcine, caprine, ovine, feline, lapine, cervine and other exotic ruminant species were found susceptible to BVDV infection and these cells may carry BVDV contaminants. Cultures from primate, equine, canine, mouse and rat (other rodents), ferret and mink, insect, reptile, amphibian, fish and avian were found to be resistant to BVDV infection and no contaminated cells in the public domain from these species were found. All attempts to infect these species' cells productively with BVDV failed.

Primary cells of bovine origin were found to vary in BVDV sensitivity from moderate (10-100 PFU/sample LOD) to high (1-10 PFU/sample LOD). Cell lines tested also fell into two categories of sensitivity with MDBK cells being moderate and EBL and BT cell lines being high. Consistently high BVDV sensitivity could be achieved with master and working cell banks of either EBL or BT between the passages of 15 and 35 (1:3 weekly).

The requirement for weekly FA tests for BVDV (CVMP) was found superior to the 9 CFR requirement for a single test after 21 days. This was due to the CVMP test detecting acute BVDV infections in cell cultures and 9 CFR usually seeing only older, persistent infections. Acute BVDV infections stain with much greater potency and clarity by FA than chronic (persistent) infections (Figs. 1-3).

Fig. 1: BVD FA stain of EBL cell culture not infected with BVD virus. 400X.

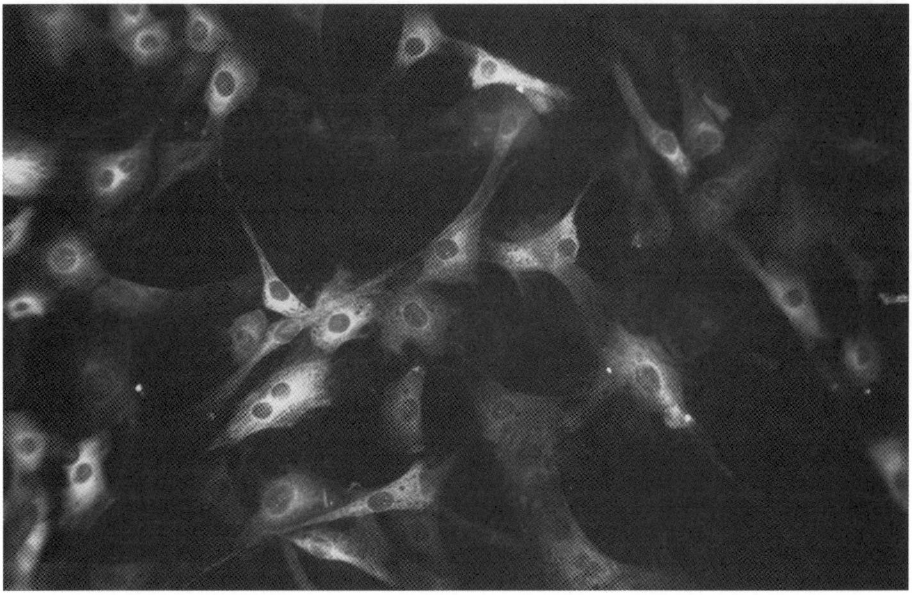

Fig. 2: BVD FA stain of EBL cell culture acutely infected with non-CPE BVD virus for 96 hours. 400X.

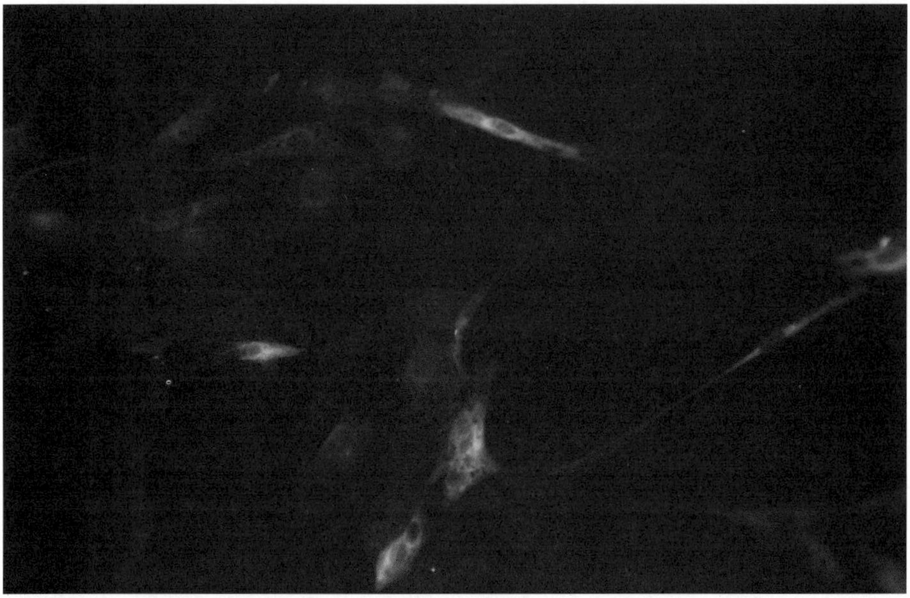

Fig. 3: BVD FA stain of EBL cell culture persistent infection with non-CPE BVD virus >three weeks. 400X.

The CVMP-mandated assay of BVDV comparative titration was found to be a highly unreliable assay for btIII BVD in cell culture with 50-60% of known FA positive CVMP assays showing no reduction in titretitre (negative interference). Laboratory validation of this method using TC-adapted strains of btIII virus worked well but many field isolates failed to inhibit the btI challenge virus.

The BVDV quantification assay required by CVMP as done by ABR proved a sensitive and reliable method of detecting BVDV in serum. Virus titres of native BVDV in TA sera ranged from undetectable to a maximum of 1000 PFU/ml sample. The majority of FBS samples ranged from 1-10 PFU/ml with the rare serum showing 100 PFU. This method removed the ambient neutralizing antibody from the virus and allowed it to propagate more freely than otherwise in the presence of unmodified TA.

The BVDV neutralizing assay requirement of CVMP (4.3.3) was relatively straightforward to perform but antibody masking of live BVDV was only seen in FBS samples where the antibody titre was relatively low. Attempts to demonstrate live, infectious BVDV in adult or calf serum where large amounts of neutralizing antibody was present were consistently unsuccessful. The maximum titre of BVDV that could be suppressed in FBS was about 3 logs, whereas calf and adult bovine serum yielded the highest neutralizing antibody level of about 4.5 logs. Attempts to recover viable BVDV from adult or calf serum by ultracentrifugation were uniformly unsuccessful.

Demonstration of btIII (non-CPE) BVDV in cell culture by FA was found to be dependent on several factors. The potency and broad reactivity of the antiserum used to make the direct FA reagent is of the greatest importance. Significant antigenic variation was seen between BVDV isolates so that the FA reagent needed to be as broadly reactive as possible. At ABR, the antiserum used to make in-house FA conjugate was made against four different BVDV isolates representing all known antigenic types. Finished FA reagent was validated on a panel of eight BVDV isolates representing a wide selection of known strains, antigenic types, genotypes and biotypes recovered from North, Central and South American as well as Australian and New Zealand sources. BVDV in cell culture was found to be somewhat fixation-sensitive for FA staining with 100% acetone being best and 75-80% acetone in water or methanol being adequate but other fixatives being unsatisfactory. Indirect FA using polyclonal antibody gave a very good signal but was somewhat prone to false positives due to background staining related to the test article. Persistently-infected cultures showed the poorest FA signal with only 5-10% of cells being positive and the intensity of fluorescence subdued to 1-2+. Acute BVDV infections in cell culture consistently gave the best FA signal with 50-100% of cells showing a 3-4+ fluorescence. Monoclonal antibody based IFA assays for BVDV, while feasible, suffered from a poor signal and the inability to react with some isolates.

Discussion

From data derived over many years at ABR, it is clear that BVDV detection is best done on acutely-infected cell cultures (<eight days) whenever possible. An acutely-infected cell culture looks and behaves normally and can be passaged to make new cultures, but after about two passages the resulting cells will mostly be negative for BVDV and only a few cells in the monolayer will be actively producing virus. These cells also show less BVDV antigen internally than acutely-infected cells resulting in a

difficult-to-read assay. The CVMP requirement for weekly FA tests guarantees the staining of at least one acutely-infected passage, thus removing some of the uncertainty in this part of the assay.

The BVDV sensitivity data and experience using all kinds of cell substrates for BVDV isolation have shown that there are cell lines of bovine origin that are adequate substitutes for primary cells in the CVMP assay. Primary cells for use in CVMP must be frozen at passage 1 to be useful for assay. It is quite difficult to perform adequate QC testing for viruses and other agents on passage 1 cells leading to possible inconsistencies in using the entire primary bank for assay purposes. At least two diploid cell lines exist (EBL and BT) that can be frozen as a working cell stock with known, stable characteristics and are useful for at least 20 passages. The MDBK cell line was found unsatisfactory for use in BVDV isolation from serum due to its relatively low sensitivity to the virus.

CVMP requires BVDV SN testing and BVDV quantification to show that the amount of measurable virus present plus the quantity that may be masked by antibody cannot exceed that shown to be inactivated by the mandated gamma irradiation dose. Results at ABR have consistently shown that BVDV may only be quantified from FBS and the maximum titre that can be found, 3 logs, plus the maximum amount of virus that can be masked, about 3 logs, equals a maximum of 3.3 logs which is far less than the reduction of 6 logs routinely seen with the mandated level of irradiation. Adult or calf serum may yield as much as 4.5 logs antibody but no countable BVDV, which is still far less than the amount of virus reduced by irradiation.

The BVDV comparative titration assay, required by CVMP, was fraught with false negative reactions in our hands, although no false positive results were seen. Although an interesting and useful research tool from the past, the BVDV interference assay has been replaced by better methods of detection in the modern QC laboratory.

Conclusions and Recommendations

In view of the previously reported characteristics of feral BVDV in serum, it is recommended by ABR that any harmonized assay accepted by the EU and the U.S. consider the following:

(i) allow the use of any suitably-validated bovine cell line in place of a primary cell;

(ii) require a specific assay (FA, IFA or IHC) every seven days or at each passage during the assay term;

(iii) drop the BVDV comparative titration requirement;

(iv) drop the BVDV SN test requirement as the amount of virus + masking antibody cannot exceed irradiation titre reduction;

(v) drop the BVDV quantification requirement as the maximum sample virus does not approach the amount removed via irradiation;

(vi) consider requiring a large volume (50-100 ml) ultracentrifuge pellet isolation for BVDV as a method of increasing the sensitivity of the assay.

It is the position of ABR Laboratory that all materials of animal origin used in the manufacture of biologicals should be irradiated as a requirement. Extraneous virus assays, however, need not be performed on any but post-irradiation material. It is also our position that the EU-mandated gamma irradiation dose is somewhat high, as serum thus treated suffers some measure of performance loss. A reduction in the required minimum dosage from 30 to 25 kGy (3 to 2.5 Mrad) will restore much lost performance while maintaining adequate virucidal activity, in view of observed loads of feral viruses.

References

1 European Medicines Agency, Committee for Medicinal Products for Veterinary Use: Guideline on requirements and controls applied to bovine serum used in the production of immunological veterinary medicinal products (EMEA/CVMP/743/00-Final) part 4.3: Tests for the presence of viral contaminants, 10 October 2001.

2 United States Code of Federal Regulations: title 9, parts 113.53c and 113.47, 2004.

3 Black JW: Performance validation for the recovery of extraneous viruses from bovine serum by EMEA/CVMP/CPMP/9 CFR: Final reports. ABR in-house dossier, October 2003.

Dr. J.W. Black, American BioResearch Laboratory, P.O. Box 6609, Sevierville, Tennessee 37964, USA.
E-mail: jblack@ambio.net

Vannier P, Espeseth D (eds): New Diagnostic Technology: Applications in Animal Health and Biologics Controls. Dev Biol (Basel). Basel, Karger, 2006, vol 126, pp 301-305.

Detection of Reticuloendotheliosis Virus in Live Virus Vaccines of Poultry

A. Fadly[1], *M.C. Garcia*[2]

[1] USDA-ARS Avian Disease and Oncology Laboratory, East Lansing, Michigan, USA

[2] Department of Avian Medicine, College of Veterinary Medicine, University of Georgia, Athens, Georgia, USA

Abstract: In vitro and in vivo assays have been used for the detection of reticuloendotheliosis virus (REV) in live virus vaccines of poultry. The presence of REV is confirmed by the demonstration of viral antigen or provirus in chicken embryo fibroblasts (CEFs) or in specific-pathogen-free chickens inoculated with vaccine. Using REV polyclonal or monoclonal antibodies, CEFs inoculated with vaccines can be examined for REV by immunofluorescence or immunoperoxidase staining methods. Cell lysates from such inoculated CEFs can also be used for detection of REV major group-specific antigen (p30) by an enzyme-linked immunoassay. Detection of proviral DNA by polymerase chain reaction (PCR) assays that amplifies the 291 base pairs product of REV LTR has been shown to be a sensitive and specific method for detection of various strains of REV in infected CEFs and in the blood of SPF chickens inoculated with contaminated fowlpox virus (FPV) vaccines. Recently, using PCR tests that amplify REV envelope and REV 3' LTR sequences provided a more accurate assessment of the insertion of REV provirus in FPV than PCR assays that amplify the REV 5' LTR. This paper reviews the most common methods used for testing live virus vaccines of poultry for contamination with REV.

INTRODUCTION

Reticuloendotheliosis virus (REV) is an avian oncornavirus that is structurally and antigenically unrelated to the leukosis-sarcoma group of viruses [1]. All REV isolates are antigenically related to each other. However, using monoclonal antibodies, REV isolates can be classed into three different subtypes, A, B and C. REV infects chickens, turkeys, ducks, geese, pheasants, quail, and probably many other avian species [1]. The most common clinical diseases induced by REV are chronic lymphomas (reticular cell tumour, B- or T-cell lymphoma) and an immunosuppressive runting disease. Although losses in REV-affected flocks can be significant due to tumour mortality and immunosuppression, the principal economic concerns of REV infection are as contaminants of live virus vaccines of poultry or as a barrier to the export of breeding stock to certain countries. Because REV can be transmitted vertically from dams to offspring, embryos and tissue cultures prepared from such embryos may harbour REV; therefore these embryos or cells could serve as a source of REV contamination of poultry and other vaccines produced in such ingredients. Accidental contamination of live virus poultry vaccines such as Marek's disease virus (MDV) and fowlpox virus (FPV) with REV [2-8] is an important source of virus infection and can cause runting disease as well as lymphomas in vaccinated chickens resulting in significant economic consequences. The fact that partial or complete insertion of REV genome in large DNA viruses such as FPV can occur also poses a problem [9-11].

Until recently, routine testing of live virus vaccine of poultry for contamination with REV was not required by regulatory agencies. However, vaccine manufacturers are required to produce vaccines in CEFs or chickens embryos obtained from flocks that are known to be free from various avian pathogens including REV. Further, there is a need to standardize methods used by vaccine manufacturers and regulatory agencies for detection of REV in live virus vaccines of poultry. The sensitivity of indirect immunofluoresence (IFA), enzyme-linked immunosorbent assay (ELISA) and polymerase chain reaction (PCR) has been compared in both in vivo and in vitro assays of REV in contaminated vaccines [4].

This paper is primarily focused on a review of methods that are currently available for detection of REV, and on the use of in vitro and in vivo assays for the detection of REV contamination in live virus vaccines of poultry.

Biological Methods for the Detection of REV

A variety of methods exists for the demonstration of REV infection in cell cultures [1,10]. Because cytopathic effects in CEFs may not be seen on primary isolation, the presence of REV is confirmed by the demonstration of viral antigens in infected cell cultures.Using polyclonal or monoclonal antibodies, the demonstration of antigens by IFA or ELISA are probably the methods of choice [1,10]. Antibodies to REV can be demonstrated in chickens inoculated with REV-contaminated vaccines by several assays, namely virus neutralization or direct ELISA. Of all the biological tests that can be used to demonstrate REV in inoculated CEF, namely IFA, ELISA, complement-fixation for REV, reverse transcriptase assay, and immuno-peroxidase plaque assay [1,10], only IFA and ELISA are considered the most common biological methods used for in vitro and in vivo assays of live virus vaccines of poultry for contamination with REV.

Immunofluorescence

Polyclonal chicken anti-REV antibody and commercially available fluorescein-conjugated anti-chicken globulin (IG) antibody are used in IFA to detect REV in CEFs inoculated with vaccines [3,4,10]. Alternatively, mouse REV monoclonal antibodies [12] and the corresponding commercial fluorescein-conjugated anti-mouse IG antibody can be used. Sample-inoculated monolayer CEFs grown on 96-well, 35 mm plates, or on glass-coverslips are processed for IFA using a standard method [1,10]. REV-infected cells have fine granular or diffuse fluorescence throughout their cytoplasm, but not in the nucleus. It is recommended that cultures inoculated with vaccines be compared with known positive and negative control cultures. The use of monoclonal antibodies facilitates interpretation of the test, because of intense specific staining and the virtual absence of background [1,10].

Enzyme-linked immunosorbent assay

The ELISA is conducted on cell lysates obtained at seven days post-inoculation (PI) from vaccine inoculated-CEFs. The ELISA for detection of REV p30 is based on the use of REVmonoclonal antibodies [12], as previously described [1,10]. The test also uses an appropriate dilution of a rabbit anti-REV antibody and peroxidase-labelled anti-rabbit IgG conjugate as the primary and secondary antibodies, respectively.

At present, the primary reagent for this test, namely REV monoclonal antibodies, is not commercially available [10], whereas the rabbit antibody may be produced by inoculation of a rabbit subcutaneously at multiple sites with 2 mg of protein from sucrose-gradient-purified strain T of REV in Freund's complete adjuvant, followed at 21, 35, and 49 days by booster immunizations, each containing 2 mg of protein in Freund's incomplete adjuvant.

Molecular Methods for the Detection of REV

Detection of proviral DNA by PCR assays that amplify the 291 base pairs product of REV LTR [13] has been shown to be a sensitive and specific method for the detection of various strains of REV in infected CEFs as well as in the blood of SPF chickens inoculated with contaminated FPV vaccines [3,4]. Proviral DNA from infected CEFs will also hybridize with specific REV probes, such as pSNV, and can be detected by dot blot or Southern blot assays [1]. Such tests have powerful specificity and should be helpful in confirming the identity of contaminant REV.

Reverse transcriptase PCR (RT-PCR) can also be used in the detection of REV in contaminated vaccines [1,10]. Most recently, using PCR tests that amplify REV envelope and REV 3' LTR sequences provided a more accurate assessment of the insertion of REV provirus in FPV than PCR assays that amplify the REV 5' LTR [9].

In Vitro Assays for the Detection of REV in Live Virus Vaccines of Poultry

Cultures of CEFs prepared from SPF chiken embryos are used for in vitro assays of live virus vaccines for contamination with REV. Inoculated CEFs are tested at various times PI for the presence of REV by IFA, ELISA or PCR [4]. To minimize cytopathic effects induced by vaccine virus, vaccines are usually pretreated with neutralizing antibody, filtration, or freeze and thaw, particularly in the case of testing vaccines containing cell-associated vaccine viruses such as Marek's disease vaccine. It is also worth noting that filtration of a known REV-contaminated FPVvaccine caused significant reduction in the frequency of detecting REV in such vaccines [14].

In Vivo Assays for the Detection of REV in Live Virus Vaccines of Poultry

Newly hatched SPF chickens are inoculated with vaccines. It is recommended that 100 bird dose of vaccine/chicken be used for in vivo assays of vaccines for contamination with REV [14]. Samples of blood are collected at various times PI from chickens inoculated with vaccines. Samples of blood from inoculated chickens are tested for the presence of REV by inoculating CEFs; inoculated CEFs are then tested for REV by IFA, ELISA or PCR (see in vitro assays above). Alternatively, DNA extracted from blood cells can also be tested for the presence of REV by PCR.

IFA and PCR were more senstive than ELISA in the detection of REV in inoculated CEFs using in vitro assays; whereas using in vivo assays, the sensitivities of IFA, ELISA and PCR were comparable [14]. If in vivo assays are used to test vaccines for contamination with REV they must include a virus test, as a negative antibody test may be misleading [14].

References

1 Witter RL, Fadly AM: Reticuloendotheliosis, in Saif YM, Barnes HJ, Fadly AM, Glisson JR, McDougald LR, Swayne DE (eds):Diseases of Poultry. Ames, IA, Iowa State University Press. 2003, pp 517-536.

2 Bendheim U: A neoplastic disease in turkey following fowlpox vaccination. Refuah Veterinaria 1973;30:35-41.

3 Fadly AM, Witter RL, Smith EJ, Silva RF, Reed WM, Hoerr FJ, Putman MR: An outbreak of lymphomas in commercial broiler breeder chickens vaccinated with a fowlpox vaccine contaminated with reticuloendotheliosis virus. Avian Pathology 1996;25:35-47.

4 Fadly AM, Witter RL: Comparative evaluation of in vitro and in vivo assays for the detection of reticuloendotheliosis virus as a contaminant in a live vaccine of poultry. Avian Diseases 1997;41:695-701.

5 Jackson CAW, Dunn SE, Smith DI, Gilchrist PT, McQueen PA: Proventriculitis "Nakanuko" and reticuloendotheliosis in chickens following vaccination with herpesvirus of turkeys (HVT). Aust Vet J 1977;53:457-459.

6 Kawamura H, Wakabayashi T, Yamaguchi S, Taniguchi T, Tarayanagi N, Sato S, Sekiya S, Horiuchi T: Inoculation experiment of Marek's disease vaccine contaminated with reticuloendotheliosis virus. Nat Inst of Animal Health Quarterly 1976;16:135-140.

7 Koyama HY, Suzuki T, Ohwada Y, Saito Y: Reticuloendotheliosis group virus pathogenic to chicken isolated from material infected with turkey herpesvirus (HVT). Avian Diseases 1976;20:429-434.

8 Yuasa N, Yoshida I, Taniguchi I: Isolation of a reticuloendotheliosis virus from chickens inoculated with Marek's disease vaccine. Natl Inst Animal Health 1976;Q16:141-151.

9 Garcia M, Narang N, Reed WM, Fadly AM: Molecular characterization of reticuloendotheliosis virus insertions in the genome of field and vaccine strains of fowlpox virus. Avian Diseases 2003;47:343-354.

10 Hertig C, Coupar BEH, Gould AR, Boyle DB: Field and vaccine strains of fowlpox virus carry integrated sequences from the avian retrovirus, reticuloendotheliosis virus. Virol 1997;235:367-376.

11 Moore KM, Davis JR, Sato T, Yasuda A: Reticuloendotheliosis virus (REV) long terminal repeats incorporated in the genomes of commercial fowlpox vaccines and pigeon poxviruses without indication of the presence of infectious REV. Avian Diseases 2000;44:827-841.

12 Cui Z, Lee LF, Smith EJ, Witter RL, Chang TS: Monoclonal-antibody- mediated enzyme-linked immunosorbent assay for detection of reticuloendotheliosis viruses. Avian Diseases 1988;32:32-40.

13 Aly MM, Smith EJ, Fadly AM: Detection of reticuloendotheliosis virus infection using the polymerase chain reaction. Avian Pathology 1993;22:543-554.

14 Fadly AM, Witter RL: Oncornaviruses: Leukosis/Sarcomas and reticuloendotheliosis, in Laboratory Manual for the Isolation and Identification of Avian Pathogens. 4th ed, Am Assoc of Avian Pathologists, 1998, vol 34, pp 185-196.

Dr. A.M. Fadly, USDA-Agricultural Research Service, Avian Disease and Oncology Laboratory, 3606 East Mount Hope Road, East Lansing, MI 48823, USA.
E-mail : fadly@msu.edu.

Vannier P, Espeseth D (eds): New Diagnostic Technology: Applications in Animal Health and Biologics Controls. Dev Biol (Basel). Basel, Karger, 2006, vol 126, p 307.

TSE Certification

P. Castle

European Directorate for the Quality of Medicines (EDQM), Council of Europe, Strasbourg, France

The TSE epidemic had a profound effect on the pharmaceutical industry in Europe. When the risks associated with raw materials of ruminant origin became apparent, the industry took stock of the situation and found that almost half of the medicines used in Europe had ingredients of ruminant origin or ingredients that were synthesised using substances of ruminant origin. In the absence of a reliable diagnostic or other test that could be used to exclude contaminated material, the industry had an obligation to reduce the risk associated with TSE by other means. The European Medicines Agency (EMEA) established a Note for Guidance on minimising TSE risk that identified a number of key factors:

- geographical origin of the ruminant material;
- source tissue;
- treatment for removal or inactivation of the TSE agent.

Compliance with the Note for Guidance became mandatory in the European Union. The practical difficulties for the industry to demonstrate compliance were such that the EMEA asked the EDQM to organise a certification scheme whereby compliance could be assessed at or near to the source of the ruminant material to streamline the evaluation of data. The pharmaceutical industry could then use sources of compliant raw materials guaranteed by an EDQM certificate. Use of EDQM certification is optional, since pharmaceutical manufacturers can choose to submit all the necessary data to the licensing authority, but the efficiency of the scheme is such that it has become the path of choice.

A panel of assessors evaluates requests for certificates, on the basis of the Note for Guidance (which is incorporated as an official general chapter of the European Pharmacopoeia). In addition to the evaluation of a data file, manufacturing sites can be inspected. The EDQM certificates are accepted by licensing authorities throughout Europe and are also recognised in a number of other countries throughout the world.

Dr. P Castle, European Directorate for the Quality of Medecines (EDQM), Council of Europe, 226 av. de Colmar, BP 907, FR-67029 Strasbourg Cedex 1, France.
E-mail : peter.castle@pheur.org

Vannier P, Espeseth D (eds): New Diagnostic Technology: Applications in Animal Health and Biologics Controls. Dev Biol (Basel). Basel, Karger, 2006, vol 126, pp 309-319.

Monitoring Veterinary Vaccines for Contaminating Viruses

H.-P. Ottiger

Institute of Virology and Immunoprophylaxis IVI, OMCL Vaccine Control, Swiss Federal Veterinary Office, Mittelhäusern, Switzerland

Key words: Avian viral vaccines, EA - extraneous agent, EP – European Pharmacopoeia, IVMPs - immunological veterinary medicinal products, PCR - polymerase chain reaction, NAT – nucleic acid testing.

Abstract: The detection of extraneous agents (EA) has always been critical for assessing the safety associated with vaccines. Vaccines and other biological medicines are derived from living substrates. This poses particular problems with regard to assuring their efficacy and safety due to the inherent variability of starting materials, the production processes, and the complex nature of the products themselves. For viral vaccines and other biological medicines, the key to effective quality control is rigorous testing and validation applied to the seed lots and also to the final batches. The quality assurance requirements for marketing products in Europe are given in the European Pharmacopoeia (EP).
With the advent of molecular biotechnology, there are increasing uses of molecular methods for the characterization and quality control of vaccines. PCR for extraneous agent testing for live and inactivated poultry vaccines has been recognized to be a valid alternative testing method and is now being used extensively as a diagnostic tool in assuring the freedom from EAs. A number of generic PCR assays for product testing have been developed and validated. The current status and issues of the application of nucleic acid testing (NAT) to the standardization and control of vaccines will be reviewed.

Quality Requirements

Certification of freedom from extraneous agents is an essential quality requirement of immunological medicinal products, like vaccines for veterinary use. Quality control is a key component in the production of medicinal products in general, and of immunological veterinary medicinal products (IVMPs) in particular. Vaccines are considered to be the most cost-effective tools in the prevention of infectious diseases; this applies in particular to avian live viral vaccines. Requirements for the purity of vaccines have increased considerably over recent years as a result of improved knowledge of the potential risks associated with the use of impure products either for the target animals themselves or for the consumers of animal products.

The biological nature of vaccines imposes a permanent risk for contamination of micro-organisms. Thus, the entire spectrum of substances of animal origin (i.e. embryonated hens' eggs, primary cell cultures etc.) used for the production of biologicals, master seed virus as well as the finished products, have to be tested [1]. Eventually, in the case of live viral vaccines, staff could transmit or contaminate infectious agents during the manufacturing process [2]. This could eventually lead to spread and/or seroconversion of flocks/livestock animals caused by such adventitious agents. For these reasons the extraneous agent (EA) testing is a crucial point in the quality control of vaccines [3].

The Regulatory Framework

The minimal requirements for the production and quality control of IVMPs are given in the European Pharmacopoeia (EP), which include general text and individual monographs on vaccines (live and inactivated) [1]. In recent years, the EP Commission and its Group of Experts No. 15V, which is responsible for veterinarian vaccines and sera, have made great efforts to revise monographs constantly and to harmonize them. Table 1 outlines the general scheme for the elaboration of monographs, publication and implementation. Principally, each draft monograph has to be first published in Pharmeuropa [4], after which members of the public are then invited to submit comments on the text to the EP Commission. The comments received are discussed among the relevant group of experts, and if considered to be appropriate, incorporated into a revised document. After final approval by the EP Commission, the new monographs will be implementated into the EP, which will then become mandatory for all member states. An additional transition time for manufacturers must be given to the new monographs. Figure 1 indicates that considering all the steps from starting until the final implementation of a new technology, this it is usually a rather lengthy and time-consuming process which may take several years.

Table 1: General scheme for elaboration of European Pharmacopoeia (EP) monographs, publication and implementation.

The outline scheme for the elaboration of new or revised monographs is the following:

- (i) a monograph is proposed for drafting by the President of the European Pharmacopoeia Commission (EPC), by a delegation, by the group of experts or by the EDQM Secretariat at the EPC;
- (ii) work on the monograph is authorized by the EPC;
- (iii) a first draft is prepared by a member of Group of Experts No. 15V;
- (iv) the first draft is submitted to the other members of the group;
- (v) comments from the members of the group are discussed by the group;
- (vi) after agreement of a draft by the group the monograph is submitted to a public enquiry by publication for comment in Pharmeuropa (quarterly information bulletin). A three-months period is allowed for comment;
- (vii) the results of the public enquiry are studied by the group and any necessary changes to the monograph are made. If extensive changes are made, steps 6 and 7 are repeated;
- (viii) the draft agreed by the group is submitted to the EPC for adoption;
- (ix) when the monograph is adopted, a "definitive" version is prepared;
- (x) monographs are published quarterly about six months after adoption and become official six months after publication.

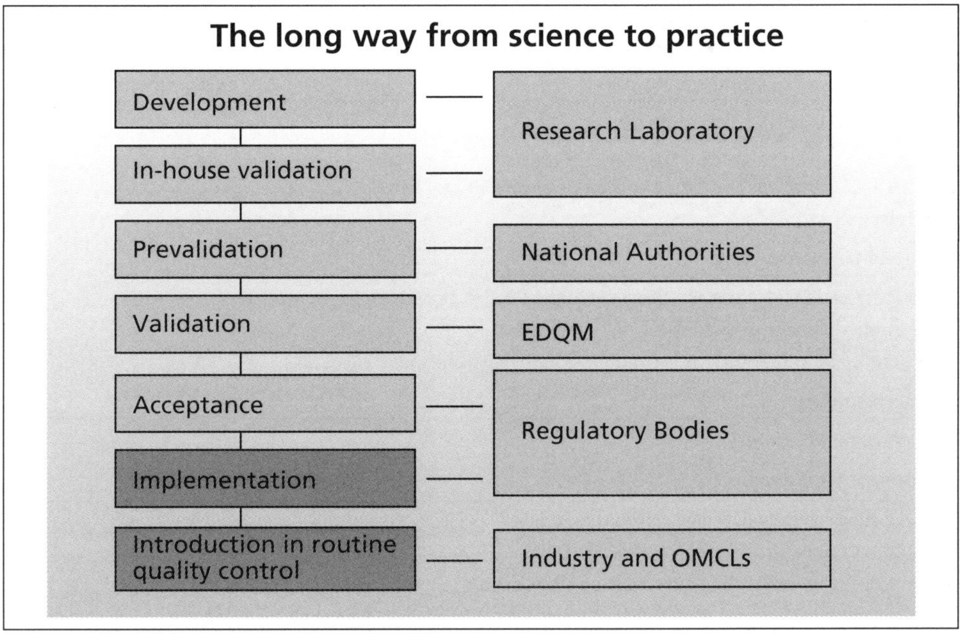

Fig. 1: Validation process of new technologies. Life cycles from technical development until final implementation.

Testing for Extraneous Agents in Avian Live Viral Vaccines

The current Fifth Edition of the European Pharmacopoeia & Supplements [1] includes several specific monographs on avian vaccines (live and inactivated) as well as the texts for extraneous agent testing in the general chapters. A selection of relevant text documents is given in Table 2. In the General Monograph "Vaccines for Veterinary Use", which is applicable to all veterinary vaccines, it is stated, that "*the methods of preparation.... are such as to maintain the identity and immunogenicity of the antigen and to ensure freedom from contamination with extraneous agents*".

- **Substances of animal origin for the production of veterinary vaccines**
  Substances of animal origin (i.e. serum, trypsin, and serum albumin) for the production of veterinary vaccines have to comply with the requirements of EP chapter 5.2.5 "Substances of animal origin for the production of veterinary vaccines" (Table 2) to minimise the risk associated with pathogens that may be present in them.

Table 2: European Pharmacopoeia (EP) texts on avian vaccines [1].

Methods of analysis/Biological tests	**Serial number**
Nucleic acid amplification techniques	2.6.21
Avian viral vaccines: Tests for extraneous agents in seed lots	2.6.24
Avian viral vaccines: Tests for extraneous agents in batches of finished products	2.6.25
General texts on vaccines	
Chicken flocks free from specified pathogens for the production and quality control of vaccines	5.2.2
Substances of animal origin for the production of veterinary vaccines	5.2.5
General monograph	
Vaccines for veterinary use	0062
Individual monographs	
Avian infectious bronchitis vaccine (inactivated)	0959
Avian infectious bursal disease vaccine (live)	0587
Avian infectious bursal disease vaccine (inactivated)	0960
Avian infectious encephalomyelitis vaccine (live)	0588
Avian infectious laryngotracheitis vaccine (live)	1068
Avian paromyxovirus 3 vaccine (inactivated)	1392
Avian viral tenosynovitis vaccine (live)	1956 (5th Edition)
Duck plague vaccine (live)	1938 (Supplement 5.3)
Duck viral hepatitis type I vaccine (live)	1315
Egg drop syndrome vaccine (inactivated)	1202
Fowl cholera vaccine (inactivated)	1945 (Supplement 5.1)
Fowl-pox vaccine (live)	0649
Infectious chicken anaemia vaccine (live)	2038 (5th Edition)
Marek's disease vaccine (live)	0589
Newcastle disease vaccine (live)	0450

- **Chicken flocks free from specified pathogens for the production and quality control of vaccines**
  Chickens, embryos or cell cultures used for the production or quality control of vaccines must be derived from eggs produced by chicken flocks free from specified pathogens (SPF). The management of maintaining such an SPF status must comply with EP chapter 5.2.2 "Chicken flocks free from specified pathogens for the production and quality control of vaccines" (Table 2).

- **Avian viral vaccines: tests for extraneous agents in seed lots and in batches of finished products**
  Testing for extraneous agents in avian live viral vaccines is regulated in the EP by the two following general documents which apply to all the individual

monographs on avian vaccines: 2.6.24 "Avian Viral Vaccines: Tests for Extraneous Agents in Seed Lots" and 2.6.25 "Avian Viral Vaccines: Tests for Extraneous Agents in Batches of finished products" (Table 2). The current scheme requires the testing for EA to be carried out with fertilized hens' eggs, cell cultures and chicks to cover a range of possible contaminants. The methods are animal and non-animal tests that have to be performed on the seed lots and on each batch of the finished product.

(i) Test for extraneous agents using embryonated hen's eggs: nine- to 11-day-old hens' eggs are used for virus cultivation, after neutralisation of the vaccine virus (at least 10 doses) with a mono-specific antiserum. One group is inoculated into the allantoic cavity and the other group on to the chorio-allantoic membrane. A third group of five- to six-day-old hens' eggs are inoculated into the yolk sac. The vaccine does not comply with the test if any deaths or abnormalities occur in the embryos, attributable to the seed lot.

(ii) Test in cell-cultures: chicken embryo kidney cells from SPF-chicks are used for virus cultivation. Cell cultures are passaged regularly and observed throughout a 21-day incubation period. Chicken erythrocytes are used to test for the presence of haemagglutinating agents. The seed lots fail with the test if any specific cytopathic effect occurs or other evidence of the presence of contaminating agents is observed.

(iii) Test for extraneous agents using chicks: Two week-old chicks (SPF-animals) are twice inoculated in a given period with multiple doses of vaccine, by the intramuscular route and eye-drops. Serum is collected from each individual before and at the end of the test period. The animals are observed and should not show signs of infection. The serum samples are then tested for antibodies against a given list of specific pathogens. The test in chicks covers 16 infectious agents (i.e. *Salmonella pullorum, Chlamydia spp.* or avian viruses; see Table 3).

Application of New Technologies: Alternative Methods to the Animal Testing Are Available

Testing for antibodies after experimental vaccination leads to a high number of animals used in the quality control of vaccines. With regard to the number of animals used, far more animals are needed for the testing of each batch than for testing seed lots. It is often not the most sensitive method but the sensitivity is partially compensated by multiple dose injections. For animal welfare reasons, animal testing must be reduced, refined or replaced whenever possible [5-9].

Efforts have been made to develop and validate nucleic acid amplification techniques (NAT). Polymerase chain reaction (PCR) is a very promising in-vitro technique which is widely used to amplify target DNA or RNA [8,10]. In the chapters EP 2.6.24 "Avian viral vaccines: test for extraneous agents in seed lots" and EP 2.6.25 "Avian viral vaccines: test for extraneous agents in batches of finished products" it is stated that "*Other types of tests than those indicated may be used provided they are at least as sensitive and with*

Table 3: Standard tests for avian seed virus (EP) [1].

Agent	Type of test
Avian adenovirus, group 1	SN, EIA, AGP
Avian encephalomyelitis virus	AGP, EIA
Avian infectious bronchitis virus	EIA, HI
Avian infectious laryngotracheitis virus	SN, EIA, IS
Avian leucosis viruses	SN, EIA
Avian nephritis virus	IS
Avian orthoreoviruses	IS, EIA
Avian reticuloendotheliosis virus	AGP, IS, EIA
Chicken anaemia virus	IS, EIA, SN
Egg drop syndrome virus	HI, EIA
Avian infectious bursal disease virus,	
Serotype 1 and 2	AGP, EIA, SN
Influenza A virus	AGP, EIA, HI
Mareks' disease virus	AGP
Newcastle disease virus	HI, EIA
Turkey rhinotracheitis virus	EIA
Salmonella pullorum	AGG
Additional tests for turkey seed virus	
Avian infectious haemorrhagic enteritis virus	AGP
Avian paramyxovirus 3	HI
Avian infectious bursal disease virus type 2	SN
Chlamydia spp.	EIA
Additional tests for duck seed virus	
Duck and goose parvoviruses	SN
Duck enteritis virus	SN, EIA
Duck hepatitis virus type 1	SN
Chlamydia spp.	EIA
Additional tests for goose seed virus	
Duck and goose parvoviruses	SN, EIA
Duck enteritis virus	SN
Goose haemorrhagic polyomavirus	Test in animals

Abbreviations: AGG = agglutination; AGP = agar gel precipitation; SN = serum neutralization; EIA = enzyme immunoassay (i.e. ELISA); HI = haemagglutination inhibition.

appropriate specificity. Nucleic acid amplification techniques (2.6.21) give specific detection for many agents and can be used after validation for sensitivity and specificity."

It is clear that in many laboratories the PCR is replacing virus isolation or bacterial cultivation for the detection of agents that are difficult or impossible to culture. There are several reasons for this trend, e.g. expensive cell culture and maintenance facilities, the presence of replicating viruses, usually a long time to complete the diagnosis, and the use of animal experimentation. Molecular biology assays have become relatively cheap, safe and user-friendly tools in diagnostic laboratories [11].

The Principles of PCR

Nucleic acid amplification techniques have to be performed and validated according to the requirements given in EP Method of Analysis 2.6.21 "Nucleic acid amplification techniques" (see Table 2). Briefly, PCR is an amplification reaction which involves several rounds of copying a specified stretch of target nucleic acid, in this case from the genome of an infectious agent. The amplified region is defined by two short complimentary oligonucleotides-primers flanking the target sequences. Using the heat-stable DNA polymerase, which is not denatured during heat cycling, it is possible to copy the sequences between the primers. By repeating up to 35 times the amplification process, the amount of target DNA gained is enough for detection or sequencing. The sensitivity is very high, considering that several million copies are produced. The specificity of the reaction is determined by the used oligonucleotide-primers. The primers must be designed to detect the selected target infectious agents. The genomes of many infectious agents contain ribonucleic acids RNA that can be amplified by including a reverse-transcription step before the PCR. Nowadays, heat-stable DNA polymerases with reverse-transcriptase activity are available enabling the RT-PCR taking place in the same tube without further handling. Serious precautions have to be taken to avoid cross-contaminations of the test samples or reagents such as (i) extractions, preparations, amplifications and handling of final products in separated compartments; (ii) frequent change of disposable gloves during testing; (iii) using positive displacement micro-pipettes and aerosol-resistant tip racks.

Nucleic Acid Testing for Adventitious Viruses

For extraneous agent (EA) testing, NAT has already proved to be a suitable tool for the identification of many of the adventitious viruses. A number of generic PCR assays for final product testing have been recently developed and validated [12-30]. Specific tests for adventitious infectious agents are conducted using various polymerase chain reaction based assays: a list of available PCR tests is given in Table 4. Some methods are well established, but others are still under development. In many cases, these assays have been validated, that is, proof-tested using spiked vaccine and vaccine substrate materials. Today, the quality standards for ISO/EN 17025 accreditated laboratories and those working under GMP/GLP require careful validation of tests. Therefore, it is not possible to introduce straightforward assays which were developed for other purposes (i.e. diagnostic tests in organs, etc.) than for quality control of vaccines. For design and development of new assays we have put an increasing emphasis on the process of standardisation, as for a complete testing scheme (i.e. all viruses should be tested simultaneously) there should not be too many variations in handling between individual assays. Special attention has therefore been put on the use of common commercially available reagents, enzymes and nucleic acid extraction kits and standardized cycling parameters. Therefore, most of the methods were or will have to be adjusted, so they can be performed in an efficient manner in-praxis. Furthermore, it is likely that matrix effects such as the presence of oil-adjuvants and other components could interfere with the extraction of nucleic acids, decrease the recovery of materials and inhibit the enzymatic activities, and therefore must be carefully ruled out. This is especially an issue for preparations such as inactivated vaccines. However, new data indicate that in-vitro tests are as sensitive as the in vivo tests [31].

Table 4: Available standardized PCR procedures for the detection of avian viruses.

Avian virus		Genome	Target genes *	Reference
Avian adenovirus	AdV	DNA	n.s.	Jiang et al, 1999 [12]
Avian encephalomyelitis virus	AEV	RNA	n.s.	Xie et al, 2005 [13]
Avian infectious bronchitis virus	IBV	RNA	Nucleocapsid (316 bp) Nucleocapsid (240 bp)	Falcone et al, 1997 [14] Falcone et al, 1997 [14]
Avian infectious laryngotracheitis virus	ILTV	DNA	P32 (588 bp)	Vögtlin et al, 1999 [15]
Avian leucosis viruses, subgroup A and B	ALV	RNA	gp85 (466 bp) gp85 (371 bp)	Häuptli et al, 1997 [16] Häuptli et al, 1997 [16]
Avian leucosis virus, subgroup J	ALV-J	RNA	Pol (545 bp)	Smith et al, 1998 [17]
Avian nephritis virus	ANV	RNA	n.s.	
Avian orthoreoviruses	ARV	RNA	S2 (399 bp) S4 (437 bp)	Bruhn et al, 2004 [18] Bruhn et al, 2004 [18]
Avian reticuloendotheliosis virus	REV	RNA	n.s.	Fadly et al, 1997 [19] Takagi et al,1996 [20]
Chicken infectious anaemia virus	CIAV	DNA	Cux-1 (419 bp) n.s.	Taylor (unpublished) Falcone et al, 2000 [21]
Egg drop syndrome virus	EDS	DNA	n.s. n.s.	Kumar et al, 2003, [22] Dhinakar Raj, 2003 [23]
Avian infectious bursal disease virus, Serotype 1 and 2	IBDV	RNA	VP1 (360 bp) VP2 (282 bp)	Timgren et al, 2002 [24] Timgren et al, 2002 [24]
Influenza A virus	AIV	RNA	Matrix gene (244bp)	Fouchier et al, 2000 [25]
Mareks disease virus, serotypes 1, 2 and HVT	MDV	DNA	gA (680 bp) UL19 (521 bp)	Lang et al, 2005 [26] Lang et al, 2005 [26]
Newcastle disease virus	NDV	RNA	F0 (310 bp)	Stäuber et al, 1995 [27]
Turkey rhinotracheitis virus	TRTV	RNA	n.s.	Bäyon-Auboyer et al, 1999 [28]
Avian infectious haemorhagic enteritis virus	HEV	DNA	n.s.	Hess et al, 1999 [29]
Duck enteritis virus	DEV	DNA	n.s.	Plummer et al, 1998 [30]

* n.s. – not standardized/method not validated according to ISO/EN17025 standards.

The experiences of the last decade indicate that the PCR techniques will eventually supersede many of the classical methods of infectious agent detection [11]. NAT is particular suitable for testing for extraneous agents. It has the potential to detect virus-specific nucleic acids of known agents and is a very sensitive, specific and fast method. For the purpose at hand, we are suggesting that the established line of specific tests now used on all modern vaccine materials be used.

Prospects

PCR for extraneous agent testing for live and inactivated poultry vaccines has been recognized to be a valid alternative testing method and is now being used extensively as a diagnostic tool in ensuring the freedom from extraneous agents.

The quality standards also have an important implication for animal welfare. However there is a need for further development of new assays and improving existing ones. Further collaborative studies and investment in training, equipment and facilities are needed to promote and enhance these developments internationally. For the safety of biologicals, and of vaccines in particular, a close cooperation between scientific base, legislation and manufacturers is needed for implementation. Only international harmonization will encourage industry-wide adoption of new technologies by vaccine manufacturers and control authorities. In particular the process of standardization and validation requires special attention.

Acknowledgements

This work was funded by the Swiss Federal Veterinary Office.

References

1 Council of Europe. European Pharmacopoeia. Editions of the Council of Europe, Strasbourg, France, 2005, Fifth Edition.

2 Elgizoli M, Vögtlin A, Bruckner L, Ottiger HP: Sequence analysis of a Newcastle disease virus contaminating avian vaccines. Union of Swiss Societies for Experimental Biology USGEB 1998. Abstract p 112.

3 van der Kamp M: Replacement, reduction or refinement of animal use in the quality control of veterinary vaccines: development, validation and implementation, in Brown F, Cussler K, Hendriksen C (eds): Replacement, Reduction and Refinement of Animal Experiments in the Development and Control of Biological Products. Dev Biol Stand. Basel, Karger, 1996 vol 86, pp 73-76.

4 http://www.pheur.org/.

5 Jungbäck C, Küchler B, Schirk EM: Poultry vaccines: an analysis of the animal trials required for vaccine testing and the way to reduce or refine these tests, in Brown F, Hendriksen C, Sesardic D (eds): Alternatives to Animals in the Development and Control of Biological Products for Human and Veterinary Use. Dev Biol Stand. Basel, Karger, 1999, vol 101, pp 79-83.

6 Bruckner L, Bongers J, Castle P, Flore PH, Guittet M, Halder M, Jungbäck C, Le Gros FX, Tollis M, Venugopal K, Wilhelm M, Zeegers J, Zigterman G: Three Rs approaches to the production and quality control of avian vaccines. The Report and Recommendations of the ECVAM Workshop 41. ATLA 2000;28:241-258.

7 Halder M: Three Rs potential in the development and quality control of immunobiologicals. ALTEX 2001;18(Suppl):13-47.

8 Ottiger HP: Use of PCR for extraneous agent testing in avian vaccines, in Brown F, Hendriksen C, Sesardic D, Cussler K (eds): Advancing Science and Elimination of the Use of Laboratory Animals for Development and Control of Vaccines and Hormones. Dev Biol. Basel, Karger, 2002, vol 111, pp 23-26.

9 Castle P: Alternatives to animal testing: achievements and recent developments in the European Pharmacopoeia, in Brown F, Cussler K, Hendriksen C (eds): Replacement, Reduction and Refinement of Animal Experiments in the Development and Control of Biological Products. Dev Biol Stand. Basel, Karger, 1996, vol 86, pp 21-29.

10 Lucken R, Stolp R: Identification of extraneous agents and other contaminants, in Pastoret PP, Blancou J, Vannier P, Verschueren C (eds): Veterinary Vaccinology. Elsevier 1997, pp 208-109.

11 Manual of diagnostic tests and vaccines for terrestrial animals, Office International des Epizooties (OIE), Fifth Edition, 2004.

12 Jiang P, Ojkic D, Tuboly T, Huber P, Nagy E: Application of the polymerase chain reaction to detect fowl adenoviruses. Can J Vet Res 1999;63:124-128.

13 Xie Z, Khan MI, Girshick T, Xie Z: Reverse transcriptase-polymerase chain reaction to detect avian encephalomyelitis virus. Avian Disease 2005;49:227-230.

14 Falcone E, D'Amore E, Di Trani L, Sili A, Tollis M: Rapid diagnosis of avian infectious bronchitis virus by the polymerase chain reaction. J Virol Methods 1997;64:125-130.

15 Vögtlin A, Bruckner L, Ottiger HP: Use of polymerase chain reaction (PCR) for the detection of vaccine contamination by infectious laryngotracheitis virus. Vaccine 1999;17(20-21):2501-2506.

16 Häuptli D, Bruckner L, Ottiger HP: Use of reverse transcriptase polymerase chain reaction for detection of vaccine contamination by avian leukosis virus. J Virol Methods 1997;66:71-81.

17 Smith LM, Brown SR, Howes K, Mcleod S, Arshad SS, Barron GS, Venugopal K, McKay JC, Payne LN: Development and application of polymerase chain reaction (PCR) tests for the detection of subgroup J avian leukosis virus. Virus Research 1998;54:87-98.

18 Bruhn S, Bruckner L, Ottiger HP: Application of RT-PCR for the detection of avian reovirus contamination in avian viral vaccines. J Virol Methods 2005;123:129-130.

19 Fadly AM, Witter RL: Comparative evaluation of in-vitro and in-vivo assays for the detection of reticuloendotheliosis virus as a contaminant in a live virus vaccine of poultry. Avian Disease 1997;41:695-701.

20 Takagi M, Ishikawa K, Nagai H, Sasaki T, Gotoh K, Koyama H: Detection of contamination of vaccines with the reticuloendotheliosis virus by reverse transcriptase polymerase chain reaction (RT-PCR) Virus Research 1996;40:113-121.

21 Falcone E, Tarantino M, Massi P: Detection of vaccine contamination by chicken anaemia virus by polymerase chain reaction. Progress in the Reduction, Refinement and Replacement of Animal Experimentation 2000;951-958.

22 Kumar NS, Kataria JM, Koti M, Dhama K, Toroghi R: Detection of egg drop syndrome 1976 virus by polymerase chain reaction and study of its persistence in experimentally infected layer birds. Acta Virol 2003;47(3):179-184.

23 Dhinakar Raj G, Sivakumar S, Matheswaran K, Chandrasekhare M, Thiagarajan V, Nachimuthu K: Detection of egg drop syndrome virus antigen or genome by ELISA or polymerase chain reaction. Avian Pathology 2003;32:545-550.

24 Timgren R: Use of RT-PCR for the detection of vaccine contamination by infectious bursitis disease (Gumboro) virus. Dissertation Faculty of Veterinary Medicines, University of Berne, 2001.

25 Fouchier RAM, Bestebroer, TM, Sander H, Van der Kemp L, Rimmelzwaan GF, Osterhaus AD: Detection of influenza A viruses from different species by PCR-amplification of conserved sequences in the matrix gene. J Clin Microbiology 2000:4096–4101.

26 Lang K: Detection of Marek's disease virus contamination in poultry vaccines using PCR. Dissertation Faculty of Veterinary Medicines Vetsuisse, University of Berne, 2005.

27 Stäuber N, Brechtbühl K, Bruckner L, Hofman MA: Detection of Newcastle disease virus in poultry vaccines using the polymerase chain reaction and direct sequencing of amplified cDNA. Vaccine 1995;13:360-364.

28 Bäyon-Auboyer MH, Jestin V, Toquin D, Cherbonnel M, Eteradossi N: Comparison of F-, G- and N-based RT-PCR protocols with conventional virological procedures for the detection and typing of turkey rhinotracheitis virus. Arch Virol 1999;144:1091-1109.

29 Hess M, Raue R, Hafez HM: PCR for specific detection of haemorrhagic enteritis virus of turkey, an avian adenovirus. J Virol Methods 1999;81:199-203

30 Plummer P, Alefantis T, Kaplan S, Oconnell P, Shawky S, Schat KA: Detection of duck enteritis virus by polymerase chain reaction. Avian Disease 1998;42:554-564.

31 Bootz F, Sieber I, Popovic D, Tischhauser M, Homberger FR: Comparison of the sensitivity of in vivo antibody production tests with in vitro PCR-based methods to detect infectious contamination of biological materials. Laboratory Animals 2003;37:342-351.

Dr. H.-P. Ottiger, Institute of Virology and Immunoprophylaxis IVI, OMCL Vaccine Control, Swiss Federal Veterinary Office, CH-3147 Mittelhäusern, Switzerland.
E-mail: hanspeter.ottiger@ivi.admin.ch.

Vannier P, Espeseth D (eds): New Diagnostic Technology: Applications in Animal Health and Biologics Controls. Dev Biol (Basel). Basel, Karger, 2006, vol 126, p 321.

Conclusions and Recommendations

Chairman: *David Espeseth,* USA

Vannier P, Espeseth D (eds): New Diagnostic Technology: Applications in Animal Health and Biologics Controls. Dev Biol (Basel). Basel, Karger, 2006, vol 126, pp 323-327.

Conclusions, Constraints, Recommendations and Solutions

D. Espeseth

Perkasie, PA, USA

Session I

PRINCIPLES OF NOVEL TECHNIQUES: TESTS DESCRIPTION AND POTENTIALITIES

Chairman: *Rick Hill,* USA

Conclusions

- Advances in technology in the past decades have resulted in PCR, rt-PCR, multiplex rt-PCR, continuous enzymatic amplification, nanotechnology-based immunoassays, and DNA micro-arrays procedures (to name a few).
- We are now looking to the use of these new chemistry-based assays in addition to traditional antigen-antibody assays.
- The use of these new technologies in the study of FMD provides a practical example of how they can contribute to improve our knowledge of biology, evolution of viruses, and epidemiology of disease.
- These new technologies present promising new tools.

Recommendations

- New technologies have great potential.
- Further development and standardization are needed when applying these technologies as diagnostic tools for international animal movement, disease surveillance, and control programs.
- New technologies should be used to complement existing tests.

Session II

Standardization, Validation and Overcoming the Problems of Molecular Diagnostic Methods

Chairman: *Alejandro Schudel,* OIE

Conclusions

- The WOAH (OIE) is commended for developing a standardized international framework for the evaluation of diagnostic test validation.
- There is a diversity of views on the preferred configuration of PCR methods for diagnosis, whether parallel or multiplex, and the need to use internal controls.
- Lessons can be learned from experiences in plant science (GMO's) and criminology (forensic use of DNA detection).

Recommendations

- Diagnostic test kit manufacturers are encouraged to submit validation dossiers for new tests for consideration by the OIE for its register.
- National Authorities should endeavor to harmonize their requirements for diagnostic test kit approval with those of the OIE.
- A standardized approach should be encouraged for the validation of PCR and other molecular diagnostic methods.
- Veterinary diagnosticians should learn from the experiences of other disciplines in developing validation systems that are appropriate for the purpose.

Constraints for transportation of reagents, reference materials, samples. Consequences for reference laboratories

Conclusions

- Stringent procedures are required for transport of select agents and associated materials, but when met, shipments are permissible.
- The complexity and cost of preparing Category A shipments is discouraging; OIE and Reference Laboratories' need assistance, awareness and training.
- Blanket Permits are needed for shipment of samples/isolates to OIE Reference Laboratories.
- OIE needs to define Standardized inactivation procedures for all substances, and shippers must verify compliance.

Recommendations

- Regulatory authorities should move from zero tolerance to risk assessment and mitigation of risk in the transport of samples and isolates as used in international animal trade.
- OIE should continue to participate in the UN's development of model guidelines for dangerous goods.

- National and International organizations should negotiate with airlines that will not accept Category B shipments.
- Regulatory agencies should provide import permit exemptions for products that meet specified inactivation procedures, and establish a list of permit exempt proficiency test organisms in lieu of requiring each site to obtain permits.
- Restrictions on use of organisms should be on the permit.
- OIE should develop movement standards (bio-safety and bio-security information), that address the requirements for possession of biological agents of concern.

Standardization and validation, etc. (QA & RT)

Conclusions

- Standardized OIE procedures and methodologies are needed to evaluate major characteristics of diagnostic kits, such as, sensitivity, specificity, repeatability, reproducibility, and predictive values.
- The need for standardization/validation of tests was a common theme of the Conference, especially for nucleic acid extraction procedures.
- The need to organize networks of laboratories to ensure standardization of proficiency tests for molecular techniques was noted.
- Quality Assurance and Accreditation of laboratories by international standards (ISO 17025/OIE guidelines) is a key factor.
- A mechanism to facilitate exchange of strains of organisms is needed, especially when crises are occurring.

Recommendations

- Reference laboratories should produce panels of reference materials for use by public and private laboratories and companies. Funding should be provided to support such activity. Reference materials should also include panels for batch release of test kits.
- Authorities should provide flexibility in requirements for the transport and handling of agents that gives priority to authorization of samples for assays and includes lists of excepted agents instead of requiring consideration on a case-by-case or agent-by-agent basis.
- We should promote negotiations with IATA to alleviate requirements for shipments between authorized laboratories.

Session III

Application in Detection of Pathogens and Screening Programmes

Chairman: *Paul Midtlyng,* Norway

Conclusions

- Molecular diagnostic methods are being developed for a wide variety of diagnostic situations across all species.

- Research for proof-of-concept is diverse, but for many applications there is sufficient progress in validation to support the transfer of these methods from research to everyday use.
- There is a possibility that few veterinary applications will be validated by private investors alone.
- The process of standardization and validation is highly dependent on the financial resources invested by research agencies and government laboratories.
- Examples from aquaculture suggest that animal production industries – in particular when disease risks and their economic consequences are severe – are ready to accept and implement these new methods on a broad scale.
- The development of new and improved pathogen detection methods has not only the potential to improve disease control tactics but also to shift entire paradigms of disease control, as was argued by the example of classical swine fever.

Recommendations

- To explore the potential of these new methods and keep disease control strategies up-to-date with technology developments, a closer and more intense dialogue should be organized between the users in veterinary administrations and animal production, and the diagnostic research community.

Session IV

Certainties and Uncertainties of Agent or Toxin Identification by Molecular Methods

Chairman: *Steve Edwards,* United Kingdom

Conclusions

- New diseases will emerge (including new variants of known diseases) and many will be zoonotic.
- There is a need for molecular tools for detection and investigation of these emerging disease agents.
- Molecular methods enable rapid, high-throughput testing, but diagnosticians should not altogether abandon the traditional methods.
- A number of new technologies offer promise, including rt-PCR, micro-arrays, NASBA, etc., but need further evaluation and standardization.
- It is likely that different technologies will be optimal for different applications.

Recommendations

- More work should be done on sample preparation and nucleic acid extraction.
- The advantages and disadvantages of multiplex versus parallel testing approaches should be assessed.
- Lower-cost approaches to micro-array methods should be developed.
- National and International reference laboratories should exchange protocols, reagents, strains, and sequences.

- Standardization projects should be funded and supported, and laboratories should endeavor to achieve consensus approaches that can be adopted by the OIE.

Session V

EXAMPLES OF APPLICATION IN QUALITY CONTROL OF HUMAN AND VETERINARY VACCINES

Chairman: *Jacques Léchenet,* France

Conclusions

- In vitro tests in conjunction with consistency of production are a way to decrease use of laboratory animals in the control of vaccines, but collaboration between all stakeholders is needed to succeed.
- Product quality must be built in during production, since potency testing of completed product alone is unable to assure required quality. In vitro tests may play a useful role as indicators for some manufacturing processes for some organisms. This approach could lead to a harmonized system, however, harmonization is a lengthy process and requires resources, time, and full collaboration toward a common goal.

Recommendations

- The evaluation of purity using new technologies that provide more sensitive and easier tests is not a great help if performed only on the finished product. Indeed the product does not change by such testing, just its acceptance after production. We should apply new technology tests at intermediate stages of production using a risk/benefit approach linked with their use; such as, the quality testing of starting material. Only such earlier testing could truly change the built in quality.
- DNA vaccines are a new tool that may pose new questions because of their novelty. Further research is required for the development of individual products and at the technology level. In addition to research, some political and educational efforts are needed toward customers to mitigate fear and possible rejection of a very powerful method of production.
- Some of the new tools presented during this session may permit us to quickly prepare vaccines against new diseases, as discussed in the European Technology Platform for Global Animal Health (http://www.ifah.be/Europe/EUPlatform/Platform.htm).

List of Participants

AGHAZADEH MESHGI Mahzad, Islamic Azad University, Science and Research Campus, Dept of Food Hygien, IR-22564 Theran, Iran.
ALBERT David, AFSSA Lerpaz, 23 av. du Général de Gaulle, FR-94706 Maisons Alfort, France.
ALONSO Sara, Bionostra S.L., C/Ronda de Poniente 4-2, ES-28760 Tres Cantos, Spain.
ALTMANN Michaela, Ages, Institute for Veterinary Disease Control, Kudlichstrasse 27, AT-4021 Linz, Austria.
AMADDEO Demetrio, Istituto Zooprofilattico Sperimentale Lazio e Toscana, Via Appia 1411, IT-00178 Rome, Italy.

BAHARI Aliasghar, Bu-Ali Sina University, Dept. of Animal Science, Hamadan, Iran.
BAIGENT Susan, Institute for Animal Health, Compton, Newbury, RG20 7NN Berkshire, United Kingdom.
BANKS Malcom, Veterinary Laboratories Agency, Woodham Lane, New Haw, Addlestone, KT15 3NB Surrey, United Kingdom.
BARDONNET Nicolas, Qiagen, 3 av. du Canada, LP 809, FR-91974 Courtaboeuf Cedex, France.
BARLIC-MAGANJA Darja, Veterinary Faculty, University of Ljubljana, Gerbiceva 60, SI-61115 Ljubljana, Slovenia.
BASOVA Daria, All Russian Research Institute for Animal Health, RU-600901 Vladimir, Yur'evets Arriach, Russia.
BERGMANN Sven, Friedrich Loeffler Institut, Boddenblick 5a, DE-17493 Insel Riems, Germany.
BERGMANN Ingrid, Pan American FMD Center, PAHO/WHO, P.O. Box 589, Av Pres. Kennedy 7778, Duque de Caxias, Cep 25040-004, Río de Janeiro, Brasil.
BERTHE Thomas, LDA 22, Zoopole le Sabot, 9 rue du Sabot, FR-22440 Ploufragan, France.
BERTHEAU Yves, INRA, 10 rte de St-Cyr, FR-78026 Versailles, France.
BLACK John, American Bioresearch, 1419 Chapman Highway, Sevierville, TN 37876, USA.
BLANCHARD Patricia, University of California Animal Health and Food Safety Laboratory, 18830 Rd 112, Tulare, CA 93274, USA.
BLANCHARD Yannick, AFSSA Ploufragan, Zoopôle, BP 53, FR-22440 Ploufragan, France.
BLANCHARD Béatrice, Adiagene, 38 rue de Paris, FR-22000 Saint Brieuc, France.
BLOND Denis, VIRBAC Laboratories, BP 27, FR-06511 Carros, France.
BOSCHIROLI Maria Laura, AFSSA Maisons Alfort, 22 rue Pierre Curie, FR-94703 Maisons-Alfort Cedex, France.
BRIDONNEAU Philippe, Scanelis, 23 ch. des Capelles, FR-31076 Toulouse, France.
BRUCKNER Lukas, Institute of Virology and Immunoprophylaxis / IVI, Sensemattstrasse 293, CH-3147 Mittelhausern, Switzerland.
BURR Paul, BioBest Laboratories Ltd, Pentlands Science Park, Penicuik Midothian, EH26 0PZ, United Kingdom.
BUSCARLET Laure, Diffchamb, 8 rue Saint Jean de Dieu, FR-69007 Lyon, France.

CAMPBELL Colin, University of Edinburgh, 49 Little France Crescent, Edinburgh, EH25 9RG, United Kingdom.
CAPORALE Vincenzo, Scientific Commission Animal Diseases OIE, Via Campo Boario, IT-64100 Teramo, Italy.
CARICATI Celso, Beraca Sabara Quimicos E, Rua Souza Melo 73, BR-3707000 Sao Paulo, Brasil.

CARRILLO Consuelo, Plum Island Animal Disease Center, USDA, PO Box 848, Greenport, NY 11944, USA.
CARTER Anthony, Guildhay Ltd, 6 Riverside Business Centre, Walnut Tree Close, Surrey GU1 4UG, United Kingdom.
CASTLE Peter, EDQM, European Directorate for the Quality of Medecine, 226 av. de Colmar, BP 907, FR-67029 Strasbourg Cedex 1, France.
CATTOLI Giovanni, Istituto Zooprofilattico Sperimentale Delle Venezie, Via Romea 14/A, IT-35020 Legnaro (Padova), Italy.
CELER Vladimir, National Institute of Public Health, CHPR, Palackeho 1-3, CZ-612 42 Brno, Czech Republic.
CERVELLATI Cecilia, Fatro SpA, Via Emilian 285, IT-40064 Ozzano Emilia, Italy.
CHARLETY Paul Julián, EPLD, Etablissement Public des Laboratoires Départementaux, Rue P. Bonnard, Cité Administrative, FR-64000 Pau, France.
CHEN Hualan, National Avian Influenza Reference Laboratory, Harbin Veterinary Research Institute CAAS, 427 Maduan Street, CN-150001 Harbin, Republic of China.
CHENUAUD Pierre, AFSSA Fougères – ANMV, B.P. 90203, FR-35302 Fougeres Cedex, France.
CIABATTI Ilaria, Istituto Zooprofilattico Sperimentale Lazio e Toscana, Via Appia 1411, IT-00178 Rome, Italy.
CLAASSEN Ivo, Departement of Bacteriological & TSE, PO Box 2004, NL-8203 AA Lelystad, The Netherlands.
CREELAN Julie, Veterinary Sciences Division, Dept of Agriculture, Stormont, Stoney Rd, Belfast, BT4 3SD, United Kingdom.
CURRIE Richard, Fort Dodge Animal Health, Flanders Road, Hedge End, Southampton, SO30 4QH, United Kingdom.

DE BATTISTI C., Istituto Zooprofilattico Sperimentale Delle Venezie, Via Romea 14/A, IT-35020 Legnaro (Padova), Italy.
DE ST VIS Blandine, LSI, Le Bois Dieu, 1 Bis Allée de la Combe, FR-69380 Lissieu, France.
DEPNER Klaus R., Bundesforschungsanstalt für Viruskrankheiten der Tiere, Federal Research Centre, Boddenblick 5a, DE-17498 Insel Riems, Germany.
DESCARSIN Véronique, AES, Route de Dol, FR-35270 Combourg, France.
DESHAIES Eric, Synbiotics Europe, 2 rue A. Fleming, FR-69367 Lyon Cedex 07, France.
DEVULDER Grégory, Biomerieux, Chemin de l'Orme, FR-69280 Marcy l'Etoile, France.
DIAZ François, OIE, 12 rue de Prony, FR-75017 Paris, France.
DIEV Viatcheslav, All Russian Research Institute for Animal Health, RU-600901 Vladimir, Yur'evets Arriach, Russia.
DINH Thi Bich Lan, Hue University, Vietman, 180 Pasteur, District 1, Ho Chi Minh, Vietnam.
DIRBAKOVA Zuzana, State Veterinary Institute, Pod Drahami 918, SK-960 86 Zvolen, Slovakia.
DORY Daniel, AFSSA Ploufragan, Zoopôle, BP 53, FR-22440 Ploufragan, France.
DRISKELL Jeremy, Iowa State University, 42 Spedding, Ames, IA 50011, USA.
DUBOURGET Philippe, Merial, 254 rue Marcel Mérieux, FR-69007 Lyon, France.
DUCAN Paul, Merck Research Laboratories, WP 17-101, Sumneytown Pike, West Point, PA 19486, USA.
DUGUE Raphael, IRM, Rue Newton, ZI Mitry Compans, FR-77290 Mitry Mory, France.

EDWARDS Steven, Veterinary Laboratories Agency, Woodham Lane, New Haw, Addlestone, Surrey KT15 3NB, United Kingdom.
ERNEST Isabelle, GSK – Biologicals, 14 rue Hubert Larock, BE-4280 Hannut, Belgium.
ERRINGTON Jane, VLA Penrith, Merrythought, Cumbria, Penrith, United Kingdom.
ESPERON Fernando, Facultad de Veterinaria, Avda Puerta de Hierro, ES-28040 Madrid, Spain.
ESPESETH David, Espeseth Consulting, 404 Longleaf Dr, Perkasie, PA 18944, USA.

FADLY Aly M., USDA-ARS Avian Disease and Oncology Laboratory, 3606 East Mount Hope Rd, East Lansing, MI 48823, USA.
FAKUR Shaahin, Faculty of Vet. Med. Islamic Azad University Sanadaj, Pasdaran Ave. 618, Sanandaj, Iran.
FELIZIANI Francesco, Istituto Zooprofilattico Sperimentale Umbria e Marche, Via Gaetano Salvemini 1, IT-06129 Perugia, Italy.
FRIMODIG Christina, Svanova Biotech AB, Science Park Glunten, SE-75183 Uppsala, Sweden.

GALLEAU Stéphanie, Merial, Parc Industriel de la Plaine de l'Ain, Allée des Cyprès, FR-01150 Saint Vulbas, France.
GARIN BASTUJI Bruno, AFSSA Lerpaz, 23 av. du Général de Gaulle, FR-94706 Maisons Alfort, France.
GAUDRY Daniel, la Boucharderie, FR-16440 Roullet, France.
GAY Cyril, USDA Agricultural Research Service, 5601 Sunnyside Avenue, Building 4-2194, Beltsville, MD 20705, USA.
GOODYEAR Mark, Pfizer Inc., 333 Portage St, Kzo 267-409, Kalamazoo, MI 49007, USA.
GORIS Nesya, Cerva Coda Var, Groeselenberg 99, BE-1180 Brussels, Belgium.
GRADINARU Dragos, Institut Pourquier, 326 rue de la Galéra, FR-34097 Montpellier, France.
GREINER Matthias, Danish Institute for Food Agency and Veterinary Research, Morkhoj Bygade 19, DK-2860 Soborg, Denmark.
GUERIN Bernard, LNCR - Laboratoire National Contrôle Reproducteurs, 13 rue Jouët, FR-94704 Maisons Alfort, Cedex, France.

HAKHVERDYAN Mikhayil, National Veterinary Institute, Swedish University of Agricultural Sciences, Ulls Vag 2B, SE-75 189 Uppsala, Sweden.
HANDBERG Kurt, Danish Veterinary Institute, Hangovej 2, DK-8200 Arhus N, Denmark.
HAREL Josée, Université de Montréal, 3200 Sicotte, CP 5000, St Hyacinthe, J2S 7C6, Canada.
HENAULT Sylvie, AFSSA Lerpaz, 23 av. du Général de Gaulle, FR-94706 Maisons Alfort, France.
HESSELINK Win, Intervet International BV, PO Box 31, NL-5830 AA Boxmeer, The Netherlands.
HIGASHI Hisako Gondo, Instituto Butantan, 1500 av. Dr Vital Brazil, Sao Paulo, 5503, Brasil.
HILL Rick, USDA Center fo Veterinary Biologics, 510 S 17th Street, Ames, IA 50010, USA.
HJERTNER Bernt, Queen's University Belfast, Stoney Road, Stormont, Belfast, BT8 8HU, United Kingdom.
HOFFMANN Bernd, Friedrich Loeffler Institut, Boddenblick 5a, DE-17493 Insel Riems, Germany.
HUETER Michael, Pfizer Inc, 333 Portage St, KZO 267-409, Kalamazoo, MI 49007, USA.
HUOVILAINEN Anita, EELA, National Veterinary and Food Research Institute, Hameentie 57, PO Box 45, FI-00581 Helsinki, Finland.

ILOTT Martin, V.M.D., Woodham Lane, New Have, Weybridge, Surrey KT15 3NB, United Kingdom.

JACOBS Liesbeth, CEDI Diagnostics, Edelhertweg 15, NL-8219 PH Lelystad, The Netherlands.
JAKAVA VILJANEN Miia, EELA, National Veterinary and Food Research Institute, Hameentie 57, PO Box 45, FI-00581 Helsinki, Finland.
JEMERSIC Lorena, Croatian Veterinary Institute, Savska cesta 143, HR-10 00 Zagreb, Croatia.
JOLY Jean Pierre, Ifremer, Av. du Mus de Loup, FR-17390 La Tremblade, France.
JUNGBAECK Carmen, Paul Ehrlich Institut, Paul Ehrlich Str. 51-59, Postfach 1740, DE-63225 Langen, Germany.

KAPLAN Claude, IP Pragmatics Ltd, 8 Edward Crescent, Wareham, United Kingdom.
KASBARI Mohamed, AFSSA Direction Générale, 27-31 av. du Général Leclerc, BP 19, FR-94701 Maisons Alfort, France.
KIM Yong Joo, NVRQS, Anyang 6 domg 480, 430 An Yang City, Gyung Gi, Republic of Korea.
KOOI Bart, CIDC Lelystad, Houtribweg 39, NL-8221 RA Lelystad, The Netherlands.
KOOLEN Johanna, Synbiotics Europe, 2 rue A. Fleming, FR-69367 Lyon Cedex 07, France.
KRAMPS Johannes A., CIDC Lelystad, PO Box 2004, NL-8203 AA Lelystad, The Netherlands.
KUNTZ-SIMON Gaelle, AFSSA Ploufragan, Zoopôle - BP 53, FR-22440 Ploufragan, France.
KUZMAK Jacek, National Veterinary Research Institute, Partyzantow 57, PL-24100 Pulawy, Poland.
KWARTA Karen, Iowa State University, 42 Spedding, Ames, IA 50011, USA.

LAAMANEN Ilona, EELA, National Veterinary and Food Research Institute, Hameentie 57, PO Box 45, FI-00581 Helsinki, Finland.
LACOURT Alain, LDA 35, 24 rue Antoine Joly, FR-35031 Rennes Cedex, France.
LAROUCAU HUET Karine, AFSSA Lerpaz, 23 av. du Général de Gaulle, FR-94706 Maisons Alfort, France.
LAU Terence L.T., Hong Kong DNA Chips, 1/F, Cosmos Centre, 108 Soy Street Mongkok, Hong Kong SAR, Kowloon, Republic of China.
LE GALL Ghislaine, LVD 29, ZA Créac'h gwen, 22 av. de la Plage des Gueux, FR-29334 Quimper Cedex 9, France.
LE POTIER Marie Frédérique, AFSSA Ploufragan, Zoopôle - BP 53, FR-22440 Ploufragan, France.

LE TALLEC Bertrand, LDV 73, 321 ch. des Moulins, FR-73024 Chambéry, France.
LECHENET Jacques, Mérial, 254, rue Marcel-Mérieux, FR-69007 Lyon, France.
LELLI Rossella, Istituto Zooprofilattico Sperimentale, Via Campo Boario, IT-64100 Terramo, Italy.
LEON Albertine, Laboratoire Départemental, Frank Duncombe, 1 rte de Rosel, FR-14280 Saint Contest, France.
LIGHTNER Donald, University of Arizona, 1117 E.Lowell, Tucson, AZ 85721, USA.
LIN Min, Canadian Food Inspection Agency, 3851 Fallowfield Rd, PO Box 11300, Station H, Ottawa, Ont. K2H 8P9, Canada.
LOISY Fabienne, Ceeram, 22 av. de la Passagère, FR-44800 Saint Herblain, France.
LOMBARD Michel, 22 rue Crillon, FR-69006 Lyon, France.

MADANI Nora, AFSSA Lerpaz, 23 av. du Général de Gaulle, FR-94706 Maisons Alfort, France.
MAGAR Ronald, Canadian Food Inspection A, 59 Camelot Drive, Ottawa, Canada.
MALDONADO Jaime, Laboratorios Hipra, Avenida la Selva s/n, ES-17170 Amer (Girona), Spain.
MALLET Laurent, Sanofi Pasteur, Building F, 1541 Av. Marcel Mérieux, FR-69280 Marcy l'Etoile, France.
MANGOLD Beverly, Tetracore Inc., 11 Firstfield Rd. Suite C, Gaithersburg, MD 20878, USA.
MARTIN RODRIGUEZ Ana Carmen, Bionostra, S.L., C/Ronda de Poniente 4-2°, ES-28760 Tres Cantos, Spain.
McCULLOUGH Sam, Veterinary Research Lab., Stoney Road, Stormont, BT4 3SD, Belfast, United Kingdom.
McKAY Paul, Fisheries Research Servic – FRS, Marine Laboratory, 375 Victoria Rd, PO Box 101, Aberdeen, AB11 9DB, United Kingdom.
McKENNA Thomas, APHIS, Foreigh Animal Disease Diag. Lab., USDA, PO Box 848, Greenport, NY 11944, USA.
McKILLEN John, Department of Agriculture and Rurale Development for Northern Ireland, Stoney Road Stormont, BT4, Belfast, United Kingdom.
MEDJITNA, T.Djamel E., Institute of Virology and Immunoprophylaxis/IVI, Sensemattstrasse 293, CH-3147 Mittelhausern, Switzerland.
MERZA Malik, Svanova Biotech AB, Science Park Glunten, SE-75183 Uppsala, Sweden.
MIDTLYNG Paul J., Veso, PO Box 8109 Dep, NO-0032 Oslo, Norway.
MILNE Catherine, EDQM, European Directorate for the Quality of Medecine, 226 av. de Colmar, BP 907, FR-67029 Strasbourg Cedex 1, France.
MINTIENS Koen, Veterinary and Agrochemical Research Centre, Groeselenberg, 99, BE-1180 Brussels, Belgium.
MOALIC Pierre Yves, Labofarm, 4 rue Théodore Botrel - BP 351, FR-22603 Loudeac, France.
MOFLE Robin, SAFC Biosciences, 13804 W. 107 th Street, Lenexa, KS 66215, USA.
MOJZIS Miroslav, State Veterinary Institute Zvolen, POD 918, SK-96086 Zvolen, Slovakia.
MOMBARG Mark, Fort Dodge Animal Health, Huizerstraatweg 117, NL-1411 GM Naarden, The Netherlands.
MONCEYRON JONASSEN Christine, National Veterinary Institute, PO Box 8156 Dep, NO-0033 Oslo, Norway.
MONNE Isabella, Istituto Zooprofilattico Sperimentale Delle Venezie, Via Romea 14/A, IT-35020 Legnaro (Padova), Italy.
MORGADO José Carlos Pinho, Merial Saude Animal Ltda, Av Carlos Grimaldi 1701, 4° andar, 13091 Campinas SP, Brasil.
MOZZER Otto, Vallee SA, Avenida Hum, 1500, 39 Montes Claros, Brasil.

NAGL Ivan, SVI Prague, PO Box 4, CZ-190 Prague, Czech Republic.
NEITZERT Erika, Sindan, Rua do Rocio, NR 313, 9 Andar, CJ 92, Sao Paulo, Brasil.
NELSON William, Tetracore Inc., 11 Firstfield Rd. Suite C, Gaithersburg, MD 20878, USA.
NOMIKOU Kiki, Ministry of Rural Development and Food, Center of Athens Veterinary Institutions, 25 Neapoleos Str, Aghia Paraskevi, GR-15310 Athens, Greece.
NUBLING Micha, Paul Ehrlich Institut, Paul Ehrlich Str. 51-59, Postfach 1740, DE-63225, Langen, Germany.

OESCH Bruno, Prionics AG, Wagistr. 27a, CH-8952 Schlieren, Switzerland.
OTTIGER Hans Peter, Institute of Virology and Immunoprophylaxis / IVI, Sensemattstrasse 293, CH-3147 Mittelhausern, Switzerland.

PALYA Vilmos, Ceva Phylaxia, Szallas Utca 5, PO Box 23, HU-1107 Budapest, Hungary.
PAPIEROK Gerard Marie, Bio Veto Test Laboratoires, ZA des Playes, 285 av. de Rome, FR-83500 La Seyne Sur Mer, France.
PARK Kyoung Sik, National Institute of Toxicological Research, Bioproduct Technical Support Division, 5 Nokbundong, Eunpyung ku 122, Seoul, South Korea.
PELLETIER Claire, LDA 71, 267 rue des Epinoches, FR-71000 Macon, France.
PEPIN Michel, AFSSA Lerpaz, 23 av. du Général de Gaulle, FR-94706 Maisons Alfort, France.
PETRINI Stefano, Fondazione Iniziative Zooprofilattiche e Zootecniche, Via Bianchi, 1, IT-25214 Brescia, Italy.
PHUNG Thang Long, Hue University, Vietman, 180 Pasteur, District 1, Ho Chi Minh, Vietnam.
PINGRET Jean Luc, Scanelis, 23 ch. des Capelles, FR-31076 Toulouse, France.
PLANCON Daniel, Virbac Laboratories, BP 27, FR-06511 Carros, France.
PORTER Marc, Iowa State University, 42 Spedding, Ames, IA 50011, USA.
POZZI Nathalie, LNCR - Laboratoire National Contrôles Reproducteurs, 13 rue Jouët, FR-94704 Maisons Alfort Cedex, France.
PRONOST Stéphane, LDA 14, 1 rte de Roussel, FR-14033 Caen, France.

QUINTING Birgit, Centre d'Economie Rurale, 1 rue Marcel Carmel, BE-6900 Marloie, Belgium.

RAMADORI Giorgio, Universita degli Studi di Milano, 36, Via C. Pascal, IT-20133 Milano, Italy.
RAYMOND Jean-Claude, Biomerieux, ch. de l'Orme, FR-69280 Marcy l'Etoile, France.
REZAZADEH Fereydon, University of Tehran, Vet. Med. Faculty, PO Box 14155-6453, Tehran, Iran.
RIBIERE Magali, AFSSA Sophia Antipolis, 105 rte des Chappes, BP 111, FR-06902 Sophia Antipolis Cedex, France.
ROSSI Jen, Qiagen, 3 av. du Canada, LP 809, FR-91974 Courtaboeuf Cedex, France.
ROTA Paul, Centers of Disease Control, MS-C-22, 1600 Clifton Road, Atlanta, GA 30333, USA.

SACHSE Konrad, Friedrich Loeffer Institut, Naumburger Str 96, DE-07743 Jena, Germany.
SADEGHI Mohammad Reza, Bu-Ali Sina University, Dept. of Animal Science, Hamadan, Iran.
SAWYER Jason, Veterinary Lab. Agency – Weybridge, New Haw, Addlestone, KT15 3NB Surrey, United Kingdom.
SCHLENCK Alexandra, CHU Nantes, Quai Moncousu, Service Génétique Médicale, FR-44093 Nantes Cedex, France.
SCHMIDT Ulrike, Intervet International BV, Wim de Korverstraat 35, NL-5830 AA Boxmeer, The Netherlands.
SCHNEIDER Thomas, Federal Ministry for Consumer Protection Food and Agricultural, Rochusstrabe 1, DE-53123 Bonn, Germany.
SCHON Joseph, Laboratoire de Médecine Veterinaire de l'Etat, 54 av. Gaston Diderich, BP 2081, LU-1020 Luxembourg, Luxembourg.
SCHUDEL Alejandro, OIE, 12 rue de Prony, FR-75017 Paris, France.
SCHYNTS Frédéric, Centre d'Economie Rurale, 1 rue Marcel Carmel, BE-6900 Marloie, Belgium.
SENNE Dennis, National Veterinary Services Laboratories, 1800 Dayton Ave, Ames, IA 50010, USA.
SIMEK Bronislav, State Veterinary Institute Jihlava, Akademicka 3, 94901 Nitra, Slovakia.
SNOW Michael, Fisheries Research Servic – FRS, Marine Laboratory, 375 Victoria Rd, PO Box 101, Aberdeen, AB11 9DB, United Kingdom.
STEPHANO Marco Antonio, Instituto Butantan, Av. Dr Vital Brazil, 1500, 5503 Sao Paulo, Brasil.
STORSET Arne, Aqua Gen AS, PO Box 1240, Pir-Senteret, NO-7462 Trondheim, Norway.
SUBASINGHE Roahna, FAO, Fisherie Department, Viale delle Terme di Caracalla, IT-00 100 Rome, Italy.
SWAM Hanny, Intervet International BV, Wim de Korverstraat 35, NL-5830 AA Boxmeer, The Netherlands.
SZCZEPANEK Thomas, Eurobio Laboratoires, 7 av. de Scandinavie, FR-91940 Les Ulis, France.

TAMPAKIS Konstantinos, Qiagen Ltd, Quagen House, Fleming Way, Crawley, West Sussex RH10 9NQ, United Kingdom.
TARRIKAS Siiri, Estonian Veterinary and Food Laboratory, Kreutzwaldi 30, EE-51006 Tartu, Estonia.
TEKES Lajos, Central Veterinary Institute Budapest, Tabornok u.2., HU-1149 Budapest, Hungary.
TERZIC Svjetlana, Croatian Veterinary Institute, Savska cesta 143, HR-10 00 Zagreb, Croatia.
THEVENON Jérôme, Ceva Phylaxia, Szallas Utca 5, PO Box 23, HU-1107 Budapest, Hungary.
THOMAS Lee Ann, USDA, APHIS, VS, 4700 River Rd, Riverdale, MD 20737, USA.

TIGNON Marylène, Veterinary & Agrochemical Research Centre, Groeselenberg 99, BE-1180 Brussels, Belgium.

TRONO Karina, Inta – Castelar, Las Cabanas y los Reseros s/n, 1712 Castelar, Argentina.

VAN DER WERF Sylvie, Institut Pasteur, 25 - 28 rue du Dr Roux, FR-75015 Paris, France.

VAN RIJN Piet A., CIDC Lelystad, Houtribweg 39, NL-8221 RA Lelystad, The Netherlands.

VAN SOLT Conny, ID Lelystad, Animal Sciences Group, Edelherweg 15, NL-8219 PH Lelystad, The Netherlands.

VANNIER Philippe, AFSSA Ploufragan, Zoopôle, BP 53, FR-22440 Ploufragan, France.

VATCHER Susan, National Research Council of Canada, Industrial Research Assistance Program, 550 University Av., Classroom Centre, Suite 106, Charlottetown, Prince Edward Island C1A 4P3, Canada.

VERMET Laurent, Bio-Rad, 3 Bd Raymond Poincaré, FR-92430 Marnes La Coquette, France.

VICENS Serge, Bio Veto Test Laboratoires, ZI des Playes, 285 av. de Rome, FR-83500 La Seyne sur Mer, France.

VILLA Riccardo, Istituto Zooprofilattico, Via A. Bianchi, 7/9, IT-25124 Brescia, Italy.

VON BANNISSEHT WIJSMULLER Thea E., Animal Health Service, In Southern Netherlands, Arnsbergstraat 7, NL-7418 EZ Deventer, The Netherlands.

WAKELEY Phil, Veterinary Laboratories Agency, 58 Beech Hill Road, Sunningdale, Berkshire SL5, United Kingdom.

WANJOHI Cathryn, Kenya Agricultural Research Institute, PO Box 58137, Kabete, Nairobi, Kenya.

WELLENBERG Gerard, Animal Health Service Institute, Arnsrerg Straat 9, NL-7400 AA Deventer, The Netherlands.

WENG Man Vong, Civic and Municipal Affairs Bureau of Macao, Av. Almeida Ribeiro, N° 163, Edf Sede do IACM, NA, Macao, Republic of China.

WILLIAMS Russell, VLA Penrith, Merrythought, Cumbria, Penrith, United Kingdom.

WISSELINK Henk, ID Lelystad, Animal Sciences Group, Edelherweg 15, NL-8219 PH Lelystad, The Netherlands.

ZIENTARA Stephan, AFSSA-Maisons Alfort, 22 rue Pierre Curie, FR-94703 Maisons-Alfort Cedex, France.

Index of Authors

Developments in Biological Standardization

Vol. 68	Geneva, 1986	Cells - Products - Safety
Vol. 69	London, 1987	Cytokines Laboratory and Clinical Evaluation
Vol. 70	Arlington, VA, 1988	Continuous Cell Lines as Substrates for Biologicals
Vol. 71	Utrecht, 1989	Monoclonal Antibodies for Therapy, Prevention and In vivo Diagnosis of Human Disease
Vol. 72	Annecy, 1989	Progress in Animal Retroviruses
Vol. 73	Shizuoka, 1990	Pertussis: Evaluation and Research on Acellular Pertussis Vaccines
Vol. 74	Bethesda, MD, 1990	Biological Product Freeze-Drying and Formulation
Vol. 75	London, 1990	Virological Aspects of the Safety of Biological Products
Vol. 76	Bethesda, MD, 1991	Continuous Cell Lines - An International Workshop on Current Issues
Vol. 77	Annecy, 1991	Standardization of the Immunopharmacology of Natural and Synthetic Immunomodulators
Vol. 78	Bethesda, MD, 1991	Poliovirus Attenuation: Molecular Mechanisms and Practical Aspects
Vol. 79	Ploufragan, 1992	The First Steps Towards an International Harmonization of Veterinary Biologicals: 1993 and Free Circulation of Vaccines within the EEC
Vol. 80	Heidelberg, 1992	Transmissible Spongiform Encephalopathies - Impact on Animal and Human Health
Vol. 81	Cannes, 1992	Virological Safety Aspects of Plasma Derivatives
Vol. 82	Albany NY, 1993	Recombinant Vectors in Vaccine Development
Vol. 83	Annecy, 1993	Genetic Stability and Recombinant Product Consistency
Vol. 84	Langen, 1993	Non-Target Effects of Live Vaccines
Vol. 85	Santa Fé, NM, 1994	Genetics of Streptococci, Enterococci and Lactococci
Vol. 86	Langen, 1994	Replacement, Reduction and Refinement of Animal Experiments in the Development and Control of Biological Products
Vol. 87	Geneva, 1995	New Approaches to Stabilisation of Vaccines Potency
Vol. 88	Bethesda, MD, 1995	Viral Safety and Evaluation of Viral Clearance from Biopharmaceutical Products
Vol. 89	Rome, 1995	Pertussis Vaccine Trials
Vol. 90	Oslo, 1996	Fish Vaccinology
Vol. 91	San Francisco, CA, 1996	Development of Specifications for Biotechnology Pharmaceuticals Products
Vol. 92	Bergen, 1996	Modulation of the Immune Response to Vaccine Antigens
Vol. 93	Annecy, 1996	Safety of Biological Products Prepared from Mammalian Cell Culture
Vol. 94	Bethesda, MD, 1997	Simian Virus 40 (SV40): A Possible Human Polyomavirus
Vol. 95	Paris, 1997	Preclinical and Clinical Development of New Vaccines
Vol. 96	Washington, DC, 1995	Characterization of Biotechnology Pharmaceutical Products

Vol. 97	London, 1997	Biological Characterization and Assay of Cytokines and Growth Factors
Vol. 98	London, 1997	Inactivated Influenza Vaccines Prepared in Cell Culture
Vol. 99	Strasbourg, 1998	Animal Sera, Animal Sera Derivatives and Substitutes Used in the Manufacture of Pharmaceuticals: Viral Safety and Regulatory Aspects
Vol. 100	Geneva, 1998	A Celebration of 50 Years of Progress in Biological Standardization and Control at WHO
Vol. 101	London, 1998	Alternatives to Animals in the Development and Control of Biological Products for Human and Veterinary Use

Developments in Biologicals

Vol. 102	San Francisco, CA, 1999	Advances in Transfusion Safety
Vol. 103	Annecy, 1999	Physico-Chemical Procedures for the Characterization of Vaccines
Vol. 104	Langen, 1999	Development and Clinical Progress of DNA Vaccines
Vol. 105	Paris, 2000	Progress in Polio Eradication: Vaccine Strategies for the End Game
Vol. 106	Rockville, MD, 2000	Evolving Scientific and Regulatory Perspectives on Cell Substrates for Vaccine Development
Vol. 107	London, 2000	The Design and Analysis of Potency Assays for Biotechnology Products
Vol. 108	Langen, 2001	Advances in Transfusion Safety
Vol. 109	Washington, DC, 2000	Biologics 2000 – Comparability of Biotechnology Products
Vol. 110	Palm Cove, 2001	Orphan Vaccines – Bridging the Gap
Vol. 111	Utrecht, 2001	Advancing Science and Elimination of the Use of Laboratory Animals for Development and Control of Vaccines and Hormones
Vol. 112	Bethesda, MD, 2001	Immunogenicity of Therapeutic Biological Products
Vol. 113	Berlin, 2001	Process Validation for Manufacturing of Biologics and Biotechnology Products
Vol. 114	Ames, IA, 2002	Vaccines for OIE List A and Emerging Animal Diseases
Vol. 115	Bergen, 2002	Laboratory Correlates of Immunity to Influenza – A Reassessment
Vol. 116	Los Angeles, CA, 2003	Development of Therapeutic Cancer Vaccines
Vol. 117	Langen, 2003	Consideration of Alternative Licensing Procedures for Vaccines for Minor Species, Minor Indications and Autogenous/Autologus Products
Vol. 118	Langen, 2003	PDA/EMEA European Virus Safety Forum
Vol. 119	Buenos Aires, 2004	Control of Infectious Animal Diseases by Vaccination
Vol. 120	Bethesda, MD, 2003	Advances in Transfusion Safety
Vol. 121	Bergen, 2003	Progress in Fish Vaccinology
Vol. 122	Bethesda, MD, 2003	State of the Art Analytical Methods for Characterization of Biological Products and Assessment of Comparability
Vol. 123	Rockville, MD, 2004	Vaccine Cell Substrates 2004
Vol. 124	Paris, France, 2005	OIE/FAO International Scientific Conference on Avian Influenza
Vol. 125	Kiev, Ukraine, 2005	First International Conference on Rabies in Europe

Benefit from a wealth of new findings

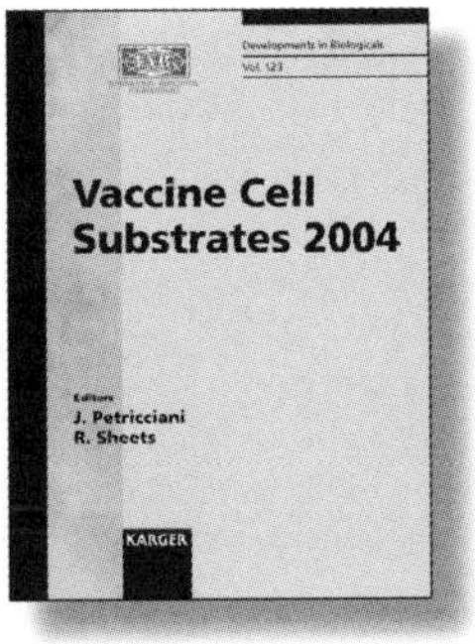

Vaccine Cell Substrates 2004

Editors
John Petricciani
Rebecca Sheets

Immunology, Infectious Diseases, Vaccination, Virology, Public Health, Biotechnology

Selected Contents

Introduction
Introduction to Vaccine Cell Substrates 2004: ***Sheets, R.L.***

Potential Oncogenicity of Cellular DNA
Cell Substrates: Where do we Stand after 50 Years of Discussion?: ***Petricciani, J.***
The Uptake and Fate of Exogenous Cellular DNA in Mammalian Cells: ***Strain, A.J.***
Lessons Learned from Gene Therapy Concerning and the Use of Integrating Vectors and the Possible Risk of Insertional Oncogenesis: ***Cichutek, K.***
Adaptation of the WHO Guideline for Residual DNA in Parenteral Vaccines Produced on Continuous Cell Lines to a Limit for Oral Vaccines: ***Lebron, J.A.; Troilo, P.J.; Pacchione, S. et al.***
Biological Activity of Residual Cell-Substrate DNA: ***Peden, K.; Sheng, L.; Pal, A.; Lewis, A.***

Potential Oncogenicity of Viruses Latent in Cells
Molecular Methods for Virus Discovery: ***Jarrett, R.F.; Johnson, D.; Wilson, K.; Gallagher, A.***
Significance of Retrovirus Contaminants of Cell Substrates (Abstract): ***Coffin, J.M.***
Practical Assay Issues with the PERT/PBRT Assay: A Highly Sensitive Reverse Transcriptase Assay: ***Chang, A.; Dusing, S.***
Antibodies to SV40 in Sera from Humans and Macaques: ***Minor, P.; Pipkin, P.A.***
Polio Vaccines, SV40 and Human Tumours, an Update on False Positive and False Negative Results: ***Elmishad, A.G.; Bocchetta, M.; Pass, H.I.; Carbone, M.***

Routine Adventitious Agent Tests – Can in Vivo Methods be Replaced?
Adventitious Agent Test Methods: ***Sheets, R.L.***
Scope and Practicality of in vivo Testing for Adventitious Agents: ***Whiteman, M.D.***
Detection of Adventitious Viruses in Biologicals – A Rare Occurrence: ***Nims, R.W.***
Approaches to the Release of a Master Cell Bank of PER.C6 Cells; A Novel Cell Substrate for the Manufacture of Human Vaccines: ***Lewis, J.A.; Brown, E.L.; Duncan, P.A.***
Newer Molecular Approaches to Detect Known Viruses and Previously Unrecognized Viruses in Cell Substrates: ***Wright, P.F.***
A Cell-Based Assay for Mouse Prions (Abstract): ***Weissmann, Ch.***

Level of Assurance Provided by Guidelines, Especially ICH Q5A and Q5D
Basis for Sample Size in International Conference on Harmonization Guidances: ***Lubiniecki, A.***
The Impact of Technology on Assurance (Abstract): ***Moore, W.A.***
Statistical Bases for Defining Level of Assurance: ***Schofield, T.***

Experience with Novel Cell Lines
Comprehensive Characterization of the 293-ORF6 Cell Line: ***Butman, B.T.; Lizonova, A. et al.***
Insect Cells as a New Substrate for Vaccine Production: ***Dalemans, W.***
Safety Assessment of Madin Darby Canine Kidney Cells as Vaccine Substrate: ***Medema, J.K. et al.***
Tumorigenicity Assessments of PER.C6® Cells and of an Ad5-Vectored HIV-1 Vaccine Produced on this Continuous Cell Line: ***Ledwith, B.J.; Lanning, C.L.; Gumprecht, L.A. et al***

Bovine (and Porcine) Viruses
Bovine Serum – Regulatory Issues: ***Robertson, J.S.***
Vaccine Cell Substrates: Bovine and Porcine Virus Considerations: ***Wessman, S.J.***
Assays for Bovine Viruses (Abstract): ***Wisher, M.***
Animal-Free Culture Process for Manufacture of Viral Vaccines: Process Validation and Regulatory Perspectives: ***Pierard, I.; Mareschal, J.-C. et al.***

Bovine Spongiform Encephalopathy
Susceptibility of Common Laboratory Fibroblasts to TSE Infection (Abstract): ***Priola, S.A.***
Diagnostics for TSE Agents: ***Raeber, A.J.; Oesch, B.***
Production Cell Substrates and vCJD: ***Minor, P.D.; Cooper, J.***
TSE Risk Assessment for Starting Materials Used During, or in the Manufacture of Vaccines for Human Use – A Consultant's View of the Commercial Approach: ***Bradley, R.***
Bovine Spongiform Encephalopathy (BSE) in the United States: an Overview and Critique of the Response with Suggestions for Additional Actions (Abstract): ***Detwiler, L.A.***

www.karger.com/debis

KI 06220

Developments in Biologicals, Vol. 123
Edited by the International Association for Biologicals IABs
ISSN 1424-6074

Vaccine Cell Substrates 2004
International Conference, Rockville, Md., USA, June 2004: Proceedings
Editors: Petricciani, J. (Calsbad, Ca.); Sheets, R. (Bethesda, Md.)
X + 366 p., 35 fig., 51 tab., soft cover, 2006
CHF 340.– / EUR 243.– / USD 309.25
Prices subject to change
EUR price for Germany, USD price for USA only
ISBN 3–8055–7999–3

KARGER

Order Form

Please send: ______ copy/ies
Postage and handling free with prepayment

Payment:
☐ Check enclosed ☐ Please bill me
Please charge to my credit card
☐ American Express ☐ Diners ☐ Eurocard
☐ MasterCard ☐ Visa

Card No.: ____________________
Exp. date: ____________________

Orders may be placed with any bookshop, subscription agency, directly with the publisher or through a Karger distributor.

Fax: +41 61 306 12 34
S. Karger AG, P.O. Box, CH–4009 Basel (Switzerland)
E-Mail orders@karger.ch, **www.karger.com**

Name/Address: ____________________

Date: ____________________
Signature: ____________________